D0758121

Practical Methods
for Design and Analysis
of Complex Surveys

Practical Methods for Design and Analysis of Complex Surveys

Revised Edition

Risto Lehtonen
The Social Insurance Institution, Finland

Erkki J. Pahkinen
University of Jyväskylä

JOHN WILEY & SONS
Chichester · New York · Brisbane · Toronto · Singapore

Other Wiley Editorial Offices

John Wiley & Sons, Inc., 605 Third Avenue,
New York, NY 10158-0012, USA

Jacaranda Wiley Ltd, 33 Park Road, Milton,
Queensland 4064, Australia

John Wiley & Sons (Canada) Ltd, 22 Worcester Road,
Rexdale, Ontario M9W 1L1, Canada

John Wiley & Sons (SEA) Pte Ltd, 37 Jalan Pemimpin #05-04,
Block B, Union Industrial Building, Singapore 20057

Library of Congress Cataloging-in-Publication Data

Lehtonen, Risto.
 Practical methods for design and analysis of complex / surveys
 Risto Lehtonen and Erkki J. Pahkinen.
   p.   cm. — (Statistics in practice)
 Includes bibliographical references and indexes.
   ISBN   0 471 93934 X
   1. Sampling (Statistics)   2. Surveys — Methodology.   I. Pahkinen,
Erkki. II. Title.   III. Series: Statistics in practice (Chichester,
England)
QA276.6.L46   1994
001.4'33 — dc20   94-20917
                                                            CIP

British Library Cataloguing in Publication Data

A catalogue record for this book is available from the British Library

ISBN 0 471 93934 X

Typeset in 10/12pt Photina by Thomson Press (India) Ltd, New Delhi
Printed and bound in Great Britain by Biddles Ltd, Guildford and King's Lynn

Contents

Preface

This book deals with both survey sampling and survey analysis. By integrating these broad topics in the same text we try to fill the gap between sampling methods and statistical analysis in complex surveys, which are often presented in separate books. We believe that from the user's point of view, an integrated text can be helpful.

For survey sampling, our aim is to provide the reader with practical tools for basic and more advanced sampling and estimation. The main focus is on demonstrating the gains from the use of available auxiliary information for efficient sampling and estimation. Measured with the design-effect statistic, this gain is shown to be substantial especially in descriptive surveys. Today, use of complex survey data for analytical purposes is also rapidly increasing. We therefore supply the reader with approximative methods for variance and covariance estimation, analysis methods for frequency tables, and multivariate methods usable in complex surveys. Special efforts are made in pointing out the importance of allowing for sampling design complexities for reliable analysis results, in such matters as stratification, clustering and weighting. Again, the design-effect statistic is extensively used.

The text is essentially applied. We concentrate on basic principles of the methodology and illustrate each method with numerical examples and case studies taken from real surveys. About 40% of the material is concerned with applications. The applications come from business surveys, educational surveys, health surveys, social surveys, and socioeconomic surveys. For practical purposes, selected commercially available software products for survey analysis are also demonstrated. The surveys described in the book are obtained from various sources; the main sources have been certain previous books published in Finnish by the authors, dealing largely with methodological and empirical work of the authors carried out at the Social Insurance Institution of Finland, the Statistics Department of the University of Jyväskylä, and Statistics Finland.

We have aimed to cover the recent methodology that is important and increasingly used in practice and is, in most cases, accessible by commercial software for survey analysis.

We have updated the text through the book for this revised edition. A new Appendix 2 is included to demonstrate the finite-population properties of the estimators used in Chapter 2. In Appendix 1, a new program for survey analysis is added.

Many individuals and organizations have supported and encouraged us during the writing of this book. Discussions with Professor Vic Barnett, Rothamsted Experimental Station, and detailed comments he has kindly given on the text, have been very helpful. We are grateful to Professor Carl-Erik Särndal, University of Montreal, for his valuable comments on a part of the text. Dr Michel Hidiroglou, Statistics Canada, Mr David Morganstein, Westat, and Dr Babubhai V. Shah, Research Triangle Institute, kindly commented on Appendix 1. We are also grateful to Professor Esko Kalimo and Docent Arpo Aromaa, Social Insurance Institution, Professor Jouko Kari, University of Jyväskylä, and Dr Eero Heikkonen and Mr Seppo Kouvonen, Statistics Finland, for giving an opportunity to utilize the survey data sets needed for this book. Esko Kalimo and Arpo Aromaa also commented on certain parts of the text.

Comments on various parts of the book have also been given by Mr Antero Malin, Mr Kari Nissinen and Ms Anja Roikonen, all of the University of Jyväskylä. Antero Malin also assisted in statistical analyses in Section 9.3, and Anja Roikonen in turn in Chapters 2–4. Further, Dr Seppo Laaksonen, EUROSTAT, commented on Chapter 4, and Mr Erkki Nenonen, Social Insurance Institution, helped in the preparation of Appendix 3. The staff of our home organizations assisted us in the preparation of the final manuscript. Mr Timo Byckling and Mr Kari Djerf helped with Chapters 2 to 4, and Ms Leila Hölttä and Mr Rauno Veijola drew the graphs. We are very grateful for all this assistance.

Ms Karen Moore and Ms Rosie Poultney of Rothamsted Experimental Station kindly edited the text to make our English more readable. Final editing and proofreading took place under the direction of Ms Helen Ramsey at John Wiley & Sons, Chichester. We are grateful to Mr Stuart Gale, John Wiley & Sons, for his patience in our preparation of the revised edition of this book.

Close cooperation between the authors has taken place in writing the manuscript for this book and in integrating the materials. However, Risto Lehtonen takes the main responsibility for Chapters 1 and 5 to 8, Sections 9.2 and 9.3 and Appendices 1 and 3, and Erkki Pahkinen for Chapters 2 to 4, Section 9.1 and Appendix 2.

We greatly appreciate the financial support given by the Academy of Finland which made possible our writing of this book.

Risto Lehtonen   **Erkki J. Pahkinen**
Helsinki   *Jyväskylä*

May 1996

Series Preface

Statistics in Practice is an important international series of texts which provide direct coverage of statistical concepts, methods and worked case studies in specific fields of investigation and study.

With sound motivation and many worked practical examples, the books show in down-to-earth terms how to select and use a specific range of statistical techniques in a particular practical field within each title's special topic area.

The books meet the need for statistical support required by professionals and research workers across a wide range of employment fields and research environments. The series covers a variety of subject areas: in medicine and pharmaceutics (e.g. in laboratory testing or clinical trials analysis); in social and administrative fields (e.g. sample surveys or data analysis); in industry, finance and commerce (e.g. for design or forecasting); in the public services (e.g. in forensic science); in the earth and environmental sciences, and so on.

But the books in the series have an even wider relevance than this. Increasingly, statistical departments in universities and colleges are realizing the need to provide at least a proportion of their course-work in highly specific areas of study to equip their graduates for the work environment. For example, it is common for courses to be given on statics applied to medicine, industry, social and administrative affairs and the books in this series provide support for such courses.

It is our aim to present judiciously chosen and well-written workbooks to meet everyday practical needs. Feedback of views from readers will be most valuable to monitor the success of this aim.

Vic Barnett
Series Editor
1994

1

Introduction

General Outline

This book deals with *sample surveys* which can be conceptually divided into two broad categories. In *descriptive surveys* certain, usually few, population characteristics need to be precisely and efficiently estimated. For example in a business survey, the average salaries for different occupational groups are to be estimated, based on a sample of business establishments. *Statistical efficiency* of the sampling design is of great importance. *Stratification* and other means of using *auxiliary information*, such as the sizes of the establishments, can be beneficial in sampling and estimation with respect to efficiency. Inference in descriptive surveys concerns exclusively a fixed population, although superpopulation and other models are often used in the estimation.

Analytical surveys, on the other hand, are often multi-purpose so that a variety of subject matters are covered. In the construction of a sampling design for an analytical survey, a feasible overall balance between statistical efficiency and *cost efficiency* is sought. For example, in a survey where personal interviews are to be carried out, a sampling design can include several stages so that in the final stage all the members in a sample household are interviewed. While this kind of clustering decreases statistical efficiency it often provides the most practical and economical method for data collection. Cost efficiency can be good, but gains from stratification and from use of other auxiliary information can be of minor concern for statistical efficiency when dealing with many diverse variables. Although in analytical surveys descriptive goals can still be important, of interest are often, for example, differences of subpopulation means and proportions, or coefficients of logit and linear models, rather than totals or means for the fixed population as in descriptive surveys. Statistical testing and modelling therefore play more important roles in analytical surveys than in descriptive surveys.

Both descriptive and analytical surveys can be *complex*, e.g. involving a complex sampling design such as multi-stage stratified cluster sampling. Accounting for the sampling complexities is essential for reliable estimation and analysis in both types of surveys. This holds especially for the clustering effect which involves *intra-cluster correlation* of the study variables. This

1

affects variance estimation and testing and modelling procedures. And if *unequal selection probabilities* of the population elements are used, appropriate weighting is necessary in order to attain estimators with desired statistical properties such as unbiasedness or consistency with respect to the sampling design. Moreover, *element weighting* may also be necessary for adjusting for nonresponse, and imputation for missing variable values may be needed, in both descriptive and analytical surveys.

Thus there are many common features in the two types of complex surveys and often in practice no real difference exists between them. A survey primarily aimed at descriptive purposes can also involve features of an analytical survey and vice versa. However, making the conceptual separation can be informative, and is a prime intention behind the structuring of the material in this book.

Topics Covered

To be useful a book on methods for both design and analysis of complex surveys should cover topics on sampling, estimation, testing, and modelling procedures. We have structured a survey process so that we first consider the principles and techniques for sample selection. The corresponding estimators for the unknown population parameters, and the related standard error estimators, are also examined so that estimation under a given sampling design can be made manageable in practice, reliable, and efficient. These topics are considered in the first part of the book (Chapters 2 and 3), mainly under the framework of descriptive surveys.

Estimation and analysis specific to analytical surveys is considered in the second part of the book (Chapters 5–8). For complex analytical surveys, more sophisticated techniques of variance estimation are needed. Our main focus in such surveys however is on testing and modelling procedures. Testing procedures for one-way and two-way tables, and multivariate analysis (including methods for categorical data and logistic and linear regression) are selected because of their importance in survey analysis practice. Topics relevant to both descriptive and analytical surveys, concerning techniques for handling missing data such as *reweighting* and *imputation*, are placed between the two main parts of the book (Chapter 4).

Fully worked examples and case studies taken from real surveys on health and social sciences are used to illustrate the various methods. Finally (Chapter 9), additional case studies are presented covering a range of different topics such as business surveys, socioeconomic surveys and educational surveys.

In Chapters 2 and 3, the basic and more advanced sampling techniques, namely *simple random sampling, systematic sampling, sampling with probability proportional to size, stratified sampling* and *cluster sampling* are examined for the estimation of three different population parameters. These parameters are the *population total, ratio* and *median*. The estimators of these parameters provide examples of linear, nonlinear and robust methods, respectively. A small fixed population is used throughout to illustrate the estimation methods, where the main focus is in the derivation of appropriate *sampling weights* under each sampling technique. Special efforts are made in comparing the relative performances of the estimators (in terms of their standard errors) and the available information on the structure of the population is increasingly utilized. Use of such *auxiliary information* is considered for two purposes: the sampling design and the estimation of parameters for a given sampling design. The use of this information varies between different sampling techniques, being minor in the basic techniques and more important and sophisticated in others, such as in stratified sampling and in cluster sampling. *Estimation using poststratification, ratio estimation and regression estimation* are considered in some detail under the framework of *model-assisted estimation*. The *design effect* is extensively used for efficiency comparisons. It is shown that proper use of auxiliary information can considerably increase the efficiency of estimation. In connection with Chapters 2 and 3, statistical properties of the total, ratio and median estimators, such as bias and consistency, are presented in Appendix 2.

In Chapter 5, we extend the variance estimation methodology of Chapters 2 and 3 by introducing additional (approximative) techniques for variance estimation. Subpopulation means and proportions are chosen to illustrate ratio-type estimators commonly used in analytical surveys. *Linearization method* and the *sample re-use techniques* including *balanced half-samples, jackknife* and *bootstrap*, are demonstrated for a two-stage stratified cluster sampling design taken from the Mini-Finland Health Survey. This survey is chosen because it represents an example of a realistic but manageable design. Approximation of variances and covariances of several ratio estimators is needed for testing and modelling procedures. Using the linearization method, various sampling complexities including clustering, stratification and weighting are accounted for (in Chapter 6) to obtain consistent variance and covariance estimates. These approximations are applied to the Occupational Health Care Survey sampling design which is slightly more complex than that of the previous survey. We also introduce certain simplified variance estimators based on the so-called *effective sample sizes*. In these variance estimators, which are intended to adjust for *extra-binomial variation*, design effects of ratio estimators are used.

The analysis of complex survey data is considered in Chapters 7 and 8. For testing procedures of goodness of fit, homogeneity and independence hypotheses in one-way and two-way tables we introduce two main approaches, the first of these using *Wald-type test statistics* and the second, *Rao–Scott-type adjustments* to standard Pearson and Neyman test statistics. The main aim in these test statistics is to adjust for the clustering effect. These testing procedures rely on assumption of an asymptotic chi-square distribution of the test statistic with appropriate degrees of freedom; this assumption presupposes a large sample and especially a large number of sample clusters. For designs where only a small number of sample clusters are available, certain degrees-of-freedom corrections to the test statistics are derived leading to F-distributed test statistics.

In Chapter 8 we turn to *multivariate survey analysis* where a binary or a continuous response variable and a set of predictor variables are assumed. In the analysis of categorical data with logit and linear models, *weighted least squares estimation* is used. Further, for logistic and linear regression in cases where some of the predictors are continuous, we use the *pseudolikelihood method*. For proper analysis using either of these methods, certain analysis options are suggested. Under the full design-based option, all the sampling complexities are properly accounted for, thus providing a generally valid approach for complex surveys. An option based on effective sample sizes can in some cases be used to adjust for extra-binomial variation. The option based on an assumption of simple random sampling is used as a reference when measuring the effects of sampling complexities on estimation and test results. Using these options, multivariate analysis is further demonstrated in the additional case studies in Chapter 9.

The *nuisance* (or *aggregated*) *approach*, where the clustering effects are regarded as disturbances to estimation and testing, is the main approach for the *design-based* analysis in this book. In this approach, the main aim is to eliminate these effects to obtain valid analysis results. In the alternative *disaggregated approach* which also provides valid analyses, clustering effects are themselves of intrinsic interest. We demonstrate this approach for *multi-level modelling* of hierarchically structured data in the last of the additional case studies in Chapter 9.

Computation

To be manageable in practice, we have in the examples and case studies demonstrated the methodology using commercially available software products for survey analysis. SUDAAN and PC CARP are chosen for this purpose. In addition, SPSS software is used to demonstrate the sampling techniques, ML3 is used for multi-level modelling in Section 9.3, and SAS software is used in demonstrating the sample re-use methods. A brief summary of selected software for the analysis of complex surveys is presented in Appendix 1. A SAS macro for bootstrap variance approximation is given in Appendix 3.

Use of the Book

This book is primarily intended for researchers, sample survey designers, and statistics consultants working on the planning, execution or analysis of descriptive or analytical sample surveys. We have aimed to supply such workers with an applied source covering in a compact form the relevant topics of recent methodology for design and analysis of complex surveys. By using real data sets with computing instructions and computerized examples the reader can also be led to a deeper understanding of the methodology.

The material in the book can also be used in university-level methodological courses. A first course in survey sampling can be based on Chapters 2–4 where the students can also be guided to real sampling and estimation using the small population provided. A more advanced course can be based on Chapters 5–8. Also, useful data sets are supplied for practising variance approximation, testing procedures and multivariate analysis in complex surveys. Chapter 4 might be included in a more advanced course.

2

Basic Sampling Techniques

Simple random sampling, systematic sampling and *sampling with probability proportional to size* are introduced as the basic sampling techniques in this chapter. We start with a discussion of sampling, and sampling errors, and estimation of a given sampling scheme. Definitions of some key concepts are given.

Sampling and Sampling Error

In survey sampling a *fixed finite population* is under consideration, where the population elements are *labelled* so that each element can be identified. *Probability sampling* provides a flexible device for the selection of a random sample, or a *sample* for short, from such a fixed population. A key property of probability sampling is that for each population element a positive probability of selection is assigned; this probability need not be equal for all the elements. A specific sampling scheme is used in drawing the sample. The term *sampling scheme* refers to the collection of techniques or rules for the selection of the sample. The composition of the sample is thus randomized according to the probabilistic definition of the sampling scheme.

In principle, a large number of different samples could be drawn from a population using a particular sampling scheme. Depending on which specific population elements happen to be drawn, different numerical estimates are obtained from the sample for an unknown population parameter such as a *total*, i.e. the sum of the population values of a variable. *Sampling error* describes the variation of the estimates calculated from the possible samples. In the design of the sample selection procedure for a specific survey, a sampling scheme is desired under which the sampling error would be as small as possible. In order to attain this goal, knowledge on the structure of the population can be helpful. Relationships between the sampling scheme and the structure of the population are considered for various specific sampling situations in this chapter and in Chapter 3. In this discussion, the *standard error* of an unbiased estimate is used as a measure of the sampling error, and the comparison of the sampling errors under various sampling schemes is carried out using the *design-effect* statistic.

Estimation from Selected Sample

When an actual sample is drawn using a specific sampling scheme, measurements are recorded from the sampled elements for some variable of interest, called a *study variable.* After data collection, statistical analyses can be carried out. For example, an estimate of the population total of the study variable and its estimated standard error are frequently calculated. In this chapter and the next, we examine practical methods for designing manageable sampling procedures and for carrying out proper estimation under a given sampling scheme. For this, let us first discuss various approaches concerning the role of the sampling scheme in the estimation process.

When a survey is analysed in practice it is emphasized that estimation should take into account the structure of the sampling scheme. To accomplish this, the analysis is carried out using the so-called *design-based approach* (abbreviated DES). An essential property of the design-based approach is that any of the complexities due to the sampling scheme can be properly accounted for in the estimation. These complexities can arise, for example, when elements have unequal selection probabilities and will be discussed further in Chapters 2 and 3. These features of a sampling scheme can be incorporated into the estimation in the design-based approach because a fixed finite population with labelled elements is being considered. By using the labels assigned to each element, appropriate *sampling design identifiers* can be included in the sample data set and used in the analysis. Making use of the sampling identifiers is examined in some detail in this and the next chapter, for estimation under various sampling schemes.

Conversely, an analysis ignoring all the sampling complexities would be based on the so-called *model-based approach,* abbreviated IID (independent identically distributed). Strictly speaking, an *infinite population* is assumed rather than a finite population, and therefore, the population elements cannot be labelled. Thus, the population total cannot be defined as a parameter since the population is assumed infinite. Moreover, there is no concept equivalent to sampling design for the model-based approach. For these reasons, the IID approach is not appropriate for estimation in complex surveys as such. However, certain features are applicable. In the design-based approach, it can sometimes be useful to assume that the finite population is a realization from some hypothetical *superpopulation.* This assumption together with appropriate *auxiliary information* can be used by postulating models for the estimation of parameters of the finite population under consideration. When auxiliary variables are included in the estimator but the inference is still design-based, we call this the *model-assisted design-based approach*, or more simply the *model-assisted approach* (Särndal *et al.* 1992). This approach is discussed in the last part of Chapter 3.

There is another important relation between the design-based and model-based approaches. For a certain sampling situation, namely where elements are selected with equal probabilities and are replaced in the population after each draw, the estimation approximates closely that under the IID approach. This sampling scheme is called *simple random sampling with replacement*, and will occasionally be used as a reference design when comparing the efficiencies of more complex sampling schemes.

Let us consider the design-based, model-assisted and model-based approaches more closely to show how a model assumption can be used to simplify the estimation for a certain sampling scheme, and to highlight the way in which the model-based approach differs from the other two. Suppose that a shipping company wants to know the approximate total weight of the passengers on a ferry. This piece of information is important for future planning. Weighing all the passengers would be too expensive and time-consuming, thus sampling would be more appropriate in this context. Suppose, therefore, that every 100th passenger is weighed and drawn into the sample. This yields a sample data set of n passengers denoted by $y_1, ..., y_k, ..., y_n$. The researcher is faced with the problem of estimating the total weight of passengers using this *sample*, and moreover, of evaluating the precision of the estimate.

In estimating the total weight of passengers, the researcher notes that the sample was drawn from a specific fixed finite population using a particular sampling scheme. Obviously, *systematic sampling* was used, and from the passenger register the total number of passengers on board, N, would be known. An *estimator* for the total weight is easily defined in the form $\hat{t} = N\bar{y}$, where $\bar{y} = \sum_{k=1}^{n} y_k/n$ is the sample mean of the n passenger weights. To assess the sampling error, the standard error of $\hat{t}$ should be estimated as the square root of the variance estimate $\hat{v}(\hat{t})$. To estimate $\hat{v}(\hat{t})$, the researcher uses the textbook variance estimator $\hat{v}_{srs}(\hat{t}) = N^2(1 - n/N)\hat{s}^2/n$, which is for *simple random sampling without replacement*, where $\hat{s}^2 = \sum_{k=1}^{n}(y_k - \bar{y})^2/(n-1)$ is the sample variance of the passenger weights.

The estimates obtained using the above formulae would usually be adequate for practical purposes. But it is instructive to progress further and examine the present estimation problem more closely. Actually, the researcher made a procedure simplifying assumption when estimating the variance of $\hat{t}$ as an estimator from simple random sampling. In fact, the variance formula for systematic sampling would be more complex, because another design parameter, the *intra-class correlation* ω, should be included. The two variance estimators are related by $\hat{v}_{sys}(\hat{t}) = \hat{v}_{srs}(\hat{t})[1 + (n-1)\hat{\omega}]$, where $\hat{v}_{sys}$ is the variance estimator under systematic sampling.

The design parameter ω measures the overall homogeneity of population elements, in the groups of 100 passengers, with respect to the values of the study variable. Unfortunately, the variance estimator $\hat{v}_{sys}(\hat{t})$ is not suitable for practical purposes, since only one element is drawn into the sample from each sampling interval. Therefore, an estimate $\hat{\omega}$ cannot be obtained from the selected sample without auxiliary information on the order in which the passengers step on board, or without making a simplifying model assumption for the process of boarding.

The simplest model assumption would be that the passengers step on board in a completely random order. In this case the intra-class correlation would be zero, because each group of 100 passengers coming on board is assumed to closely mirror the composition of the total passenger population. Then, the variance of $\hat{t}$ estimated from systematic sampling would coincide with that from simple random sampling. By using this simplifying model assumption, we thus implicitly make use of auxiliary information in the design-based analysis, in the form of a superpopulation assumption. For systematic sampling, the alternative ways of making use of auxiliary information, or a model assumption, are examined in Section 2.4. There, it will be shown that proper use of auxiliary information not only simplifies the estimation but can also make the estimation more efficient.

It should be noted that the estimation of the total weight of passengers would not be possible by solely relying on the model-based approach. The main reason is that under the IID approach, a fixed finite population (the passengers on the ferry) is not defined and, therefore, the parameter for the total weight does not exist. But the sample mean $\bar{y}$ can be calculated, and if we use the population size N as auxiliary information we may write the variance estimator of $\hat{t} = N\bar{y}$ in the form $\hat{v}_{iid}(\hat{t}) = N^2 \hat{s}^2/n$, which is also the variance estimator from simple random sampling with replacement.

It is obvious that the target parameters in design-based and model-assisted design-based estimation concern the finite population and that in model-based estimation infinite population parameters are of interest. This conceptual duality appears also in discussion on testing procedures and multivariate analysis in Chapters 7 and 8. There, it will be emphasized that for large finite populations, the difference between a finite-population parameter, such as a regression coefficient, and the corresponding superpopulation parameter is small, and thus inference on a finite-population parameter also constitutes inference on the appropriate infinite-population parameter.

In this and the next chapter, five different sampling techniques are introduced and selected population parameters are estimated with corresponding standard errors under the design-based approach. It will become evident that it is essential to derive appropriate *element weights* specific to each sampling scheme. In the example above, the weights would be equal to N/n for all passengers, i.e. the inverse of the probability of selecting a passenger in the sample. This weight derivation holds, for example, for both simple random and systematic sampling; for more complex schemes, the weights are not necessarily equal for all elements. The estimators and standard error estimators are derived for a given sampling scheme so that the correct weights are incorporated into the equations. Moreover, it will be pointed out to what extent, and how, auxiliary information available on the population can be used with a specific sampling scheme. In addition to the use of auxiliary information in sampling, such information will also be used for certain model-assisted estimators applied to a selected sample for reducing standard errors and to obtain estimates close to the corresponding population values. There, a new type of a weight is derived called the *g weight*. Its value depends on both the selected sample and the chosen model-assisted estimator.

2.1 BASIC DEFINITIONS

The formal framework and basic definitions are now given for Chapters 2–4, and the various sampling schemes are briefly described in relation to their use of auxiliary information.

Population and Variables

A *finite population* $\{u_1, \ldots, u_k, \ldots, u_N\}$ of N elements is considered with elements labelled from 1 to N. For simplicity, let the kth element of the population be represented by its label k, so that the finite population can be denoted by

$$U = \{1, \ldots, k, \ldots, N\}.$$

We denote by y the *study variable* with unknown population values $Y_1, \ldots, Y_k, \ldots, Y_N$. In some cases an additional study variable, x, and an auxiliary variable, z, are also used. The unknown population values of x are denoted by $X_1, \ldots, X_k, \ldots, X_N$. The *auxiliary variable* z represents additional information on the finite population and is usually assumed known for all the N population elements. The known population values of the auxiliary variable are denoted by $Z_1, \ldots, Z_k, \ldots, Z_N$.

Population Parameters

A *parameter* of the finite population U is a function of the population values Y_k of the study variable y; in some cases, the function includes population values X_k of the study variable x. Typical parameters are the **total**, the **ratio** and the **median**. They are defined as follows:

Total $T = \sum_{k=1}^{N} Y_k = Y_1 + Y_2 + \cdots + Y_N$
Ratio $R = T/T_x$, where T_x is the population total of the study variable x
Median $M = F^{-1}(0.5)$, where F is the population distribution function of y

The population total has been chosen because of its importance in survey sampling, most notably by descriptive surveys carried out by statistical agencies publishing *official statistics*. Much of the literature on survey sampling deals with the estimation of population totals. Because the population mean $\bar{Y}$ is a simple transformation of the total, i.e. $\bar{Y} = T/N$, the estimators presented below for totals are equally applicable to means with a few minor changes. Instead of the mean, the median is considered since it is often a more appropriate measure of location, as is the case for the demonstration data used later. The ratio is chosen as a more complicated parameter to estimate, and because it is frequently used in practice. Ratio-type estimators will be extensively used in the survey analyses considered in Chapters 5 to 9.

Sampling Design and Sample

The aim of a sample survey is to estimate the unknown population parameters T, R or M based on a sample from the population U. A sample is a subset of U. There are many different samples that could be drawn. We denote by S the set of all possible samples of size n ($n < N$) from the population U. The actual sample is denoted by $s = \{1, \ldots, k, \ldots, n\}$, so that s is one of the possible samples in the set S. To draw a sample from U a specific sample selection scheme is used. Under a sampling scheme it is possible to state the *selection probability* for a sample s. This probability is denoted as $p(s)$. Formally, the function $p(\cdot)$ is called a *sampling design*. The sampling design determines the statistical properties (expectation and sampling error) of random quantities such as the sample total, sample ratio and sample median calculated from the sample drawn under the actual sampling scheme. In what follows we will use interchangeably the terms sampling scheme and sampling design, although somewhat different definitions have been given for these concepts in the literature. For the purposes of this book the terms are taken to refer roughly to the way in which we draw a sample from the fixed population.

Under a fixed sampling design $p(\cdot)$ an *inclusion probability* is assigned for each population element to indicate the probability of inclusion of the element in the sample. For a population element k the inclusion probability is denoted by π_k. It is also called the *first-order* inclusion probability. Such inclusion probabilities will be extensively used when we introduce the various sampling techniques.

A population element can appear more than once in a sample s if sampling involves *replacement* of the selected element in the population after each draw. Such a sampling design is of a *with-replacement*-type (WR). On the contrary, under *without-replacement*-type sampling (WOR), a population element can appear in a sample s only once. The with-replacement assumption simplifies the estimation under complex sampling designs and is often adopted, although in practice sampling is usually carried out under a without-replacement-type scheme. Obviously, the difference between with-replacement and without-replacement sampling becomes less important when the population size is large and the sample size is noticeably smaller than it.

The study variable y is measured for the elements belonging to the sample s. The n sample values of y are denoted by lower-case letters $y_1, \ldots, y_k, \ldots, y_n$. In some cases, as for the estimation of the ratio R, the data set also includes the measurements x_k, $k = 1, \ldots, n$, of a study variable x. We assume for simplicity that the measurements are free from measurement errors. In addition to the study variables, the data set should include appropriate information on the sampling design, i.e. the *design identifiers* such as stratum and cluster identifiers and a weight variable. These variables are described in detail under each sampling technique to be introduced.

Estimator

An *estimator* of a population parameter refers to a specific computational formula or algorithm which is used to calculate the sample statistics for the selected sample. Estimators that are *unbiased* or *consistent* with respect to the sampling design are usually desired so that the expectation of an estimator equals, or approximates more closely, the population parameter, with increasing sample size n. The following three estimators will be considered:

Total $\hat{t} = \sum_{k=1}^{n} w_k y_k$, where w_k is the element weight
Ratio $\hat{r} = \hat{t}/\hat{t}_x$, where $\hat{t}_x$ is the estimated total of x
Median $\hat{m} = \hat{F}^{-1}(0.5)$, where $\hat{F}$ is the estimated distribution function of y

The observed numerical value obtained by using an estimator for the actual sample is called an *estimate*.

A combination of a sampling design $p(\cdot)$ and an estimator is a *strategy*. This concept will be used especially in the last part of Chapter 3 when discussing model-assisted estimation.

Variance of Estimator

The estimates for a population parameter vary from sample to sample. This sampling error describes the uncertainty of inference based on a particular sample. The sampling error is measured by the variance $V_{p(s)}$ of an estimator. Because $V_{p(s)}$ depends on the sampling design, it is also called the *design variance*. Its value can be estimated from the actual sample by using an appropriate *variance estimator*, which will be denoted by $\hat{v}_{p(s)}$. The square root of a variance estimator is the *estimated standard error* (s.e.) of an estimator.

Strictly speaking, the design variance is only appropriate for unbiased estimators; for biased estimators, a more general measure of sampling error called the *mean square error*, MSE, should be used. The MSE can be expressed as the sum of the design variance and the squared bias of an estimator, where the bias is the deviation of the expected value of an estimator from the corresponding parameter. Generally, in survey estimation, unbiased or approximately unbiased estimators are preferred, so that use of design variances can be justified. This holds also for *finite-population consistent estimators* whose bias decreases with increasing sample size.

Design Effect

Different sampling designs use different design variances of an estimator of a population parameter. A convenient way to evaluate a sampling design is to compare that design variance of an estimator to the design variance from a references sampling scheme of the same (expected) sample size. Usually, simple random sampling with or without replacement is chosen as the reference. For example, for an estimator $\hat{t}$ of the total T, the ratio of the two design variances, called the *design effect* and abbreviated DEFF, is defined by

$$\text{DEFF}_{p(s)}(\hat{t}) = \frac{V_{p(s)}(\hat{t})}{V_{srs}(\hat{t})},$$

where $p(\cdot)$ refers to the actual sampling design. Obviously, obtaining a DEFF requires the values of both design variances. These are rarely available in practice. However, in some instances we will calculate such figures. In practice, an estimate of the design effect is calculated using the corresponding variance estimators for the sample data set. An estimator of the design effect is thus

$$\text{deff}_{p(s)}(\hat{t}) = \frac{\hat{v}_{p(s)}(\hat{t})}{\hat{v}_{srs}(\hat{t})}.$$

More generally, the design effect can be defined for a strategy $\{p(\cdot), \hat{t}^*\}$, where $p(\cdot)$ denotes the sampling design and $\hat{t}^*$ denotes a specified estimator for the total T:

$$\text{DEFF}_{p(s)}(\hat{t}^*) = \frac{V_{p(s)}(\hat{t}^*)}{V_{srs}(N\bar{y})},$$

where $\bar{y} = \sum_{k=1}^{n} y_k/n$ is the sample mean of y. In this DEFF, $\hat{t}^*$ is a design-based or model-assisted estimator of T under $p(s)$ and $N\bar{y} = \hat{t}$ is the SRS (simple random sampling) estimator, and $V_{p(s)}$ and V_{srs} are the corresponding variances. For example, auxiliary variables can be included in the estimator $\hat{t}^*$ of the total.

As a rule, a sampling design is equally as efficient as simple random sampling if DEFF is equal to one, more efficient if DEFF is less than one and less efficient if DEFF is greater than one. The efficiencies of different sampling designs or strategies will be compared by using a design-effect statistic based on either of the definitions given above.

Use of Auxiliary Information in Sampling and Estimation

A *sampling frame*, i.e. a list or register of the population elements from which the sample is drawn, often includes additional information on the structure of the population. This information can be useful in the construction of the sampling design and in improving the efficiency of the estimation for the actual sample. Auxiliary information can also be taken from other sources such as registers, databases and official statistics. To be useful, auxiliary information should be related to the variation of the study variable. Therefore, during the sampling phase, auxiliary information is used primarily in descriptive surveys where a small number of study variables are considered. In addition to descriptive surveys, increasing use is being made of auxiliary information in analytical surveys.

In Chapter 2 simple random sampling, systematic sampling and sampling with probability proportional to size are considered. The use of auxiliary information in these schemes is as follows.

Simple random sampling (SRS)   The sample is drawn without using auxiliary information on the population. Therefore, a simple random sampling scheme provides a reference when assessing the gain from the use of auxiliary information in more complex designs or in improving the estimation.

Systematic sampling (SYS)   Auxiliary information is used in the form of the list order of population elements in the sampling frame. For example, if the values of the study variable increase with the list order, then systematic sampling appears to be more efficient than simple random sampling. *Intra-class correlation,* an additional design parameter in the design variance of an estimator, provides a measure of the correlation between list order and the values of the study variable.

Sampling with probability proportional to size (PPS)   An auxiliary variable z is assumed to be a measure of the size of a population element. Varying inclusion probabilities can be assigned using this auxiliary variable. The magnitude of sampling error depends on the relationship between the study variable y and the auxiliary variable z.

In Chapter 3 auxiliary information will be used in two ways. In stratified sampling and in cluster sampling, auxiliary information is used in the construction of the sampling design. Further, in a given sample, auxiliary information can be used during the estimation phase for poststratification, ratio estimation and regression estimation. The use of auxiliary information in stratified sampling and in cluster sampling is as follows.

Stratified sampling (STR)   The population is first divided into non-overlapping subpopulations called strata, and sampling is executed independently within each stratum. The total sampling error is the sum of the stratum-wise sampling errors. If a large share of the total variation of the study variable is captured by the variation between the strata, then stratified sampling can be more efficient than simple random sampling.

Cluster sampling (CLU)   The population is assumed to be readily divided into naturally formed subgroups called clusters. A sample of clusters is drawn from the population of clusters. If the clusters are internally homogeneous, which is usually the case, then cluster sampling is less efficient than simple random sampling. The *intra-cluster correlation coefficient* is the important design parameter in cluster sampling and it measures the internal homogeneity of the clusters.

These five sampling techniques can be used to construct a manageable sampling design for a complex sample survey, either using a particular method or more usually a combination of methods. In all the schemes, excluding simple random sampling, auxiliary information on the structure of the population is required. Auxiliary information can be used in the construction of the sampling design and, for a given sample, to improve the efficiency. As a rule, sampling error can be decreased by the proper use of auxiliary information. Thus, it is worthwhile to make an effort to collect this type of data.

Further Reading

The following books are suggested as further reading for Chapters 2–4. A classical general overview on sampling techniques can be found in Cochran (1977). The principles of survey sampling are discussed at an intermediate level in Barnett (1991). Särndal *et al.* (1992) considers topics on model-assisted survey sampling by extensively using auxiliary information. Thompson (1992) discusses sampling issues specific to biological, ecological and geological surveys. Methods for handling missing data are summarized in Little and Rubin (1987). Background information on computational algorithms used in the examples in this and the next chapter can be found in Fuller *et al.* (1989).

2.2   THE *PROVINCE'91* POPULATION

In practical survey sampling we are interested in finite populations which are limited in size. Indeed, real populations are generally very large as will be seen later in this book when practical survey samples are analysed. For example, in the Finnish Occupational Health Care Survey to be used in Chapters 6–8, the target population was about 2 million employees in over 100 000 establishments. In a sample of 1542 locations the total sample size was about 17 000 employees. It is obvious that different kinds of errors such as measurement, nonresponse and sampling errors can be present in such a large-scale sample survey.

In the case of real surveys it is not easy to see how the sampling error arises and how the properties of the estimators depend on it. For this reason, we have chosen a more restricted problem and a small finite population in order to demonstrate different sampling schemes and their influence on sampling error. For example, the parameters total, ratio and median of the target population can be calculated exactly and compared to their estimates computed from the appropriate sample. This allows a view of the whole target population. This finite population consists of only 32 population elements from which a sample of fixed size of 8 units is drawn. It is immediately obvious that there is an enormous gap between this demonstration survey and a real sample survey such as the Occupational Health Care Survey. But the demonstration data set can help to clarify such important concepts and issues as how to determine the sampling distribution and how a sampling design affects estimators and their design variances.

To illustrate the main ideas, a small data set under the title *Province'91* has been taken from the official statistics of Finland. This data set will be used as a sampling frame in Chapters 2 to 4. Finland is divided into 14 provinces from which one has been selected for demonstration. This province comprises 32 municipalities and had a total population of 254 584 inhabitants on 31 December 1991. The data set is presented in Table 2.1.

The *Province'91* population contains three kinds of information categorized according to their purpose throughout the survey process. The first phase is sampling design in which identification variables, such as labels, and the ability to identify important subgroups of the population such as strata and clusters, are needed. Here, as the population of elements are municipalities, the name or register number serves as an identifier of an population element. The other two types of information define the study and auxiliary variables.

In the official statistics of Finland municipalities are listed in alphabetical order with urban municipalities in the first group and rural municipalities in the second group. This gives a natural order for a certain sampling technique called systematic sampling and further, allows the population of municipalities to be divided into non-overlapping subpopulations called strata. Another type of population subgroup is formed by combining four

Table 2.1   The *Province'91* population. Percentage unemployment (%UE) and totals of unemployed persons (UE91), labour force (LAB91), population in 1991 (POP91) and number of households (HOU85) by municipality in the province of Central Finland in 1985. Source: Statistics Finland 1991.

ID	LABEL	STR	CLU	%UE	UE91	LAB91	POP91	HOU85
	Urban			**12.67**	**8 022**	**63 314**	**129 460**	**49 842**
1	Jyväskylä	1	1	12.20	4 123	33 786	67 200	26 881
2	Jämsä	1	2	11.07	666	6 016	12 907	4 663
3	Jämsänkoski	1	2	13.83	528	3 818	8 118	3 019
4	Keuruu	1	2	12.84	760	5 919	12 707	4 896
5	Saarijärvi	1	3	14.62	721	4 930	10 774	3 730
6	Suolahti	1	5	15.12	457	3 022	6 159	2 389
7	Äänekoski	1	3	13.17	767	5 823	11 595	4 264
	Rural			**12.63**	**7 076**	**56 011**	**125 124**	**41 911**
8	Hankasalmi	2	5	15.07	391	2 594	6 080	2 179
9	Joutsa	2	6	9.38	194	2 069	4 594	1 823
10	Jyväskylän mlk.	2	7	11.82	1 623	13 727	29 349	9 230
11	Kannonkoski	2	4	18.64	153	821	1 919	726
12	Karstula	2	4	13.53	341	2 521	5 594	1 868
13	Kinnula	2	8	13.92	129	927	2 324	675
14	Kivijärvi	2	8	15.63	128	819	1 972	634
15	Konginkangas	2	3	21.04	142	675	1 636	556
16	Konnevesi	2	5	12.91	201	1 557	3 453	1 215
17	Korpilahti	2	1	11.15	239	2 144	5 181	1 793
18	Kuhmoinen	2	2	12.91	187	1 448	3 357	1 463
19	Kyyjärvi	2	4	11.31	94	831	1 977	672
20	Laukaa	2	5	12.11	874	7 218	16 042	4 952
21	Leivonmäki	2	6	10.65	61	573	1 370	545
22	Luhanka	2	6	10.34	54	522	1 153	435
23	Multia	2	7	11.24	119	1 059	2 375	925
24	Muurame	2	1	9.79	296	3 024	6 830	1 853
25	Petäjävesi	2	7	15.08	262	1 737	3 800	1 352
26	Pihtipudas	2	8	13.02	331	2 543	5 654	1 946
27	Pylkönmäki	2	4	17.98	98	545	1 266	473
28	Sumiainen	2	3	12.80	79	617	1 426	485
29	Säynätsalo	2	1	10.28	166	1 615	3 628	1 226
30	Toivakka	2	6	11.72	127	1 084	2 499	834
31	Uurainen	2	7	16.47	219	1 330	3 004	932
32	Viitasaari	2	8	14.16	568	4 011	8 641	3 119
	Whole province			**12.65**	**15 098**	**119 325**	**254 584**	**91 753**

neighbouring municipalities in a cluster. Thus the total number of clusters is eight. The identification variables STR (stratum) and CLU (cluster) correspond to the urban vs rural and neighbouring municipalities, respectively.

For the following calculations, the *total number of unemployed persons* on 30 November 1991, abbreviated as UE91, is taken as the study variable. Technically, the process is as follows: using a certain sampling technique a fixed-size sample of eight municipalities is selected. From this observed sample a design-based estimate of a parameter of UE91 is calculated, and its efficiency studied, by means of the design-effect statistic. For model-assisted estimation and for sampling proportional to size (PPS), an auxiliary variable from the Finnish Population Census (1985) is selected. This is the *number of households*, abbreviated as HOU85. The reason for taking HOU85 as an auxiliary variable is that it is available from the population register and is highly correlated with the study variable UE91. The frequency histogram for UE91 is displayed in Figure 2.1. Since the distribution is skewed, the mean is not the most appropriate statistic for location and the median has been chosen for further analysis.

Three different types of population parameters are considered: total *T*, ratio *R* and median *M*. The total of UE91 is the number of unemployed persons. The population total is given by

$$T_{ue91} = \sum_{k=1}^{32} Y_k = 15\,098.$$

Another population total is the *size of labour force* LAB91, which can also be calculated from the figures in Table 2.1. This total is given as

$$T_{lab91} = \sum_{k=1}^{32} X_k = 119\,325.$$

Finally, *the total population size* in the *Province'91* population data is 254 584 inhabitants. Totals have long been the main parameter of interest in classical sampling theory and official statistical agencies often produce survey estimates of population totals.

In what follows the total T_{ue91} remains the target parameter which will be estimated under the various sampling techniques. It provides in a single figure the information on how many persons are unemployed in the province under consideration. Because an estimator $\hat{t}$ of the total is a *linear estimator* on the observations, its design variance and the corresponding variance estimator are simple and tractable.

Another interesting population parameter is the unemployment rate in this province. It can be given as the ratio of two totals

$$R = \frac{T_{ue91}}{T_{lab91}} = (15\,098/119\,325) = 0.1265.$$

A more practical expression of the ratio is to express it as an unemployment percentage given by %UE $= 100R = 100 \times 0.1265 = 12.65\%$.

Although the parameter R is simple, the design variance of an estimator $\hat{r}$ of the ratio can be complicated even if the sampling design is not complex. This is because the estimator of the ratio is of a *nonlinear* type and calls for approximations in the derivation of the design variance. In classical sampling theory a ratio estimator refers to ratio estimation; this will be considered in Section 3.3.

The third parameter of interest is the median or 50th percentile of the distribution of municipalities according to the number of unemployed persons. It is obtained by first deriving the population cumulative distribution function (c.d.f.) given by

$$F(y) = \sum_{k=1}^{N} I(y_k \le y)/N$$

where $I(y_k \le y) = 1$ if $y_k \le y$ and zero otherwise. From the c.d.f. of UE91 the population median M is calculated as

$$M = F^{-1}(0.5) = 229.$$

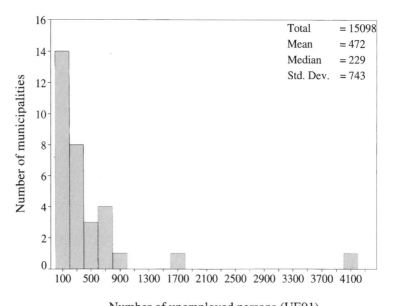

Figure 2.1 Frequency histogram for the number of unemployed persons in 1991 (the *Province'91* population; $N = 32$).

Here, the median has been chosen instead of the mean since the distribution of the number of unemployed persons is very skewed; the mean is $\bar{Y} = 472$. The median estimator $\hat{m}$ belongs to the family of *robust estimators*. These estimators are reasonably unaffected by extreme or outlying observations. However, the derivation of the design variance of the median estimator and the corresponding variance estimator can be cumbersome and requires approximations.

We have defined three population parameters: the total T, the ratio R and the median M. In the *Province'91* population these parameters have clear interpretations. The parameter T measures the total number of unemployed persons in the whole province and the parameter R, multiplied by 100, gives the province's unemployment percentage. The parameter M, the median, gives information on the location of the distribution of unemployed persons and is more appropriate than the mean because of the strongly skewed distribution of UE91.

In the following examples, we will take a sample of $n = 8$ elements from the *Province'91* population using five different sampling techniques. These are simple random sampling (SRS), systematic sampling (SYS), stratified sampling (STR), sampling proportional to size (PPS) and cluster sampling (CLU). Sampling causes sampling error which varies according to the sampling design but the computationally manageable size of the demonstration population will provide an opportunity to analyse the behaviour of the sampling distributions.

2.3 SIMPLE RANDOM SAMPLING AND DESIGN EFFECT

Simple random sampling can be regarded as the basic form of probability sampling applicable to situations where there is no previous information available on the population structure. This sampling technique ensures that each population element has an equal probability of selection, and thus the resulting sample constitutes a 'representative microcosm' of the population.

Simple random sampling serves two functions. Firstly, it sets a baseline for comparing the relative efficiency of other sampling methods. Secondly, amongst the more advanced sampling techniques such as stratified sampling and cluster sampling, simple random sampling can be used as the final method for selecting the elementary or *primary sampling units* and for working out randomization.

Simple random sampling is seen in this section from the viewpoint that sampling a subset from a population always gives rise to sampling variation in computations. A parameter for a fixed and finite population, as for example in the *Province'91* population, the total number of unemployed in the province, is a fixed number ($T = 15\,098$), which is a constant. However, if a sample of 8 municipalities is selected out of this population of 32 municipalities, then naturally the sample estimate of the total number of unemployed in that province is also a constant, but will vary among different samples depending on sample structure. This variation leads to uncertainty in statistical inference, and the way it comes into being is the reason for labelling it a *sampling error.*

However, in actual practice there is only one sample to be analysed. The random variation due to sampling needs to be kept under control in statistical inference, and consequently one has to be familiar with the *sampling distributions* of the estimators of the unknown population parameters. In the case of a simple random sample the sampling distribution is relatively easy to determine, but generally the magnitude of the sampling error is such that it is rarely resorted to exclusively but more advanced sampling designs are used as well.

In the following, simple random sampling is introduced by looking at three sampling techniques: *Bernoulli sampling* (SRSBE), *simple random sampling with replacement* (SRSWR) and *simple random sampling without replacement* (SRS). Instructions for these sampling techniques are given using SPSS for Windows, Release 6.1 syntax. SPSS is chosen because of its popularity in many applications. These instructions have been used to select a sample of eight elements from the *Province'91* population for further analysis. On this basis, sample estimates for three parameters are supplied: total T, ratio R and median M. The estimates are obtained by using the PC CARP program which produces point estimates and appropriate standard error and design-effect estimates.

Finally, the behaviour of the sampling error is examined by simulating 1000 Monte Carlo samples from the *Province'91* population and calculating the mean and variance of this sampling distribution. In the case of an unbiased estimator, the mean of the sampling distribution of the estimator should be equal to the parameter under consideration and the variance of the simulated distribution is expected to be close to the design variance of the estimator. A design variance can be calculated exactly in a fixed and known population as exemplified by the *Province'91* population. The examination of simple random sampling is concluded by presenting design-effect parameters and the corresponding estimates obtained from the actual sample.

Sample Selection

Simple random sampling can be executed by three specific selection techniques: Bernoulli sampling, simple random sampling with replacement and simple random sampling without replacement. In the first method, the sample size is not fixed in advance; in the two other methods it is fixed. Sample selection in both the Bernoulli and without-replacement types of random sampling can be conveniently carried out by a *list-sequential* procedure applied to a database. In the with-replacement type of selection, on the other hand, each separate instance of sampling has to be done by lottery or a *draw-sequential* procedure. All these techniques belong to the class of *equal-probability sampling designs* where the inclusion probabilities are $\pi_k = \pi$, i.e. a constant for all population elements.

Bernoulli sampling (SRSBE)   The selection probability is set first, which in this case is the constant π with regard to all elements so that $0 < \pi < 1$. The value of the constant π is fixed so that the expected or mean sample size is $E(n_s) = \pi N$. In practice the selection is done by appending two variables to the population register; let one variable be PI with the same value or a chosen π for each observation and the other variable EPSN takes a value drawn from a uniform distribution over the range (0,1). The kth population element is included in the sample if EPSN $< \pi$. Following this procedure all the population elements are treated sequentially. This method leads to a variation in sample size with the expected value $E(n_s) = N\pi$ and the variance $V(n_s) = N(1 - \pi)\pi$. This creates estimation problems in small samples, but varying sample size is relatively unimportant in large samples. The relevant SPSS code to select a 25% size Bernoulli sample from the *Province'91* population is given below. Note that an additional variable PI is first appended to comparison with the number EPSN produced by the random-number generator. The selection of an element is achieved by using the command SELECT IF, and the selected sample is saved by the command SAVE. The Bernoulli sample is of a without-replacement type and has expected sample size of eight elements.

```
TITLE Bernoulli sampling SRSBE; PI=0.25, N=32.
GET FILE='input dataset'.
COMPUTE PI=0.25.
COMPUTE EPSN=UNIF(1).
SELECT IF (EPSN LE PI).
SAVE OUTFILE='output dataset'.
```

Simple random sampling with replacement (SRSWR)   Simple random sampling with replacement is based on a selection by lottery from the population by replacing the chosen element in the population after each draw. The probability of the selection of an element remains unchanged after each draw, and any two separately selected samples are independent of each other. This property also explains why this method is used as the default sampling technique in many theoretical statistical studies; the selection method leads to a sample which is close to a pure model-based or IID approach situation in large data sets. Thus, a simple random sample drawn with replacement and an IID data set are often seen as parallel cases. Moreover, because the SRS assumption considerably simplifies the formulae for estimators, especially variance estimators, it is often used as a reference, e.g. in design-effect calculations when working with more complex sampling designs.

In the following SPSS code the lottery is achieved by applying the random-number generator UNIF(32) eight times on a group of natural numbers $1, 2, \ldots, 32$. *Sampling fraction* or *sampling rate* is thus $n/N = 0.25$. Note that a population element can be present several times in the actual sample; these occurrences are recorded in the weight variable W.

```
TITLE Simple random sampling with replacement SRSWR; n=8, N=32.
GET FILE='input dataset'.
COMPUTE L=L+ID.
LEAVE L.
COMPUTE E=L-ID.
NUMERIC W (F2).
COMPUTE W=0.
DO REPEAT A=A1 to A8.
IF (ID=1) A=UNIF (32).
LEAVE A.
IF (E<A AND A(=L) W=W+1.
END REPEAT.
SELECT IF (W GT 0).
SAVE OUTFILE='output dataset'.
```

Simple random sampling without replacement (SRS)   The most common simple random sampling method used in practice is that of simple random sampling without replacement, abbreviated SRS. The probability of the selection of a single element is a constant, but this is related to how far the sampling has progressed, since the probability of selecting an element still present in the population increases with each draw. This causes difficulties in calculating the variance estimators; with-replacement sampling, dealt with earlier, is easier in this respect. A subcommand SAMPLE is available in SPSS for without-replacement sampling. The survey analyst has to know either (a) the population size N and the sample size n or (b) only the sampling rate n/N. Using (a), the SPSS code for SRS from the *Province'91* population is as follows:

```
TITLE Simple random sampling without replacement SRS; n=8, N=32.
GET FILE='input dataset'.
SAMPLE 8 FROM 32.
SAVE OUTFILE='output dataset'.
```

Using this simple code the SRS sample in Table 2.2 was obtained. The sampling rate is again $n/N = 0.25$.

It is noteworthy that this sample could have been produced by any of the three SRS methods, namely, Bernoulli, with replacement or without replace-

Table 2.2 A simple random sample drawn without replacement ($n = 8$) from the *Province'91* population.

Element	Study variables	
LABEL	UE91	LAB91
Jyväskylä	4 123	33 786
Keuruu	760	5 919
Saarijärvi	721	4 930
Konginkangas	142	675
Kuhmoinen	187	1 448
Pihtipudas	331	2 543
Toivakka	127	1 084
Uurainen	219	1 330

Sampling rate $= 8/32 = 0.25$

ment. Even under complex designs, the assumption can be made that the actual sample would be a realization of one of these basic selection techniques. This being the case, simple random sampling without replacement can also be used as the reference when dealing with actual complex designs. The sample just drawn will now be subjected to design-based estimation.

Estimation

Statistical inference generalizes from the sample to the target population, by calculating point and interval estimates for parameters and, further, by performing tests of statistical hypotheses. For the *Province'91* population, the interest focuses on the population total T, the relative proportion $100R\%$ and the median M, with the calculations including point estimates and their standard error estimates reflecting sampling errors. In the case of simple random sampling the design is not complex but can still be used to highlight the essential features when developing design-based estimators, design variances and the estimators for these variances.

When the corresponding estimates have been computed from the sample, the desired confidence intervals can be obtained. Moreover, a statistical test can be performed on the percentage of unemployed in the province. For example, we can test whether the percentage has stayed the same since last year, i.e. $H_0 : 100R\% = 100R_0\% = 9\%$.

Let us introduce the formulae for the estimators $\hat{t}$, $\hat{r}$ and $\hat{m}$ of the total T, the ratio R and the median M, and the corresponding design variance and standard error estimators under simple random sampling without replacement. For the total T we have an estimator $\hat{t}$ given in the standard form by

$$\hat{t} = N\bar{y} = N \sum_{k=1}^{n} y_k/n \qquad (2.1)$$

or the sample mean $\bar{y}$ multiplied by the population size N. The estimator can be expressed as $\hat{t} = \sum_{k=1}^{n} w_k y_k = (N/n) \sum_{k=1}^{n} y_k$, where $w_k = N/n$. The consant N/n is the *sampling weight*, and is the inverse of the sampling fraction n/N. Alternatively, an estimator for the total can be written by first defining the inclusion probability of an population element. Under SRS, the *inclusion probability* of a population element k is $\pi_k = n/N$ or the same constant for every population element. Based on the inclusion probabilities, an estimator of the total can be expressed as a more general *Horvitz–Thompson*-type estimator:

$$\hat{t}_{ht} = \sum_{k=1}^{n} w_k y_k = \sum_{k=1}^{n} \frac{1}{\pi_k} y_k = \frac{N}{n} \sum_{k=1}^{n} y_k. \qquad (2.2)$$

In this case, the estimators $\hat{t}$ and $\hat{t}_{ht}$ obviously coincide, because the inclusion probabilities $\pi_k = n/N$ are equal for each k. The *Horvitz–Thompson*-type estimator is often used, for example, with probability-proportional-to-size sampling where inclusion probabilities vary. The estimator has the statistical property of unbiasedness in relation to the sampling design.

The estimator of the ratio R is the ratio of the estimators of two totals or

$$\hat{r} = \hat{t}/\hat{t}_x, \qquad (2.3)$$

where $\hat{t}_x$ denotes the total of the study variable x. Although both the estimators for totals are unbiased, the estimator $\hat{r}$ of a ratio nonetheless belongs to the class of *biased estimators*. Let us consider more closely the bias of $\hat{r}$.

The bias of $\hat{r}$ is related to the linear regression existing between the two variables, y and x, which takes the form $y = A + Bx$. If the intercept is $A = 0$, then the regression line goes through the origin, which means that the ratio Y_k/X_k is constant among the elements of the population. In this instance the ratio estimator $\hat{r}$ is unbiased, whereas if $A > 0$ the bias amounts to

$$\mathrm{BIAS}(\hat{r}) = E(\hat{r}) - R \dot{=} V_{srs}(\bar{y}) \frac{A}{\bar{Y}^2 \bar{X}},$$

where $V_{srs}(\bar{y})$ denotes the design variance of $\bar{y}$ under the SRS design and $\bar{Y}$ and $\bar{X}$ are the population means of the study variables y and x.

The formula shows that if the constant A is large, the bias is also considerable. On the other hand, with increasing sample size the variance $V_{srs}(\bar{y})$ declines leading to a reduced bias. Therefore $\hat{r}$ is a consistent estimator of R and can be considered more reliable as the sample size increases.

An estimator of the median M can be constructed by first estimating the cumulative distribution function of the study variable at the point y. The *Horvitz–Thompson*-type estimator of the c.d.f. is given by

$$\hat{F}(y) = \sum_{k=1}^{n} w_k I(y_k \leq y)/\hat{N}, \qquad (2.4)$$

where w_k denotes the weight for the kth sample element and $I(y_k \leq y)$ is one if $y_k \leq y$ and zero otherwise. The sum of the weights is $\hat{N} = \sum_{k=1}^{n} w_k$. The estimated c.d.f. is a step function which should first be smoothed to form an estimate $\hat{m}$ of the median M. Since we use PC CARP as the calculating device, the procedure is described only briefly. The smoothed distribution function is constructed by connecting the points $\hat{F}(y)$ with straight lines and the estimated quantiles, including the median, are computed from this. The procedure provides an unbiased estimator for the median.

To determine confidence intervals and test statistics the design variances, or rather the estimators of these variances, are required for the estimators $\hat{t}$, $\hat{r}$ and $\hat{m}$. They are used to estimate the sampling error brought about by the random selection of a sample from the population. Here we derive those variance estimators which are suitable for the single-sample situation. The behaviour of sampling error in more general terms is taken up separately in the context of design variances and sampling distributions of estimators.

An unbiased estimator of the design variance $V_{srs}(\hat{t})$ (see equation (2.7)) of the estimator $\hat{t}$ of the total is given by

$$\hat{v}_{srs}(\hat{t}) = N^2\left(1 - \frac{n}{N}\right)\sum_{k=1}^{n}(y_k - \bar{y})^2/n(n-1) = N^2\left(1 - \frac{n}{N}\right)\hat{s}^2/n, \qquad (2.5)$$

where $\bar{y} = \sum_{k=1}^{n} y_k/n$ is the sample mean and $\hat{s}^2$ is an estimator of the element variance $S^2 = \sum_{k=1}^{N}(Y_k - \bar{Y})^2/(N-1)$.

Variance estimators for the ratio $\hat{r}$ and the median $\hat{m}$ are considerably more complicated since both must be regarded as nonlinear estimators. The approximate variance estimator for the estimator $\hat{r}$ of the ratio is

$$\hat{v}_{srs}(\hat{r}) = \left(1 - \frac{n}{N}\right)\left(\frac{1}{\bar{x}^2}\right)\sum_{k=1}^{n}(y_k - \hat{r}x_k)^2/n(n-1). \qquad (2.6)$$

In developing this variance estimator, the ratio estimator has been *linearized* with the *Taylor series expansion*, and therefore the above equation gives an approximate estimator of the design variance. This technique will be considered in more detail in Chapter 5. The variance estimator of $\hat{m}$ also requires use of the linearization method. This implies that the variance estimator of the median cannot be expected to be very stable, especially for small samples.

Another approximative variance estimator for the median is found in PC CARP. The square root of the variance estimator, i.e. the standard error, for a median is determined as follows. A lower 0.975-level and an upper 0.025-level bound for the smoothed cumulative distribution function are created. The standard error for the pth quantile is a quarter of the horizontal distance at level p between the upper and lower bounds of the smoothed distribution function.

Analysing an SRS Sample

The computation of design-based estimates and their standard errors have been performed in this and the next chapter by using the PC CARP program. PC CARP has been chosen because of its convenience with small samples including a small number of study variables, and in part because its algorithm for the estimation of the variance of a median appeared to be well suited to a small sample. In survey analysis towards the end of the book, the SUDAAN software has been applied since the sampled data sets cover thousands of observations and a large number of study variables, and SUDAAN is well suited to data sets of this kind. With respect to the range and properties of statistical algorithms the two software products are comparable.

In PC CARP, the following *sampling-design identifiers* are required for design-based estimation: stratum identification variable (STR), cluster identification variable (CLU) and sampling weight variable (WGHT). The corresponding variables must be included in the data set to be analysed. In addition to these variables, sampling rates must be supplied under without-replacement sampling. Use of this design information is illustrated under all the sampling techniques to be considered. Our first example is of design-based estimation under simple random sampling without replacement.

Example 2.1

Analysing an SRS sample from the *Province'91* population. We will use PC CARP to produce the estimates of the total, the ratio and the median, and their standard error estimates, from the sample selected earlier under simple random sampling without replacement. First, the design identifiers are appended to the sampled data set. These include the stratum identifier STR which in the case of a simple random sample is a constant for all sample elements, i.e. STR=1. Next, we need to know whether an element belongs to a group of elements or a cluster. In element sampling, each element is a cluster of its own; therefore CLU equals the ID number of the observation. Finally, we enter the weight variable, which in this case is the inverse of the sampling rate. It is used to weight the sample observations in the estimation of the total so that the weights sum to N. In general for the estimation of a total, the weight variable should be scaled such that the sum of the weights equals the population size. In this example the population size is 32 municipalities ($N = 32$) and the selected sample includes 8 municipalities ($n = 8$); therefore, the weight variable is given the value WGHT $= 32/8 = 4$.

For the estimation using PC CARP, the design identifiers have to be the first variables in the data set in the following order: STR, CLU, WGHT. The pre-processing of the data set has to be carried out prior to entry into PC CARP. Moreover, possible missing variable values should be attributed; otherwise the observations which include missing data should be dropped out. As soon as these preliminary steps are completed the data set should resemble Table 2.3. To make the output more readable an alphanumeric variable LABEL has been included and the rest of the variables have been divided into two headlines: 'Sample design identifiers' and 'Study variables'.

It is important under without-replacement-type sampling to provide the sampling rate to account for the *finite population correction* (f.p.c.) in the variance estimators when dealing with small populations. In this example, the sampling rate is $8/32 = 0.25$, and thus the f.p.c. equals $(1 - n/N) = 0.75$. In addition to the point estimates, the output contains standard error estimates, *coefficients of variation*, which is for example for the total c.v.$(\hat{t}) = $ s.e.$(\hat{t})/\hat{t}$, and the deff estimates expressing the design effect. In this case the deff estimates are equal to unity, since the SRS design is also the reference scheme.

The output from PC CARP comprises two parts. The first part is obtained when the specification phase is completed, and provides information on the sample design and on input data. The second one includes output from the actual analysis.

Table 2.3 A simple random sample drawn without replacement from the *Province'91* population ($n = 8$) provided with the sample design identifiers.

Sample design identifiers			Element	Study variables	
STR	CLU	WGHT	LABEL	UE91	LAB91
1	1	4	Jyväskylä	4123	33786
1	4	4	Keuruu	760	5919
1	5	4	Saarijärvi	721	4930
1	15	4	Konginkangas	142	675
1	18	4	Kuhmoinen	187	1448
1	26	4	Pihtipudas	331	2543
1	30	4	Toivakka	127	1084
1	31	4	Uurainen	219	1330

Sampling rate $= n/N = 8/32 = 0.25$

PC CARP output from the problem - specification phase:

```
------------------------------------------------------------------
PC CARP    Version 1.3  Iowa State  University
                Date:   2\21\1993   Time: 17: 10
------------------------------------------------------------------
 Problem Identification
 ------------------------
 SRS
 Variables
 -----------
 Number input is 3
 Intercept generated: YES
 Sample Design Information
 --------------------------
 Stratum ID
 Cluster ID
 Weight
 Input Data
 ------------
 1. C:\<input dataset>
 Input Data Format
 ----------------------
 List directed
 Sampling Rates
 -----------------
 C:\<dataset for sampling rates>
 (F4.2)
 Output to disk file: C:\<output file>
 ------------------------------------------
```

PC CARP output from the analysis phase:

```
------------------------------------------------------------------
 TOTALS     Number of Observations is    8
 Variable Estimate S.E.                 C.V.      DEFF
 UE91   2.64400D+04  1.32823D+04  5.0235D-01  1.0000D+00
 —

 RATIOS          Number of Observations is     8
 Num.\Denom. Estimate      S.E.            C.V.        DEFF
 UE91\LAB91   1.27816D-01   4.08726D-03  3.1978D-02  1.0000D+00
 —

 UNIVARIATE 1
 Quantiles
         Estimate          S.E.       95% Confidence Interval
 0.50  2.2621505D+02  1.4951565D+02 ( 1.41385D+02, 7.39447D+02)
 Empirical Distribution Function F(x)
     x               F(x)            S.E. of F(x)        DEFF
 2.374224D+02   5.0000000D-01   1.6366342D-01   1.0000000D+00
------------------------------------------------------------------
```

It can be noted from the PC CARP output of the problem-specification phase that all three sampling identifiers are given, namely stratum, cluster and weight. The user can thus check that this information matches with the data before running the analysis.

The results of the estimation are interpreted as follows. The point estimate of the total number T of unemployed persons UE91 for the whole province is $\hat{t} = 26\,440$ and the corresponding standard error estimate is s.e.$(\hat{t}) = 13\,282$. On the basis of these two estimates, and by using the standard normal distribution $N(0,1)$ as an approximate distribution for the estimated total, the following 95% confidence interval is obtained for the total number of unemployed persons in the province:

$$\hat{t} - 1.96 \times \text{s.e.}(\hat{t}) < T < \hat{t} + 1.96 \times \text{s.e.}(\hat{t})$$

i.e. $407 < T < 52\,472$, which is so wide as to lack any significance for administrative purposes. We shall see later on how this confidence interval is affected by selecting a more effective sampling scheme in such a way as to produce a smaller sampling error.

On the basis of the output we can conclude that the estimate $\hat{r}$ for percentage unemployment in the province is $100(0.1278)\% = 12.78\%$. Since the standard error estimate s.e. of $\hat{r}$ is available, we can test statistically whether the current unemployment rate R is different from that estimated a year ago: it was then 9%, thus $H_0 : R = R_0 = 0.09$. Using again the normal approximation we have

$$Z = \frac{\hat{r} - R_0}{\text{s.e.}(\hat{r})} = \frac{0.127\,82 - 0.09}{0.004\,09} = 9.25^{***},$$

and we reject the H_0 hypothesis and conclude that the unemployment percentage of the province has gone up significantly during the past year. The significance level is denoted as *** referring to the rejection probability, i.e. the p-value of the test which in this case is less than 0.001.

Selected figures from the PC CARP output are displayed in Table 2.4. In addition to the estimates, the values of the corresponding population parameters T, R and M are supplied for comparison. Differences are evident when comparing the point estimates with their corresponding parameters. The estimate $\hat{t} = 26\,440$ for the total deviates markedly from its parameter value.

Table 2.4 Estimates from a simple random sample drawn without replacement ($n=8$); the *Province'91* population.

Variable	Parameter	Estimate	s.e.	c.v.	deff
Total UE91	15 098	26 440	13 282	0.50	1.00
Ratio (%) UE91/LAB91	12.65	12.78	0.41	0.03	1.00
Median UE91	229	226	150	. . .	1.00

On the other hand, the point estimates for the ratio and the median are near to the corresponding parameters. The deff estimates are ones, because the actual design is also the reference design.

Next, we study the design variances and sampling distributions of the estimators $\hat{t}$, $\hat{r}$ and $\hat{m}$ in greater detail.

Design Variance and Sampling Distribution

Simple random sampling is convenient for demonstrating how different estimators and their variances behave under a certain sampling design and how the sampling error is influenced by the randomization. We examine this behaviour by first calculating the design variances of $\hat{t}$, $\hat{r}$ and $\hat{m}$, denoted V_{srs}, under the SRS design. These variances can be exactly determined for the small fixed population under consideration. However, the *design variance* does not contain all the information on the sampling error; derivation of the sampling distributions of the estimators allows closer examination of the behaviour of the estimators.

Sampling distributions of estimators are often derived by simulating a large number of samples from the population using the given sampling scheme. We have simulated by the *Monte Carlo* method a total of 1000 samples of size eight ($n = 8$) elements from the *Province'91* population under SRS. From each of these samples the estimates $\hat{t}$, $\hat{r}$ and $\hat{m}$ are calculated. The distribution of each estimator constitutes an experimental sampling distribution for that estimator, i.e. the total, the ratio and the median. These distributions provide information about the location and shape of the sampling distribution.

Design variance formulae and the corresponding observed values for the total, ratio and median estimators under SRS using the *Province'91* population are:

Total T: A design variance for $\hat{t}$ is

$$V_{srs}(\hat{t}) = \frac{N^2}{n}\left(1 - \frac{n}{N}\right)\sum_{k=1}^{N}(Y_k - \bar{Y})^2/(N-1) = N^2\left(1 - \frac{n}{N}\right)S^2/n, \qquad (2.7)$$

where $\bar{Y} = \sum_{k=1}^{N} Y_k/N$ is the population mean and $S^2 = \sum_{k=1}^{N}(Y_k - \bar{Y})^2/(N-1)$ is the population variance. The observed design variance is

$$V_{srs}(\hat{t}) = \frac{32^2}{8}\left(1 - \frac{8}{32}\right)743.36^2 = 7283^2.$$

Ratio R: An approximate design variance for $\hat{r}$ is

$$V_{srs}(\hat{r}) \doteq \frac{1}{\bar{X}^2}\frac{1}{n}\left(1 - \frac{n}{N}\right)\sum_{k=1}^{N}(Y_k - R \times X_k)^2/(N-1), \tag{2.8}$$

which gives the observed value

$$V_{srs}(\hat{r}) = \frac{1}{3729^2}\frac{1}{8}\left(1 - \frac{8}{32}\right)315.91^2/(32-1) = 0.005^2.$$

Median M: There are several approximative variances available for the design variance of the median $\hat{m}$. One possibility is to approximate the variance from the cumulative distribution function as follows:

$$V_{srs}[\hat{F}(\hat{m})] = \frac{N-n}{N-1}\frac{1}{n}F(M)(1-F(M)) \doteq \frac{1-n/N}{n}0.25, \tag{2.9}$$

which is very simple because no unknowns are included. It gives

$$V_{srs}[\hat{F}(\hat{m})] \doteq \frac{1-0.25}{8}0.25 = 0.02,$$

which should be rescaled to obtain the design variance of $\hat{m}$ on the ordinary study variable scale. In the *Province'91* population, however, we use the approximate design variance from the Monte Carlo simulations (see Figure 2.4); hence we obtain

$$V_{srs}(\hat{m}) \doteq \hat{v}(\hat{m}_{mc}) = 107^2.$$

Note that the design variances are displayed in terms of squared standard errors to facilitate comparison with the standard error estimates (s.e.) exhibited in the PC CARP output. When comparing the design variance, or standard error, of an estimator to the corresponding estimate from the actual sample, it can be seen that they differ due to sample-to-sample variation. For example, the variance estimate for the total was $\hat{v}_{srs}(\hat{t}) = 13\,282^2$, and the corresponding design variance was calculated as $V_{srs}(\hat{t}) = 7283^2$. The sample estimate considerably overestimates the design variance in this case. For the ratio estimator these figures are $\hat{v}_{srs}(\hat{r}) = 0.004^2$ and $V_{srs}(\hat{r}) = 0.005^2$ which are quite close. Finally, for the median we have $\hat{v}_{srs}(\hat{m}) = 150^2$ and $V_{srs}(\hat{m}) = 107^2$; the sample estimate is again noticeably larger than the corresponding design variance.

For a closer examination of the behaviour of the estimators under simple random sampling without replacement, estimates for the total, ratio and median from Monte Carlo simulations are displayed as histograms in Figures 2.2 to 2.4. The mean of the distribution of a Monte Carlo estimator is expected to coincide with the corresponding population parameter, and the variance should approximate the design variance of the estimator.

The mean of the total estimates is $\hat{t}_{mc} = 15\,049$ which fits well with the corresponding parameter $T = 15\,098$. The variance of the total estimates is $\hat{v}(\hat{t}_{mc}) = 7278^2$, which is close to the design variance $V_{srs}(\hat{t}) = 7283^2$. In this respect the estimator $\hat{t}$ works well.

In a closer examination two peaks are noted in the histogram. The distribution does not seem bell-shaped when referred to the normal distribution which can be used as the reference (the values from the corresponding normal distribution are displayed as dots in the figure). Great discrepancies are noted between the observed and theoretical distributions. This cautions us against basing our inferences on an assumption of a normal distribution. The causes are obvious. The sampling distribution of $\hat{t}$ strongly depends on the distribution of UE91 in the *Province'91* population which is highly skewed in favour of one municipality (provincial capital), where one-third of the total population of the province lives (see Figure 2.1). The population and sample sizes are not large enough to meet the requirements of a normal approximation. Consequently, simple random sampling might not be an appropriate technique for the estimation of the total in this population.

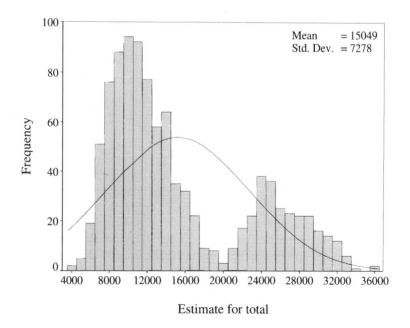

Estimate for total

Figure 2.2   Sampling distribution of the total estimator $\hat{t}$ for UE91 from 1000 Monte Carlo samples taken from the *Province'91* population under an SRS design ($N = 32$ and $n = 8$).

The simulated distribution indicates that the estimator $\hat{r}$ for the ratio UE91/LAB91 works well. The mean of the ratio estimates is $\hat{r}_{mc} = 0.128$, which is almost equal to the population parameter $R = 0.1265$. The variance $\hat{v}(\hat{r}_{mc}) = 0.006^2$ coincides with the design variance, $V_{srs}(\hat{r}) = 0.005^2$. Moreover, the distribution is reasonably bell-shaped, indicating that the normal approximation is better motivated than for the total estimator.

The median M was defined as the 50th percentile of the cumulative distribution function (c.d.f.) of the study variable y. Usually, the c.d.f is unknown and the median should be approximated. Generally used procedure for a median estimate is to arrange the sample values in ascending order $y_{(1)} < \cdots < y_{(k)} < \cdots < y_{(n)}$ and to take the middle value as the median if the sample size is odd, otherwise the median is taken as the mean of the two middle values or $\hat{m} = \frac{1}{2}[y_{(n/2)} + y_{(n/2+1)}]$. This kind of an estimator of a median is often called *50% trimmed mean*. Note that this procedure is used in common statistical packages such as SPSS and SAS.

For a symmetric population the mean and median coincide. The *Province'91* population is heavily skewed as can be seen in Figure 2.1, and therefore the difference between the population mean and median is as great as $\bar{Y} - M = 472 - 229 = 223$. This causes the mean $\hat{m}_{mc}$ of Monte Carlo medians to be

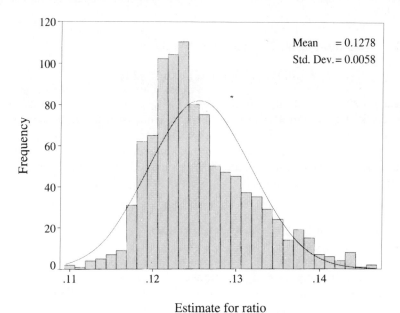

Estimate for ratio

Figure 2.3 Sampling distribution of the ratio estimator $\hat{r}$ for the ratio UE91/LAB91 from 1000 Monte Carlo samples taken from the *Province'91* population under an SRS design ($N = 32$ and $n = 8$).

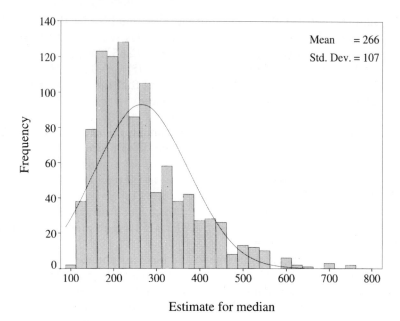

Mean = 266
Std. Dev. = 107

Estimate for median

Figure 2.4 Sampling distribution of the median estimator $\hat{m}$ for UE91 from 1000 Monte Carlo samples taken from the *Province'91* population under an SRS design (N =32 and n =8).

$\hat{m}_{mc} = 266$ indicating moderate bias whose value is

$$\mathrm{BIAS}(\hat{m}) \doteq E(\hat{m}_{mc}) - M = 266 - 229 = 37.$$

This approximative estimator is consistent in the case of a finite population as will be demonstrated in Appendix 2 where we show that, when the sample size increases, the estimator will approach the value of its target parameter.

In PC CARP another algorithm for a median is used in which the unknown c.d.f. is first estimated and then the 50th percentile is determined according to the ordinary definition of a median. This estimator is unbiased.

DEFF and Efficiency of Sampling Design

The design effect was previously defined as the ratio of two design variances where the numerator is design variance of an estimator under the actual sampling design and the denominator is the corresponding design variance under simple random sampling. This definition was originally given by Kish (1965) in which simple random sampling without replacement was taken as the reference. More formally, let the design variance of an estimator, e.g. for the total estimator $\hat{t}$, be $V_{p(s)}(\hat{t})$ under the actual design. The DEFF parameter is obtained as

$$\text{DEFF}(\hat{t}) = \frac{V_{p(s)}(\hat{t})}{V_{srs}(\hat{t})}. \tag{2.10}$$

In the design effect (2.10) it is assumed that the estimator $\hat{t}$ applies to both the actual and reference designs. For more complex actual designs, the DEFF was in Section 2.1 given also by a more general formula that allows a design-based estimator, denoted by $\hat{t}^*$, which differs from the SRS counterpart $\hat{t}$. Moreover, in the Kish definition SRS acts as the reference. In practice this definition is often interpreted more loosely. The reason for this is simply that simple random sampling either with or without replacement tends to lead to the same results if the target population is large and the sampling fraction n/N is small. This is generally the case with large-scale survey sampling. Variance estimators under SRSWR are algebraically simpler than those under SRS, so SRSWR is in this respect more convenient as the reference design. This is emphasized also in software applications for survey analysis. For example, in SUDAAN software an SRSWR variance estimator, weighted if appropriate, is used as the reference in design-effect calculations. In PC CARP, an SRS variance estimator is used if sampling rates are supplied, otherwise SRSWR is used; both are weighted if appropriate.

Obviously, if actual sampling is SRS then DEFF $= 1$. But when using SRSWR as the reference, the DEFF for the total estimator $\hat{t}$ under actual SRS is

$$\text{DEFF}(\hat{t}) = \frac{V_{srs}(\hat{t})}{V_{srswr}(\hat{t})} = \frac{N^2\left(1 - \dfrac{n}{N}\right)S^2/n}{N^2\left(1 - \dfrac{1}{N}\right)S^2/n} = \frac{N - n}{N - 1}$$

which is less than one when $n \geq 2$ and decreases with increasing sample size, indicating that SRS is more efficient than SRSWR.

In practice, the design variance $V_{p(s)}$ and the corresponding SRS (or SRSWR) reference variance of an estimator are estimated from the selected sample. Thus the DEFF must be estimated from the sampled data and for this, the estimates of the variances are used.

It should be noticed that the DEFF and its estimator deff give identical results in the special case where actual sampling is SRS, and the reference is SRSWR. For example, for the total estimate $\hat{t}$ calculated from the sample in Table 2.3, DEFF $= (32 - 8)/(32 - 1) = 0.77$, and the deff estimate is

$$\text{deff}(\hat{t}) = \frac{\hat{v}_{srs}(\hat{t})}{\hat{v}_{srswr}(\hat{t})} = \frac{13\,282^2}{15\,095^2} = 0.77.$$

However, it should be noticed that this equality of DEFF and deff does not hold for other sampling designs.

Design variances and variance estimators of the total, ratio and median were considered under simple random sampling without replacement. For the linear estimator $\hat{t}$ of the total, an analytical design variance was derived yielding basically equal formula for the corresponding variance estimator. For the ratio $\hat{r}$ as a nonlinear estimator, an approximative design variance was derived by the linearization method; the variance estimator also mirrored the design variance. And for the design variance of the robust estimator, the median $\hat{m}$, alternative approximative estimators are available whose suitability, however, varies at least for small samples.

Summary

Simple random sampling was introduced in order to promote familiarity with the most important concepts of estimation under a specific sample-selection scheme. The key statistical concepts appeared at three levels. At the first level are the unknown population parameters of the study variable, such as the total T, the ratio R and the median M, which are to be estimated from a selected sample. At the second level are the estimators of the population parameters, and the design variances of these estimators, including also the design parameters and other characteristics of the sampling distribution of an estimator. The randomization produced by the sampling involves variation in the observed values of the estimators calculated from repeated samples from the population. The design variance is intended to capture this variation which is also reflected in the sampling distribution of an estimator. It appeared that it is beneficial to be aware of the properties of the sampling distribution as a basis for appropriate point and interval estimation and for hypothesis testing. The average leverage of a sampling design is reflected in the design effect DEFF of an estimator which can be calculated if the value of the design variance is available.

In practice, only the sample actually drawn is available for the estimation. Thus, at the third level are the sample estimates of the population parameters, and the estimators of the design variances for obtaining standard error estimates and the corresponding confidence intervals. An important figure is the deff estimate calculated from the sample by using the estimated design variance and the respective variance estimate from the assumed simple random sample.

Covering all three levels, the properties of the estimators of the total, ratio and median were studied in detail for a simple random sample drawn without replacement from the *Province'91* population. The estimator $\hat{t}$ was for the total number T of unemployed persons UE91 in the province, the ratio estimator $100\hat{r}\%$ was for the percentage unemployment rate $100R\%$ in the province, and the median estimator $\hat{m}$ was for the average value M of unemployed persons per municipality. These estimators cover three important families of estimators, namely linear, nonlinear and robust estimators. In this case all the DEFF figures and deff estimates were ones because SRS was also the reference in the design-effect calculations. Under other sampling schemes we will see in later chapters how efficiency varies according to both the estimator and the sampling design, and in many cases deff estimates differing from unity will be obtained.

Finally, it is fair to notice that simple random sampling cannot be taken solely as a simple device for the demonstration of sampling error and other key concepts when discussing the basics of survey sampling, nor as the reference in efficiency comparisons. Simple random sampling can also be included as an inherent part of sampling designs in complex sample surveys; thus it is of a practical value as well.

2.4   SYSTEMATIC SAMPLING AND INTRA-CLASS CORRELATION

Systematic sampling is one of the most frequently used sample selection techniques. A list of population elements or a register serves as the selection frame from which every qth element can be systematically selected. For example, many population registers are alphabetically ordered by family name. The first member is selected at random among the first q elements. The rest of the sample is selected by taking every qth element thereafter down to the end of the list. We have devoted a great deal of space to discussing estimation in a systematic sample, since it presents a good example of the complexities encountered when estimating under a design which involves a certain design parameter in the design variance of an estimator. Here the design parameter is the intra-class correlation (ω). A further complexity arises in the estimation of the design variance; as there is no known analytical variance estimator even for such a simple estimator as the total, we shall derive several approximate variance estimators. In choosing between them, further information on the structure of the target population would be helpful.

Systematic sampling may in some cases be more effective than simple random sampling. This will occur, for example, if there is a certain relationship between the ordering of the frame population and the values of the study variable. The most common cases are those where the population is already stratified or a trend exists that follows the population ordering, or there is a periodic trend; all these situations can also be reached by appropriate sorting procedures. Periodicity may be harmful in some cases, especially if harmonic variation coincides with the sampling interval. Good *a priori* knowledge of the structure of the population is thus beneficial to gaining efficient estimation.

Sample Selection

Let us suppose that a systematic sample of size n elements is desired from a fixed population of N elements. There are two basic ways of selecting the sample. The most common is to draw a single sample of size n with a *sampling interval* of $q = N/n$. Alternatively, two, or more generally m, replicated systematic samples can be taken, each of size n/m elements, the length of the sampling interval being $m \times q$. This method is suitable if variance estimation is to be carried out by the so-called replication techniques.

Let us consider systematic sampling with *one random start*. The first task is to number the elements of the frame population consecutively by $1, 2, \ldots, q, q + 1, \ldots, N - 1, N$, where $q = N/n$ refers to the sampling interval. If q is not an integer, all sampling intervals can be defined as of equal length except one. The selection proceeds as follows. Select a random integer with an equal probability of $1/q$ between 1 and q. Let it be q_o. The sample will be composed of elements numbered $q_o, q_o + q, q_o + 2q, \ldots, q_o + (n - 1)q$, so that one member from each sampling interval is included.

Another selection with one random start can be executed by taking a random integer from the interval $[1, N]$. Let it be Q_o. Starting from Q_o, the selection proceeds forward and backward with steps of the length of the sampling interval q. The composition of the systematic sample will be $\ldots, Q_o - 2q, Q_o - q, Q_o, Q_o + q, Q_o + 2q, \ldots$. Both random start methods lead to the selection of a systematic sample size of n elements, and the methods are equivalent with respect to the estimation.

In replicated systematic sampling *multiple random starts* are used. The intended sample size n is first allocated to the m subsamples so that the sampling interval for each subsample of equal size n/q is $m \times q$. For every subsample, an integer for random start is chosen without replacement from the first sampling interval, and the selection is performed according to the first of the methods introduced above. This procedure gives a set of equal-sized replicate systematic samples comprising n distinct elements in the combined sample.

Example 2.2

Selection of a systematic sample from the *Province'91* population. The following SPSS code selects a systematic sample of size eight ($n = 8$) with the sampling interval $q = 4$ by using one random start. The algorithm includes the SPSS subroutine named MATRIX by which a random integer (INT) between one and four is drawn. In this case INT turned out to be one, and the selected sample is composed of the municipalities whose ordinal numbers are $1, 5, 9, \ldots, 29$. This systematic sample is analysed in Example 2.3.

```
TITLE Systematic sampling SYS.
MATRIX.
COMPUTE RAND=TRUNC(4*UNIFORM(1,1)).
```

```
COMPUTE INT=RAND*MAKE(32,1,1).
SAVE INT/OUTFILE=*/VAR=INT.
END MATRIX.
MATCH FILES FILE='input dataset'/FILE=*.
COMPUTE INDEX=MOD($CASENUM,4).
EXECUTE.
SELECT IF (INDEX=INT).
SAVE OUTFILE='output dataset'.
```

Inclusion Probability

In systematic sampling the number of different samples is quite small. If the sampling interval is $q = N/n$, there will be q separate systematic samples in total. Thus, the selection probability for a sample s is $p(s) = 1/q$. When one element from each sampling interval is included the inclusion probability for kth population element is $\pi_k = 1/q = n/N$, which is the same as the selection probability. The inclusion probability is also equal to that under simple random sampling without replacement. So systematic sampling is also an equal-selection-probability design.

Estimation

The ease of selection of a systematic sample does not continue into the estimation phase. Point estimates for total T, ratio R and median M are still easily calculated by using the corresponding estimators from simple random sampling. But it is not possible to estimate the design variance analytically from the selected sample; approximations have to be used for this purpose. This is the consequence of only one population member being drawn from each sampling interval. Thus no information is available in the sample on the variation within a sampling interval required to analytically estimate the variance. The problem can be illustrated in the estimation of the total T by using the estimator

$$\hat{t} = N \sum_{k=1}^{n} y_k/n, \qquad (2.11)$$

which is the same as equation (2.1) for SRS. Under systematic sampling, the design variance of $\hat{t}$ is given by

$$V_{sys}(\hat{t}) = \sum_{j=1}^{q} (\hat{t}_j - T)^2/q, \qquad (2.12)$$

where $\hat{t}_j$ is the estimator of the total in the jth systematic sample.

The amount of variance depends on the extent to which the q estimates $\hat{t}_j$ of the total vary around the population total T. If each sample closely mirrors the composition of the population, the design variance would be small and thus the estimation of the total would be efficient. But if the sample totals vary, a large variance would be obtained. The problem arises from the division of the total variation between and within the systematic samples. This will be discussed further under intra-class correlation.

The population total T is an unknown in the design variance $V_{sys}(\hat{t})$ of $\hat{t}$ as well as the q systematic sample total estimators $\hat{t}_j$. Because only one sample is selected, this variance is approximated by using one of the alternative, but more or less biased, variance estimators $\hat{v}_{sys}(\hat{t})$. The choice of the approximate variance estimator should be based either on auxiliary information available in the frame population or the use of certain methodological solutions such as sample re-use or selection of replicated systematic samples. Five approximative variance estimators are introduced in equations (2.13) to (2.17).

1. Randomly ordered population   It is often natural to assume that the values of the study variable are in random order in the frame population. If this model is correct, the variance estimator of simple random sampling without replacement, given by

$$\hat{v}_{1.sys}(\hat{t}) \doteq \hat{v}_{srs}(\hat{t}) = N^2\left(1 - \frac{n}{N}\right)\hat{s}^2/n, \qquad (2.13)$$

is unbiased under the actual systematic sample. Although seldom exactly correct, this model seems to be realistic, for example, for population registers if the elements appear alphabetically within it.

2. Implicitely stratified population   The population elements are sorted according to certain criteria. For example, in a population register persons can be listed according to sex so that females occur first listed alphabetically followed by males, also alphabetically. This kind of stratification is called *implicit stratification*. The corresponding approximate variance estimator is based on successive differences $a_i = y_i - y_{i-1}$ and is given by

$$\hat{v}_{2.sys}(\hat{t}) \doteq N^2\left(1 - \frac{n}{N}\right)(1/n)\sum_{i=2}^{n} a_i^2/2(n-1). \qquad (2.14)$$

Alternatively, it is possible to make direct use of the variance estimator of stratified random sampling with proportional allocation by using the equation (2.5) from simple random sampling without replacement in each implicit stratum; hence we get an estimator denoted by $\hat{v}_{2.str}(\hat{t})$.

3. Autocorrelated population   This possibility arises under the superpopulation mechanism which is assumed to generate a correlation ρ_q between each pair of elements of the population that are q units apart. This correlation is similar to the autocorrelation familiar from the analysis of time-series. It is expected that this correlation is positive; if not, some of the other approximations should be used. The autocorrelation coefficient can be estimated from the selected sample and used as a correction factor for the variance estimator $\hat{v}_{srs}$ as follows:

$$\hat{v}_{3.sys}(\hat{t}) \doteq N^2 \left(1 - \frac{n}{N}\right)(\hat{s}^2/n)[1 + 2/\log(\hat{\rho}_q) + 2/(\hat{\rho}_q^{-1} - 1)], \qquad (2.15)$$

where $0 < \hat{\rho}_q < 1$ is the estimated value of the autocorrelation. When the autocorrelation is greater than zero, the term in brackets is less than one and decreases towards zero with increasing $\hat{\rho}_q$. Thus, strong autocorrelation increases the efficiency.

4. Sample re-use   The parent sample is split into two or more equally sized distinct systematic subsamples. The design variance is estimated from the observed variation between the m subsamples as follows:

$$\hat{v}_{4.sys}(\hat{t}) \doteq N^2 \left(1 - \frac{n}{N}\right) \sum_{l=1}^{m} (\bar{y}_l - \bar{\bar{y}})^2/m(m-1), \qquad (2.16)$$

where $\bar{\bar{y}} = \sum_{l=1}^{m} \bar{y}_l/m$ is the mean of the m subsample means. In place of $\bar{\bar{y}}$, the estimate $\bar{y}$ can be used in (2.16). Other sample re-use methods such as bootstrap, jackknife and balanced half-samples are other possible candidates for variance estimation. Sample re-use methods will be discussed in more detail in Chapter 5.

5. Replicated systematic sample   This method resembles the one above where the parent sample is split into two or more subsamples, but here this is done before the sample selection. Selection is performed by drawing without replacement two or more replicated systematic subsamples. The variation between the m subsamples gives an opportunity to estimate the design variance. The formula for the approximate variance is the same as for the previous method, i.e.

$$\hat{v}_{5.sys}(\hat{t}) = \hat{v}_{4.sys}(\hat{t}). \qquad (2.17)$$

All the five variance estimators are approximate and thus their statistical properties depend on the validity of the respective model assumption or on the success of the splitting of parent samples. In the real world there is, of course, no assurance of this. We can, however, evaluate the validity of these variance estimators for the *Province'91* population, since it is possible to calculate the value of the design variance V_{sys} and, therefore, also the intra-class correlation ω as the design parameter.

Example 2.3

Variance approximations under systematic sampling from the *Province'91* populaton. A systematic sample of 8 municipalities $(n = 8)$ from the total of 32 municipalities in the *Province'91* population can be selected in two alternative ways:

(A) The province is divided into eight sampling intervals, each containing four municipalities. A single sample is selected, including, for example, the first municipality from each sampling interval. Thus, the sample size will be eight elements.

(B) The province is divided into four sampling intervals, each containing eight municipalities. Two parallel systematic samples are selected without replacement, one of which includes, for example, the first municipality of each sampling interval and the other, the fifth. The sample is thus composed of two distinct replicated systematic samples of four municipalities, and the total sample size is again eight municipalities.

Both methods produce in this case, the same actual sample. The sampled data is displayed in Table 2.5. Recall from Table 2.1 that the implicit stratification is based on the ordering of the municipalities in the municipality register: densely populated towns are given first, followed by rural municipalities. Systematic sampling through such a register selects municipalities from each stratum in the same proportion that they are found in the stratum. The result of this sampling is the same as stratified sampling using proportional allocation. Stratified sampling will be discussed in more detail in Section 3.1.

All the five approximate variance estimators have been calculated on the basis of the sampled data set. To compute the variance estimate under the stratification assumption the stratum identifiers receive the value STR = 1 if

Table 2.5 A systematic sample from the *Province'91* population (sample design identifiers are given for implicit stratification).

| Sample design identifiers | | | Element | Study variables | |
STR	CLU	WGHT	LABEL	UE91	LAB91
1	1	4	Jyväskylä	4 123	33 786
1	5	4	Saarijärvi	721	4 930
2	9	4	Joutsa	194	2 069
2	13	4	Kinnula	129	927
2	17	4	Korpilahti	239	2 144
2	21	4	Leivonmäki	61	573
2	25	4	Petäjävesi	262	1 737
2	29	4	Säynätsalo	166	1 615

Sampling rates: Stratum 1 = 0.25
Stratum 2 = 0.25

the municipality is a town, or STR $= 2$ for a rural municipality. Similarly, as under simple random sampling, the cluster identifier (CLU) receives the corresponding element-identification value. In proportionally stratified sampling the element weights are constants or, as here, the expansion factor equals WGHT $= 4$ like under simple random sampling. The sampling rate is given for each stratum separately, but even then it is the same figure, 0.25.

The estimation results under implicit stratification are displayed in Table 2.6 in addition to the values of the corresponding parameters. The estimates $\hat{t}$, $\hat{r}$ and $\hat{m}$ are equal to those obtained if simple random sampling estimators is used, but the variance estimates differ. Here, the variance estimator $\hat{v}_{2.str}(\hat{t})$ is used. The deff estimates for the total and the median are considerably smaller than one. Thus the use of implicit stratification in variance approximation under systematic sampling makes these estimates more precise when compared to variance estimators calculated under simple random sampling without replacement. The deff estimate of the ratio, however, is greater than one, indicating, no gain was reached from the approximation.

Let us consider more closely the variance approximations for the total $\hat{t}$. The point estimate for the total T of course remains the same under all the approximations and is $\hat{t} = 23\,580$. There are two variance estimators under the stratification assumption: the one ($\hat{v}_{2.str}$) just computed by PC CARP and the other, $\hat{v}_{2.sys}$, based on successive differences. Put together, the following approximate variance estimates are obtained:

$$\hat{v}_{1.sys}(\hat{t}) \doteq N^2\left(1 - \frac{n}{N}\right)\hat{s}^2/n = 13\,549^2 \qquad\qquad \text{deff} = 1.00$$

$$\hat{v}_{2.sys}(\hat{t}) \doteq N^2\left(1 - \frac{n}{N}\right)(1/n)\sum_{i=2}^{n} a_i^2/2(n-1) = 13\,220^2 \qquad\qquad \text{deff} = 0.95$$

$$\hat{v}_{2.str}(\hat{t}) \doteq \sum_{h=1}^{2} \hat{v}(\hat{t}_h) = 11802^2 \qquad\qquad \text{deff} = 0.76$$

$$\hat{v}_{3.sys}(\hat{t}) \doteq N^2\left(1 - \frac{n}{N}\right)(\hat{s}^2/n)[1 + 2/\log(\hat{\rho}_q) + 2/(\hat{\rho}_q^{-1} - 1)] = 8224^2$$

$$\text{deff} = 0.35$$

$$\hat{v}_{4.sys}(\hat{t}) = \hat{v}_{5.sys}(\hat{t}) \doteq N^2\left(1 - \frac{n}{N}\right)\sum_{l=1}^{m}(\bar{y}_l - \bar{\bar{y}})^2/m(m-1) = 12\,959^2$$

$$\text{deff} = 0.87.$$

Of the approximate variance estimates, the value of $\hat{v}_{1.sys}$, being based upon an assumption of simple random sampling without replacement, is the

Table 2.6 Estimates from a systematic sample drawn from the *Province'91* population using implicit stratification.

Variable	Parameter	Estimate	s.e.	c.v.	deff
Total UE91	15 098	23 580	11802	0.50	0.76
Ratio (%) UE91/LAB91	12.65	12.34	0.33	0.03	1.29
Median UE91	229	198	27	. . .	0.70

largest. The others fall more or less below it. This could indicate that, in this case, systematic sampling is more efficient than simple random sampling. In any case, the estimates clearly show that it is worth modelling the available auxiliary information from the population in order to increase the efficiency. The most efficient approximation method turns out to be autocorrelative modelling which gave the value deff $= 0.35$. This model is based on the assumption of an autocorrelated superpopulation, of which the fixed population constitutes one realization.

The results on variance estimation can be evaluated by studying the properties of the intra-class correlation coefficient ω, which is the single design parameter under systematic sampling, and the efficiency of this sampling scheme. Moreover, it is illustrated how the order in the frame register is related to the value of the intra-class correlation coefficient.

Intra-class Correlation

Systematic sampling is our first example of a design where a design parameter exists. This parameter, called the *intra-class correlation* (ω), will be included in in the design variance V_{sys} of an estimator. The magnitude of the intra-class correlation, and consequently its effect on variance estimates, depends partly on the selected sampling interval and partly upon whether there is a successive system of ordering the study variable's values in the population frame. For simplicity, we consider the estimator $\bar{y}$ of the mean $\bar{Y} = T/N$ and the corresponding design variance $V_{sys}(\bar{y})$. Because the mean $\bar{y}$ is a linear function of the total $\hat{t}$, the estimators $\bar{y}$ and $\hat{t}$ share the same design effect.

Let us begin by writing the population variance of the study variable y in the form

$$\sigma^2 = \sum_{k=1}^{N}(Y_k - \bar{Y})^2/N$$

where $\bar{Y}$ is the population mean. We then decompose the total variance σ^2 into the variance σ_b^2 between the q systematic samples and σ_w^2, the within-sample variance:

$$\sigma^2 = \sigma_b^2 + \sigma_w^2.$$

Under systematic sampling the variance between the sample means is

$$\sigma_b^2 = \sum_{j=1}^{q} (\bar{Y}_j - \bar{Y})^2 / q,$$

where $\bar{Y}_j$ is the mean of the jth sample. The variance within the systematic samples is

$$\sigma_w^2 = \sum_{j=1}^{q} \sum_{k=1}^{n} (Y_{jk} - \bar{Y}_j)^2 / N.$$

By using these variances the intra-class correlation is defined as

$$\omega = 1 - \frac{n}{n-1} \frac{\sigma_w^2}{\sigma^2},$$

or, substituting $\sigma_w^2 = \sigma^2 - \sigma_b^2$,

$$\omega = \frac{\sigma_b^2 - \sigma^2/n}{\sigma^2(n-1)/n}.$$

If the variance between the means is zero, or $\sigma_b^2 = 0$, then the intra-class correlation reaches its minimum of $-1/(n-1)$ and, correspondingly, where $\sigma_w^2 = 0$ it reaches its maximum, or $\omega = 1$.

Further, we can write the variance of the mean estimator in the form

$$V_{sys}(\bar{y}) = \sigma_b^2 = \frac{\sigma^2}{n}[1 + (n-1)\omega]. \qquad (2.18)$$

Because the variance σ^2/n represents the design variance $V_{srswr}(\bar{y})$ of simple random sampling with replacement, only the intra-class correlation needs to be considered to determine when systematic sampling is more efficient than with-replacement-type simple random sampling. For this, we study how the value of the intra-class correlation is involved in the design effect DEFF. From (2.18), the design variance of $\bar{y}$ under systematic sampling can be written as

$$V_{sys}(\bar{y}) = V_{srswr}(\bar{y})[1 + (n-1)\omega].$$

Hence, the design effect is

$$\mathrm{DEFF}_{sys}(\bar{y}) = \frac{V_{sys}(\bar{y})}{V_{srswr}(\bar{y})} = [1 + (n-1)\omega].\qquad(2.19)$$

Systematic sampling compared with simple random sampling with replacement is

(a) more efficient, if $-1/(n-1) < \omega < 0$,
(b) equally efficient, if $\omega = 0$, or
(c) less efficient, if $0 < \omega < 1$.

This can be interpreted to mean that the more heterogeneous the sampling intervals (i.e. negative intra-class correlation), the more efficient systematic sampling will be. Therefore, in systematic sampling the connection between the design parameter ω and the ordering of the frame population should be studied, or at least guessed.

Example 2.4

Intra-class correlation (ω) in the *Province'91* population. We will now calculate the intra-class correlation under systematic sampling from the *Province'91* population, where the mean $\bar{Y}$ of UE91 is to be estimated. The intra-class correlation is calculated in two cases: (a) under systematic sampling involving a single systematic sample of eight elements, and (b) under replicated systematic sampling involving two real replicate samples of four elements. Thus, in case (a) the sampling interval is $q = 4$ and in case (b) the interval is $q = 8$.

(a) A single systematic sample. By drawing the four possible systematic samples the total variation in the population

$$\sigma^2 = \sum_{k=1}^{32}(Y_k - 472)^2/32 = 732^2$$

decomposes into

$$\sigma_b^2 = \sum_{j=1}^{4}(\bar{Y}_j - 472)^2/4 = 169^2$$

and

$$\sigma_w^2 = \sigma^2 - \sigma_b^2 = 732^2 - 169^2 = 712^2.$$

Hence the intra-class correlation is

$$\omega = 1 - \frac{n}{n-1}\frac{\sigma_w^2}{\sigma^2} = 1 - \left(\frac{8}{8-1}\right)\left(\frac{712^2}{732^2}\right) = -0.082.$$

Because the intra-class correlation is negative, systematic sampling will be more efficient in this case than simple random sampling. Thus, the design effect is

$$\text{DEFF}_{sys}(\bar{y}) = 1 + (n-1)\omega = 1 + (8-1)(-0.082) = 0.426.$$

(b) Two replicated systematic samples. By drawing the eight possible systematic samples the variance components are

$$\sigma_b^2 = \sum_{j=1}^{8}(\bar{Y}_j - 471)^2/8 = 319^2$$

and

$$\sigma_w^2 = \sigma^2 - \sigma_b^2 = 732^2 - 319^2 = 659^2.$$

Hence the intra-class correlation is

$$\omega = 1 - \frac{n}{n-1}\frac{\sigma_w^2}{\sigma^2} = 1 - \frac{4}{4-1}\frac{659^2}{732^2} = -0.080.$$

Again, because the intra-class correlation is negative, systematic sampling will be more efficient than simple random sampling. Correspondingly, the design effect is

$$\text{DEFF}_{sys,rep}(\bar{y}) = 1 + (n-1)\omega = 1 + (4-1)(-0.080) = 0.676.$$

The two design effects indicate that the replicated systematic sampling scheme, in which two replicate systematic samples are drawn, is not as efficient as the systematic sampling scheme where a single systematic sample is taken. This result is related both to the length of the sampling interval and the size of a single sample. Thus, the length of the sampling interval affects the performance of systematic sampling. Moreover, similar influence can be found by studying the relationship between the values of the study variable in the frame population and the frame ordering. In practice, the most common situations are: completely random order, implicit stratification, trend, and autocorrelative dependence. In practice, these should be considered separately in each case, as is done below for systematic sampling from the *Province'91* population.

Example 2.5

Implicit stratification and DEFF. In the *Province'91* population, the urban municipalities in the province occur first, followed by the rural municipalities, both in alphabetic order. Thus, the order of the list involves two implicit strata. In the first stratum, there are the urban municipalities which are relatively large in terms of population and, thus, also in number of unemployed. Consequently, there will be a slightly declining trend with order of ID-numbers. It is not unusual to find this kind of implicit stratification and trend in official statistics. The corresponding scatterplot (Figure 2.5) shows the dependence of the study variable UE91 on the register order (ID NUMBER).

The dependence of the values of UE91 on the list order has certain implications for selecting a proper variance estimator.

1. The dispersion figure clearly shows that the successive order is not random, and thus it is not fair to consider this sample as a simple random sample. We found this out earlier when calculating $\text{DEFF}_{sys}(\bar{y}) = 0.426 < 1$. Thus, the SRS design variance V_{srs} would distinctly over-approximate the design variance V_{sys}.

2. The population is ordered successively by stratum in the register. The following stratum sizes and means of UE91 can be calculated for the implicit strata:

Stratum	ID	Size	Mean
1. Urban	1–7	7	1146
2. Rural	8–32	25	283
Whole population	1–32	32	472

Systematic sampling reveals these implicit strata and draws a sample that corresponds to a proportionally stratified sample (STR). If the stratum weights are known, the sample can be analysed as a poststratified sample, as considered in Section 3.3. The design effect under stratified sampling would be

$$\text{DEFF}_{sys,pos}(\bar{y}) = 206^2/224^2 = 0.85,$$

hence this stratification makes estimation efficient, and this can be taken into account by using the approximative variance estimator $\hat{v}_{2.sys}(\hat{t})$ based on successive differences.

3. A linear trend exists between the study variable and identification number that can be modelled by a simple linear regression

$$Y_k = 1070.72 - 36.30 \times \text{ID}_k.$$

The squared multiple correlation coefficient for this model is $R^2 = 0.21$. Using this regression model as auxiliary information in the actual estimation, we could use regression estimation (see Section 3.3). For example, the design effect

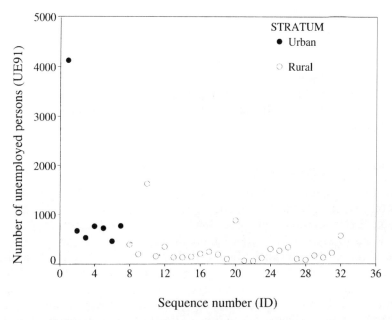

Figure 2.5 Plot of UE91 versus sequence number (ID) for the *Province'91* population. Implicit stratification to two strata is indicated.

under regression estimation would be

$$\text{DEFF}_{srs,reg}(\bar{y}) \doteq 1 - R^2 = 1 - 0.21 = 0.79,$$

which falls in the interval $0.426 < 0.79 < 1$, where 0.426 is the exact design effect for $\bar{y}$ under systematic sampling.

4. The listing order of the municipalities also includes autocorrelative dependence between the successive municipalities. Using the sampling interval $q = 4$ as the lag, the coefficient of autocorrelation turns out to be $\rho_4 = 0.09085$, so that the design effect under this autocorrelation would be

$$\text{DEFF}_{srs,autocor}(\bar{y}) \doteq 157^2/224^2 = 0.49,$$

which is very close to the exact design effect 0.426 under systematic sampling. In the case of an autocorrelated situation, the only disadvantage appears if the frame population contains harmonic variation with a period corresponding to the sampling interval. This was not the situation here.

5. Pre-sorting of the register and efficiency of systematic sampling. Frame registers are usually presented as computer databases which can be sorted by desired variables or by size order. A sorting procedure affects the contents of the sampling intervals, but is not so damaging to efficiency of estimation as might be expected. For example, the *Province'91* population was sorted by the number of unemployed in decreasing order in order to achieve a monotonic

trend. Further, the internal order of the sampling intervals was alternated so that the number of unemployed was decreasing in every second sampling interval and at every other increasing. In this way, we achieved an optimal order of the frame population with respect to systematic sampling. The corresponding design variance is $V_{sys,opt}(\bar{y}) = 165^2$ and DEFF $= 0.405$, which indicates that the advantage of sorting is fairly small in this case. Nonetheless, sorting to achieve certain implicit stratification is often used in large-scale surveys.

Summary

Systematic sampling is easy to accomplish from a frame register and therefore it is very commonly used in practice. The problem, however, is the estimation of the design variance of an estimator under systematic sampling. One solution is to use auxiliary information already available in the frame population. If reasonable, it can be assumed that the population elements are in completely random order in the register and then the estimators of simple random sampling can be used. However, if certain structure such as implicit stratification, trend or periodicity of the study variable is present in the register, it is more efficient to use this information in the estimation, by using the corresponding approximative variance estimator. In our case, the estimates obtained by using these approximative estimators were closer to the exact design variance than those produced by the estimator from simple random sampling, because a certain structure was present in the population. Particularly when working with a large systematic sample, it is worth trying out techniques based on the re-use of the selected sample, leading to other approximative variance estimators. Wolter (1985) offers a more comprehensive study of variance estimation under systematic sampling; he points out that it is worthwhile to try alternative variance estimators in order to select the most appropriate for the situation at hand.

We have dealt rather broadly with systematic sampling because of its popularity in practice, and because it involves an interesting design parameter, i.e. intra-class correlation. The design parameter is not essential as such, but has a particular effect on variance estimation, and thus on the specification of sampling error, confidence limits and sizes of tests. Consequently, the main lines of approximative variance estimation were provided and supplemented by an excursion to model-assisted estimation.

2.5   SELECTION WITH PROBABILITY PROPORTIONAL TO SIZE

Sampling error is partly due to population variance of the study variable which in turn is dependent upon population size. The effect of a large population variance on the efficiency can be reduced by choosing an appropriate sampling technique such as stratified sampling, where separate subpopulations are formed by grouping similar population elements. Sometimes, the variance between elements is so large that stratification is not enough to keep the sampling error within modest limits. Perhaps the population contains a number of elements that have an extremely large value for the study variable, causing this great variance. A more suitable sampling technique in such a case, especially for the estimation of a total, is one in which the inclusion probability depends on the size of the population element. Reduction in variance can then be expected if the size measure and the study variable are closely related. Because this sampling technique is based on inclusion probabilities proportional to relative sizes of the population elements, it is called *sampling with probability proportional to size* (PPS).

In PPS sampling, inclusion probabilities will vary according to the relative sizes of the elements. The size of a population element is measured by an auxiliary positive-valued variable z. It is assumed that the value Z_k of the auxiliary variable is known for each population element k, since the relative size equals the quotient $p_k = Z_k/T_z$, where T_z is the population total of the auxiliary variable or more precisely $T_z = \sum_{k=1}^{N} Z_k$. Commonly used size measures are variables which physically measure the size of a population element. In business surveys, for example, the number of employees in a business firm is a convenient measure of size, and in a school survey the total number of pupils in a school is also a good size measure.

The auxiliary variable z is selected such that its own variability resembles that of the study variable y. More precisely, a size measure z is sought for whose ratio to the value of the study variable is, as close as possible, a constant. That is because the efficiency under PPS depends on the extent that the ratio Y_k/Z_k remains a constant C, for all the population elements. If the ratio remains nearly a constant, then the design variance of an estimator will be small.

In PPS sampling the inclusion probabilities π_k are proportional to the relative sizes $p_k = Z_k/T_z$ of the elements, and the individual weighting of the sampled elements is based on the inverse values of these relative sizes. It is possible to draw a PPS sample either without or with replacement. Calculation of the inclusion probabilities is easier to manage under with-replacement-type sampling. Obtaining these probabilities can be complicated in without-replacement-type PPS sampling because when the first element is sampled, the relative size of the remaining $(N - 1)$ elements is changed and then new inclusion probabilities should be calculated. Various techniques have been developed to overcome this difficulty, and PPS sampling can be very efficient, especially for the estimation of the total, if a good size measure is available.

Sample Selection

A number of sampling schemes have been proposed for selecting a sample with probability proportional to size. The starting point is knowledge of the values of the *auxiliary variable z* for each population element so that probabilities of selection can be calculated. The inclusion probability π_k for a population element k is proportional to the relative size Z_k/T_z. For example, in the trivial case of simple random sampling with replacement, the relative sizes are $p_k = 1/N$ for each k. The quantity $1/N$ is also called the *single-draw selection probability* of a population element k. The inclusion probability of an element for a sample of size n would be $\pi_k = n \times p_k = n/N$. But in PPS sampling, the inclusion probabilities π_k vary and, thus, it is not an equal-probability sampling design in contrast to simple random sampling and systematic sampling.

In practice, the selection of a PPS sample can be based on the relative sizes of the population elements or, alternatively, on the cumulative sum of size measures. The cumulative total for the kth element is

$$G_j = \sum_{k=1}^{j} Z_k, \qquad j = 1, \ldots, N, \quad G_N = T_z.$$

The natural numbers $[1, G_1]$ are associated with the first population element, and the numbers $[G_1 + 1, G_2]$ to the second element; generally, the kth element receives the numbers belonging to the interval $[G_{k-1} + 1, G_k]$. The sample selection process is based on these figures.

We consider five specific selection schemes for PPS sampling. These are *Poisson sampling* which resembles Bernoulli sampling, the *cumulative total method* with replacement or without replacement, *systematic sampling with unequal probabilities* and the *Rao–Hartley–Cochran* method (RHC method; Rao *et al.* 1962). Of these, the cumulative total method with replacement and systematic sampling with unequal probabilities are considered in more detail. In the examples, the variable HOU85 measures the size of a population element. It is register-based and gives the number of households in each population municipality.

Poisson sampling This sampling scheme uses a list-sequential selection procedure. First the inclusion probabilities $\pi_k = n \times Z_k/T_z$ are calculated. Then, let $\epsilon_1, \ldots, \epsilon_k, \ldots, \epsilon_N$ be independent random numbers drawn from the uniform(0,1) distribution. If $\epsilon_k < \pi_k$, then the element k is selected. This procedure is applied to all population elements $k = 1, \ldots, N$, in turn.

Obviously, under Poisson sampling, the sample size is not fixed in advance but is a random variable. The expectation of the sample size is $E(n_s) = \sum_{k=1}^{N} \pi_k$. If a fixed-size sample is not essential for the aim of the investigation, then Poisson sampling can be a reasonable alternative to other

PPS sampling schemes, and it gives good possibilities for calculating the desired survey estimates.

The following SPSS code selects a Poisson sample from the *Province'91* population. The average sample size is eight ($n = 8$) elements, and the sample is essentially of without-replacement type. The random number from the uniform distribution is denoted by EPSN and the inclusion probability of a population element by PI.

```
TITLE Poisson sampling with expected size of 8.
GET FILE='input dataset'.
COMPUTE PI=8*HOU85/91753.
COMPUTE EPSN=UNIF(1).
SELECT IF (EPSN LE PI).
SAVE OUTFILE='output dataset'.
```

PPS sampling with replacement (PPSWR) Sample selection with replacement has its own value in the evaluation of the statistical properties of estimators, since the corresponding design variance formulae are tractable. PPS sampling with replacement is rather like simple random sampling with replacement. The difference between these two methods is due to the way that selection numbers are assigned to population elements. In simple random sampling, a single number from the set of natural numbers $1, \ldots, k, \ldots, N$ is assigned to a population element. In PPS sampling, on the other hand, a corresponding interval from the set of numbers $1, \ldots, G_k, \ldots, G_N$ is assigned to an element, where G_k are cumulative totals.

PPS sampling with replacement is performed by first producing a single random number from the interval $[1, G_N]$. This number is then compared to the numbers associated with the population elements. An element whose selection interval includes this random number will be drawn. The single-draw selection probability of an element is thus $p_k = Z_k/T_z$. The procedure is repeated until the desired number n of draws are completed. Over all the draws, the selection probability is $\pi_k = n \times p_k$. It should be noted that under with-replacement sampling the same population element may be selected several times. This is especially true for those population elements whose size is large because their selection probabilities will also be large.

A selection algorithm for PPS sampling with replacement is given below. The key subroutine in the code is LEAVE, which calculates the cumulative total. The statement A=UNIF(91753) generates a random number from the closed interval $[1, G_N]$. A weight variable W is created to count the repeated occurrences of a population element in the sample.

```
TITLE PPS sampling with replacement (n=8).
GET FILE='input dataset'.
COMPUTE L=L+HOU85.
LEAVE L.
COMPUTE E=L-HOU85.
NUMERIC W(F2).
COMPUTE W=0.
DO REPEAT A=A1 TO A8.
IF (ID=1) A=UNIF(91753).
LEAVE A.
IF (E<A AND A<=L) W=W+1.
END REPEAT.
SELECT IF W>0.
SAVE OUTFILE='output dataset'.
```

PPS sampling without replacement (PPSWOR)   When selecting without replacement a new problem arises concerning the computation of inclusion probabilities. With the selection of the first element, the single-draw probability is exactly $\pi_k = p_k$. When the first sample element has been selected, the single draw probability changes because the total T_z of the remaining $N-1$ elements in the population decreases. Particularly for large samples, the calculation of inclusion probabilities becomes tedious. For that reason numerous alternative sample selection techniques have developed to overcome this difficulty. Some techniques are restricted to the selection of only two elements when the sample size is itself two elements. We will discuss in greater detail a method that is often used in practice and enables the selection of a PPS sample of size two or more elements without replacement.

Systematic PPS sampling (PPSSYS)   This method is the easiest to operate under without-replacement-type selection with probability proportional to size. In this method, the properties of systematic sampling and sampling proportional to size are combined into a single sampling scheme. If the population is randomly preordered, this scheme offers a convenient opportunity to perform the estimation as if the sample were selected as a with-replacement-type PPS sample.

In ordinary systematic sampling, the sampling interval is determined by the quotient $q = N/n$. In systematic PPS sampling, the sampling interval is given by $q = T_z/n$. As in the ordinary one-random-start systematic sampling, we first select a random number from the closed interval $[1, q]$. Let it be q_0. The n selection numbers for inclusion in the sample are hence

$$q_0, \quad q_0 + q, \quad q_0 + 2q, \quad q_0 + 3q, \ldots, q_0 + (n-1)q.$$

The population element identified for the sample from each selection is the first unit in the list for which the cumulative size G_k is greater than or equal to the selection number. Given this method, the inclusion probability of the kth element in the sample is again $\pi_k = n \times p_k$.

In the SPSS code for systematic PPS sampling, the *size measure* is the variable HOU85 whose population total is $T_z = 91\,753$. The sample size is fixed to eight ($n = 8$) elements, and the sampling interval is $q = 91\,753/8 = 11\,469$. For the sample selection, the cumulative sum of HOU85 has been added to the population data set as the variable CUMHOU85. The sample obtained by using this code is displayed in Table 2.7. There, the municipalities are sorted by size.

```
TITLE Systematic PPS sampling PPSSYS (n=8).
GET FILE='input dataset'.
COMPUTE #CHOU85 = #CHOU85 + HOU85.
COMPUTE CUMHOU85 = #CHOU85.
COMPUTE #C=#C+1.
COMPUTE CASE=#C.
COMPUTE #SN=8.
COMPUTE #PN=91753.
COMPUTE #INT=TRUNC(#PN/#SN).
COMPUTE #RAN=TRUNC(UNIFORM(#INT)+1).
DO IF CASE=1.
COMPUTE #COMP=#RAN.
COMPUTE RAN=#RAN.
END IF.
COMPUTE SAMIND=0.
LOOP IF #COMP LE #CHOU85.
+ COMPUTE SAMIND=SAMIND+1.
+ COMPUTE #COMP=#COMP+#INT.
END LOOP.
SAVE OUTFILE ='output dataset'.
```

PPS under the Rao–Hartley-Cochran method (RHC method) The population is first divided into n subpopulations $N_1, N_2, \ldots, N_g, \ldots, N_n$ by using the size measure z so that in subpopulation g the sum T_g of the size measure will be close to T_z/n. There can be varying numbers of elements in the subgroups. Next, one element is drawn from each subpopulation with selection probabilities proportional to size so that for an element k the selection probability is $p_k = Z_k/T_g$. The RHC method is easily managed and suitable for various PPS sampling situations, and is often used in practice. For the demonstration data set, when the grouping into subpopulations has been completed, the previous SPSS code for PPS sampling with replacement can be used to give a sample selected by the RHC method.

Estimation

The estimation should be considered separately under the with-replacement and without-replacement options. Under with-replacement sampling, the single-draw selection probability of an element remains constant (i.e. equal to the relative size p_k of the element). But under without-replacement sampling the selection probabilities of the remaining population elements change after each draw and this causes difficulties, especially in variance estimation. To introduce the basic principles of estimation under PPS sampling, we shall consider the with-replacement case only. And as an approximation, systematic PPS sampling, which will be extensively used in the examples, is also simplified to the with-replacement case.

To construct the estimators, the relative size p_k of population element k is required; by using the size measure Z_k the relative size is

$$p_k = \frac{Z_k}{\sum_{k=1}^{N} Z_k} = \frac{Z_k}{T_z}.$$

The quantity p_k is also the single-draw selection probability for the kth element. The inclusion probability π_k of the element k in an n-element sample is, in turn, written as

$$\pi_k = n \times p_k = n \times \frac{Z_k}{T_z}.$$

The inclusion probabilities should fulfil the requirement $\pi_k \leq 1$. In the trivial case of $n = 1$, this holds true for each population element. When $n > 1$ and some population values Z_k are exceptionally large, the inclusion probabilities for some of these elements may be greater than one, $n \times Z_k / \sum_{k=1}^{N} Z_k > 1$. This conflict can be encountered in practice but fortunately it is solvable. One possibility is to set $\pi_k = 1$ for all those values of k for which $nZ_k > \sum_{k=1}^{N} Z_k$, i.e. to take these elements with certainty. For the remaining elements, π_k is set proportional to the size measure. For example, if only one of the population elements, say the element k', is overly large in this sense, set $\pi_{k'} = 1$, and the inclusion probabilities of the $N - 1$ remaining population elements are

$$\pi_k = (n - 1) \frac{Z_k}{\sum_{k=1}^{N} Z_k - Z_{k'}}, \qquad k \neq k',$$

which assures that the condition $\pi_k \leq 1$ holds. An application of this is shown in Example 2.8.

The two well-known estimators of the total for PPS samples, namely the *Horvitz–Thompson* or the HT estimator, and the *Hansen–Hurwitz* or the HH estimator, are essentially based on these probability quantities. Let us derive these estimators of the total T. Under PPS sampling without replacement, an unbiased *Horvitz–Thompson* estimator of T is given by

$$\hat{t}_{ht} = \sum_{k=1}^{n} \frac{y_k}{\pi_k}, \tag{2.20}$$

where π_k denotes the inclusion probability. For a with-replacement PPS scheme the corresponding HH estimator is given by

$$\hat{t}_{hh} = \frac{1}{n}\sum_{k=1}^{n}\frac{y_k}{p_k} = \frac{1}{n}(\hat{t}_1 + \cdots + \hat{t}_k + \cdots + \hat{t}_n), \qquad (2.21)$$

where each $\hat{t}_k = y_k/p_k$ estimates the total T. An estimator $\hat{r}$ of the ratio R can be derived as a ratio of two *Horvitz–Thompson* estimators, or as a ratio of two *Hansen–Hurwitz* estimators. Further, in the estimation of the median M, the empirical cumulative distribution function is constructed with the inverse inclusion probabilities $1/\pi_k$ as the element weights.

The with-replacement assumption also simplifies the estimation of the design variances. For the estimator $\hat{t}_{hh}$ of the total, the design variance under PPS with replacement is

$$V_{ppswr}(\hat{t}_{hh}) = \frac{N^2}{n}\sum_{k=1}^{N}p_k\left(\frac{Y_k}{Np_k} - \bar{Y}\right)^2 = \frac{1}{n}\sum_{k=1}^{N}p_k(T_k - T)^2, \qquad (2.22)$$

where $T_k = Y_k/p_k$. From (2.22) it can be inferred that if Y_k is strictly proportional to Z_k such that $Y_k/Z_k = C$ holds for each k, then the design variance would be zero–an ideal case rarely met in practice. An unbiased estimator of the variance is given by

$$\hat{v}_{ppswr}(\hat{t}_{hh}) = \frac{N^2}{n(n-1)}\sum_{k=1}^{n}\left(\frac{y_k}{Np_k} - \bar{y}\right)^2 = \frac{1}{n(n-1)}\sum_{k=1}^{n}(\hat{t}_k - \hat{t}_{hh})^2 \qquad (2.23)$$

where $\bar{Y}$ and $\bar{y}$ are the population mean and sample mean of the study variable y, respectively.

We use this variance estimator as an approximation under systematic PPS sampling. Approximative variance estimators can be derived also for the without-replacement case and for the Rao–Hartley–Cochran method but we omit the details here and refer the reader to Wolter (1985). Approximative estimators can also be used with software for survey analysis such as PC CARP by first deriving an appropriate weight variable. This is discussed in the following example.

Example 2.6

Estimation under systematic PPS sampling. A sample of eight $(n = 8)$ municipalities is drawn with systematic PPS sampling from the *Province'91* population such that the number of households HOU85 is used as the size measure z. The cumulative sum over the population is $T_z = 91\,753$, and under systematic PPS sampling the sampling interval would be $q = 91\,753/8 = 11\,469$.

Table 2.7   A systematic PPS sample ($n = 8$) from the *Province'91* population.

Sample design identifiers			Element	Size measure	Study variables	
STR	CLU	WGHT	LABEL	HOU85	UE91	LAB91
2	1	1.000	Jyväskylä	26 881	4 123	33 786
1	10	1.004	Jyväsk.mlk.	9 230	1 623	13 727
1	4	1.893	Keuruu	4 896	760	5 919
1	7	2.173	Äänekoski	4 264	767	5 823
1	32	2.971	Viitasaari	3 119	568	4 011
1	26	4.762	Pihtipudas	1 946	331	2 543
1	18	6.335	Kuhmoinen	1 463	187	1 448
1	13	13.730	Kinnula	675	129	927

Sampling rate: (not used here)

The largest single element 'Jyväskylä' has the value 26 881 for the variable HOU85, which is more than twice the sampling interval. Therefore, the element 'Jyväskylä' would be drawn twice, and the remaining 6 elements would be drawn from the remaining population elements (31). Such a situation is commonly managed in the following way. An element which has a size measure larger than the selection interval is drawn with certainty (but only once). For such a certainty element, the weight and the inclusion probability are one by definition. In this case, therefore, we first take 'Jyväskylä' with certainty, and then draw 7 elements from the remaining 31 population elements by systematic PPS sampling. This results in the following sample of eight ($n = 8$) municipalities. Note that the sample is sorted by the size measure HOU85.

It is important for the estimation under a systematic PPS design to construct a proper weight variable. For a population element k, the weight w_k is calculated using the formula

$$w_k = \frac{1}{p_k \times n} = 91\,753/(Z_k \times n),$$

where Z_k is the value of HOU85 for element k. However, in this case 'Jyväskylä' is an element drawn with certainty, whose weight gets the value one. The element weights of the remaining seven municipalities are calculated by

$$w_k = \frac{1}{p_k \times n} = (91\,753 - 26\,881)/(Z_k \times 7).$$

In the estimation using PC CARP, the other required design identifiers are the stratum identifier STR, which is two for the certainty element and one for the remaining elements. The element identifier is used for CLU, because each

Table 2.8 Estimates under a PPSSYS design ($n=8$); the *Province'91* population.

Variable	Parameter	Estimate	s.e.	c.v.	deff
Total UE91	15 098	15 077	521	0.03	0.0003
Ratio (%) UE91/LAB91	12.65	12.85	0.2	0.02	0.1854
Median UE91	229	134	188	...	2.017

element is taken to be a separate cluster. In addition, the finite-population correction $(1 - \sum_{k=1}^{n} p_k)$ could also be used to make it resemble without-replacement sampling. The estimates in Table 2.8 are produced for the total $\hat{t}_{ht}$, ratio $\hat{r}_{ht}$ and median $\hat{m}_{ht}$ of UE91. For comparison, the values of the corresponding parameters T, R and M are also displayed.

As expected, systematic PPS sampling is very efficient for the estimation of the total. The design-effect estimate for $\hat{t}_{ht}$ is close to zero (deff = 0.03). This results from the strong linear correlation of the size measure HOU85 and the study variable UE91, and is also due to the linearity of the estimator itself. For the estimator $\hat{r}_{ht}$ of the ratio, which is a nonlinear estimator, PPSSYS is still quite efficient but much less so, however, than for the total. And for the robust estimator $\hat{m}_{ht}$ for the median, the design is less efficient than simple random sampling. This is in part caused by the property of PPS sampling that the larger elements tend to be drawn, and these represent the margin rather than the middle part of the distribution of UE91.

Efficiency of PPS Sampling

We discuss the efficiency of PPS sampling in more detail for the estimation of the total T. It can be shown that the PPS design variance $V_{pps}(\hat{t}_{ht})$ of the estimator $\hat{t}_{ht}$ is related to the regression

$$Y_k = \alpha + \beta Z_k + \epsilon_k$$

of the size measure z and the study variable y where ϵ_k is the residual term. The relationship between the residual sum of squares and the population variance is given by

$$\frac{1}{N-1} \sum_{k=1}^{N} (Y_k - \alpha - \beta Z_k)^2 \doteq S^2 (1 - \rho_{yz}^2),$$

where S^2 is the population variance of y and ρ_{yz}^2 is the squared correlation coefficient of the variables y and z. The residual variation is small if the correlation is close to ± 1. Actually, this variance coincides with that considered later under regression estimation. The efficiency of PPS sampling should thus be examined under the above regression model, but strong correlation ρ_{yz} does not alone guarantee efficient estimation, as will become evident.

A simple condition for the efficiency of PPS sampling can be looked for by comparing the variances of the total estimators from SRSWR and PPSWR. It can been shown that

$$V_{srswr}(\hat{t}) - V_{ppswr}(\hat{t}_{ht}) = N^2\text{Cov}(z, y^2/z)/n.$$

Thus, PPS sampling is more efficient than simple random sampling if the correlation of the variable pair $(z, y^2/z)$ is positive. On the other hand, it was previously noted that most efficient PPS sampling occurs if the ratio Y_k/Z_k is a constant, say C for each population element. Then the design variance $V_{ppswr}(\hat{t}_{ht})$ attains its minimum, zero. If we insert $C = Y_k/Z_k$ in the previous covariance term, it is noted that $\text{Cov}(z, y^2/z)$ reduces to the covariance of z and y. Thus, the correlation of z and y^2/z is equal to that of the original variables z and y in this case. We conclude that a necessary condition of PPS sampling being more efficient than simple random sampling with replacement is that the study variable y and the auxiliary variable z are positively correlated in the population. But for a sufficient condition, the ratio Y_k/Z_k should remain constant over the population. These two conditions will be examined more closely in the next example.

Example 2.7

Efficiency of PPS sampling in the *Province'91* population. To evaluate the efficiency of PPS sampling two conditions should be examined. These are the

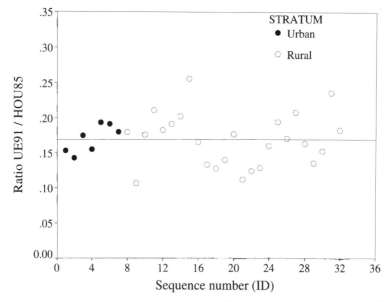

Figure 2.6   Scatterplot of the ratio UE91/HOU85 against sequence number (ID); the *Province'91* population.

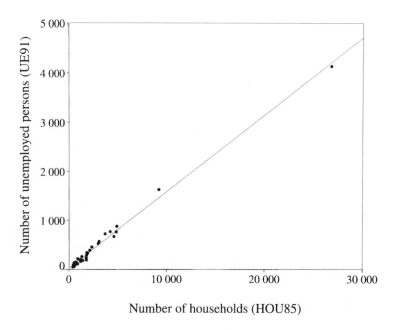

Figure 2.7   Scatterplot of UE91 against HOU85; the *Province'91* population.

stability of the ratio Y_k/Z_k across the population and the regression $Y_k = \alpha + \beta Z_k$ which, for good efficiency, should intercept the y-axis near to the origin. For these purposes two scatterplots from the *Province'91* population are displayed and appropriate coefficients are calculated.

The variation of the ratio Y_k/Z_k in the population is displayed in Figure 2.6. PPS sampling is efficient if the ratio is close to a constant over the population, as is the case here. It can be seen that the towns in the leftmost part (ID ≤ 7) are the largest, and especially amongst these, the ratio Y_k/Z_k is nearly a constant. Under PPS sampling the largest elements tend to be drawn, which means efficient estimation of the total. The same property holds also for the ratio Y_k/X_k if the ratio Y_k/Z_k and the ratio X_k/Z_k are constants.

The correlation of y and z is $\rho_{yz} = 0.997$ (see Figure 2.7). Strong correlation, however, is not sufficient for efficient estimation in a PPS sample. Let us consider the extreme case where this correlation is perfect, i.e. the regression $y = \alpha + \beta z$ holds exactly. Using the usual interpretation of regression coefficients, it can be shown that if α is large, i.e. the regression line intercepts the y-axis far from the origin, then SRSWR is more efficient than PPS. In the *Province'91* population the number of households HOU85 explains 99% of the variation in the number of unemployed UE91 and, moreover, the coefficient α is approximately zero, as can be seen from Figure 2.7.

Summary

Sampling with probability proportional to size provides a practical technique when sampling from populations with large variation in the values of the study variable and often gives a considerable gain in efficiency. The efficiency of PPS sampling depends upon two things. First, efficiency varies considerably according to the type of parameter to be estimated; here these were the total, the ratio and the median. The estimation of the total appeared to be the most efficient. Under PPS sampling an auxiliary size measure (z) must be available and for efficient estimation the size measure should be strongly related to the study variable y. A condition for this is that the variable pair $(z, y^2/z)$ is positively correlated. In the *Province'91* population this condition was satisfied, but this alone cannot guarantee efficient estimation. The ratio Y_k/Z_k must also remain constant over the population. Because this condition was satisfied in the *Province'91* population, PPS provided efficient estimation of the total. The reader who is more interested in PPS sampling is recommended to consult the books of Brewer and Hanif (1983) or Hedayat and Sinha (1991, Chapter 5).

3

Further Use Of Auxiliary Information

Auxiliary information recorded from the population elements can be success-fully used in designing a manageable and efficient sampling design and, after sample selection, to further improve the efficiency of estimators. We previously employed auxiliary information in systematic sampling to select an appropri-ate variance estimator under various assumptions about the listing order of the population frame. In PPS sampling auxiliary information was used in the sampling phase; an appropriate choice of an auxiliary size measure tended to considerably improve efficiency. In Sections 3.1 and 3.2 auxiliary information will be used for stratified sampling and cluster sampling. In both of these techniques, auxiliary information is used to design the sampling scheme; under stratified sampling, the primary goal is to improve the efficiency, whilst in cluster sampling, the practical aspects on sampling and data collection are the main motivation for the use of auxiliary information.

Auxiliary information can be used to improve the efficiency of estimation under the sample already drawn, independently of the sampling design used. A categorical auxiliary variable could be used for poststratification, i.e. stratifi-cation of the sample after selection. If a continuous auxiliary variable is available that is strongly correlated with the study variable, it is possible to improve the efficiency by using ratio estimation or regression estimation. These model-assisted techniques are introduced in Section 3.3. Model-assisted estimators are considered in the context that the finite population under study is a realization from a superpopulation. The use of these techniques can considerably improve the accuracy of estimates, i.e. produce estimates that are close to the corresponding population values and, in addition, decrease the design variances of the estimators.

Auxiliary Information in Stratified Sampling

In *stratified sampling* the target population is divided into non-overlapping subpopulations called *strata.* These are conceptually regarded as separate populations in which sampling can be performed independently. To carry out stratification, appropriate auxiliary information is required in the sampling frame. Regional, demographic and socioeconomic variables are often used as the stratifying auxiliary variables. The efficiency can benefit from stratification, because the strata are usually formed such that similar population elements, with respect to the expected variation in the values of the study variable, are collected together within a stratum. Hence, the within-stratum variation is small.

Information for the stratification can sometimes be inherent in the population. For example, strata are clearly identified if a country is divided into regional administrative areas which are non-overlapping. Separate sampling from each area guarantees the proper representation of different parts of the country in the sample. Auxiliary information of such an administrative type can be used in designing the sampling. Moreover, for example, regional comparisons or comparisons between the strata can also be conducted. Thus, in addition to functioning as a tool for making internally homogeneous sub-populations, stratification can also serve as a classifying variable in the estimation and testing procedures.

Auxiliary Information in Cluster Sampling

In *cluster sampling*, auxiliary information is used to form groups from naturally occurring *clusters* of population elements. Instead of drawing the sample directly from the element population, a sample from the population of clusters is drawn by using the clusters as the primary sampling units (PSU). Subgroups often used in practice are, for example, clusters of employees in establishments, clusters of pupils in schools, and clusters of people in house-holds. For sampling purposes, a frame of the population clusters is needed; however, it is not necessary to have a complete frame covering all the population elements, but only those elements from the sampled clusters. Recognizing the structure of the population reveals the existence of the primary sampling units. Educational surveys, in which the primary sampling unit is usually a school, and a sample of schools is first drawn from a register of schools, are good examples of the use of such a structure. Moreover, the population clusters can be stratified before sample selection. Auxiliary information in cluster sampling therefore concerns not only the grouping of the population elements into clusters but also the properties of the clusters needed if stratification is desired.

In forming clusters of population elements, groups of elements are collected together which often tend to be cluster-wise similar in the various respects relevant to the survey. This intra-cluster homogeneity involves a certain design parameter called the intra-cluster correlation. There are two main approaches which take proper account of the intra-cluster correlation necessary for valid estimation. Firstly, intra-cluster correlation can be taken as a nuisance effect in the estimation with the aim being to remove this disturbance effect from the estimation and testing results. Alternatively, the clustering can be regarded as a structural phenomenon of the population to be modelled. The population is thus seen as having a hierarchical or multi-level structure. In educational surveys, for example, the first level of the structure contains the schools, the second the teaching groups, and the third or lowest level the pupils. Pupils' measured achievements are conditioned by this hierarchical structure. Modelling methods using the multi-level structure share this approach and also presuppose that the corresponding information exists in the data set. Both the nuisance approach and the multi-level approach are discussed: respectively in Chapter 8 and in Section 9.3.

Increasing Precision of Estimates by using Auxiliary Information

There are various approaches available for improving the efficiency of the estimation that take advantage of auxiliary information. For these approaches to be effective, auxiliary variables must be strongly correlated with study variables. Auxiliary variables can be readily included in the frame population register, or can, for this particular purpose, be gathered from other registers. In the simplest case, an improvement in efficiency can be achieved by knowing the population total of the auxiliary variable which can come from a source such as official statistics. By further assuming that the ratio of the values of the study variable and the auxiliary variable is nearly a constant over the population, estimation is considerably more precise than from just simple random sampling. In fact, this kind of ratio is a special case of a linear regression model used in regression estimation. These kinds of use of auxiliary information lead to *ratio estimation* and *regression estimation*, discussed in Section 3.3. There, *poststratification* will also be considered; in that method a categorical auxiliary variable is used in the estimation phase for a selected sample, whereas in stratified sampling such auxiliary variables are used in the sampling phase.

3.1 STRATIFIED SAMPLING

Stratification of the population into non-overlapping subpopulations is another popular technique where auxiliary information can be used to improve efficiency. Such auxiliary information is often available in registers or data bases that provide sampling frames. Typical variables used in stratification are regional (e.g. county), demographic (sex, age group) and socioeconomic (e.g. income group) variables gathered in a census. To fully benefit from the gains in efficiency of stratified sampling, it is important not only to be careful when selecting stratification variables but also to appropriately allocate the total sample to the strata.

There are several reasons for the popularity of stratified sampling:

(1) For administrative reasons many frame populations are readily divided into natural subpopulations that can be used in stratification.
(2) Stratification allows for flexible stratum-wise use of auxiliary information for both sampling and estimation.
(3) Stratification can enhance the precision of estimates if each stratum is homogeneous.
(4) Stratification can guarantee representation of small subpopulations in the sample if desired.

Estimation and Design Effect

In stratified sampling, auxiliary information is used to divide the population into H non-overlapping subpopulations of size $N_1, N_2, ..., N_h, ..., N_H$ elements such that their sum is equal to N. A sample is selected independently from each stratum, where the stratum sample sizes are $n_1, ..., n_h, ..., n_H$ elements, respectively. In stratified sampling, the estimators are usually weighted sums of individual stratum estimators where the weights are stratum weights $W_h = N_h/N$. The strata can be thus regarded as mutually independent subpopulations. An estimator $\hat{t}_{str}$ for a population total T, is a weighted sum of stratum means $\bar{y}_h = \sum_{k=1}^{n_h} y_k/n_h$:

$$\hat{t}_{str} = N \sum_{h=1}^{H} W_h \bar{y}_h = \sum_{h=1}^{H} \hat{t}_h = \hat{t}_1 + \cdots + \hat{t}_h + \cdots + \hat{t}_H, \tag{3.1}$$

where $\hat{t}_h = N_h \bar{y}_h$ is the total estimator in stratum h. If all the stratum totals are unbiased estimates, then the estimator of the population total is also unbiased. Because the samples are drawn independently from each stratum, the design variance $V_{str}(\hat{t}_{str})$ of $\hat{t}_{str}$ is simply the sum of stratum variances $V(\hat{t}_h)$. For example, if simple random sampling without replacement is used in each stratum, the design variance of the estimator $\hat{t}_{str}$ is

$$V_{str}(\hat{t}_{str}) = \sum_{h=1}^{H} V_{srs}(\hat{t}_h), \tag{3.2}$$

whose unbiased estimator is correspondingly

$$\hat{v}_{str}(\hat{t}_{str}) = \sum_{h=1}^{H} \hat{v}_{srs}(\hat{t}_h). \tag{3.3}$$

The design effect DEFF of $\hat{t}_{str}$ heavily depends on the proportion of the total variation given by the division into between- and within-stratum variance components. From the variance equation (3.2) it can be inferred that to benefit from a small design variance, internally homogeneous strata should be constructed which have small within-stratum variances. The efficiency is also affected by the allocation scheme, since the individual stratum variances depend on the respective stratum sample sizes. Let us consider the calculation of DEFF with the estimation of the total T using stratified sampling with *proportional allocation* where stratum sample sizes are $n_h = n \times W_h$ and $n = \sum_{h=1}^{H} n_h$. If the elements are selected with SRS without replacement within each stratum, the estimator $\hat{t}_{str}$ is unbiased for T and

$$V_{str}(\hat{t}_{str}) = N^2(1 - n/N)\sum_{h=1}^{H} W_h S_h^2/n$$

is the design variance of $\hat{t}_{str}$, where S_h^2 is the variance of y in stratum h. Alternatively, the SRS variance $V_{srs}(\hat{t}) = N^2(1 - n/N)S^2/n$ of $\hat{t} = (N/n)\sum_{k=1}^{n} y_k$ can be written in terms of stratified sampling as follows. Assuming large n, we get

$$V_{srs}(\hat{t}) \doteq N^2(1 - n/N)\left[\sum_{h=1}^{H} W_h S_h^2 + \sum_{h=1}^{H} W_h(\bar{Y}_h - \bar{Y})^2\right]/n,$$

where $\bar{Y}_h$ is the population mean in stratum h, and the first term in brackets measures the within-stratum variation and the squared differences $(\bar{Y}_h - \bar{Y})^2$ measure the variation of the stratum means around the grand mean $\bar{Y}$, i.e. the between-stratum variation. The total variance is thus split into the within-stratum and between-stratum variance components. Therefore, the DEFF of $\hat{t}_{str}$ is given by

$$\text{DEFF}_{str}(\hat{t}_{str}) \doteq \frac{\sum_{h=1}^{H} W_h S_h^2}{\sum_{h=1}^{H} W_h[S_h^2 + (\bar{Y}_h - \bar{Y})^2]}, \tag{3.4}$$

or by analogy with analysis of variance:

$$\text{DEFF}_{str}(\hat{t}_{str}) \doteq \frac{\text{WITHIN-STRATUM VARIANCE}}{\text{TOTAL VARIANCE}},$$

where total variance = within-stratum variance + between-stratum variance. Thus the DEFF of $\hat{t}$ under STR with proportional allocation is always less than or equal to 1. Disproportional allocation, which is often more appropriate in practice, can further reduce the DEFF, and this will be demonstrated.

Allocation of Sample

Allocation provides a tool for determining the number of sample units to be taken from each stratum under the constraint that the total number of units to be sampled is n. The modest target is to find an allocation scheme which enables efficient and unbiased estimation. Outlines will be given for achieving optimal allocation under the rather restricted situation of a descriptive survey with one study variable. It should be noted, however, that in a large-scale analytical survey it is impossible to reach global optimality for the allocation with a stratified sampling design, because, generally, numerous study variables are present.

Optimality of the allocation depends on the stratum sizes and, more generally, on the share of the total variance of the study variable to the between-stratum and within-stratum variances. Of the many methods of allocation suggested in the literature, *optimal* or *Neyman allocation*, and *power* or *Bankier allocation* will be considered, in addition to proportional allocation.

1. *Proportional allocation*   This is the simplest allocation scheme and, this as such, widely, used in practice. It presupposes a knowledge of the stratum sizes only, since the sampling fraction n_h/N_h is constant for each stratum. The number of sample elements n_h in stratum h is given by

$$n_{h,pro} = n \times \frac{N_h}{N} = n \times W_h.$$

Proportional allocation guarantees an equal share of the sample in all the starta, but can produce less efficient estimates than generally expected.

As the sampling fraction is a constant n/N in each stratum, the inclusion probability of any population element k is also a constant $\pi_k = \pi = n/N$. The scheme therefore provides an equal-probability sampling design equivalent to that of SRS. This property simplifies the estimation because then

$$\hat{t}_{str} = \hat{t} = N \sum_{h=1}^{H} \sum_{k=1}^{n_h} y_{hk}/n,$$

so the within-stratum means need not be calculated. For this reason a proportionally allocated sample has the property of *self-weighting*. This property is not present in the other allocation schemes where the inclusion probabilities vary between strata.

2. *Optimal or Neyman allocation*   This can be used if S_h, the standard deviations for individual strata of the study variable, are known. The number of

sample units n_h in stratum h under optimal allocation is given by

$$n_{h,opt} = n \frac{N_h S_h}{\sum_{h=1}^{H} N_h S_h}.$$

In practice, S_h is rarely known, but from experience gained in past surveys close approximations to the true standard deviations may be made. In optimal allocation, a stratum which is large or has a large within-stratum variance gets more sampling units than a smaller or more internally homogeneous stratum. This type of allocation provides the most efficient estimates under stratified sampling.

3. *Power allocation*   This is suggested for surveys where there are numerous small strata and also a need for precise estimates at each stratum level. For example, under power allocation the n_h that are required to efficiently estimate stratum totals are given by

$$n_{h,pow} = n \frac{(T_{zh})^a \text{c.v.} (y_h)}{\sum_{h=1}^{H} (T_{zh})^a \text{c.v.} (y_h)},$$

where T_{zh} is the stratum total of an auxiliary variable z and c.v. $(y_h) = S_h/\bar{Y}_h$ is the coefficient of variation of y in stratum h. The constant a is called the power of allocation and in practice, a suitable choice of a may be $\frac{1}{2}$ or $\frac{1}{3}$. This choice can be viewed as a compromise between the Neyman allocation and an allocation which leads to approximately constant precision for all strata.

Example 3.1

Optimal allocation under stratified simple random sampling in the *Province'91* population. The population is first divided into two strata, one urban and the other rural. Of all the municipalities, seven $(N_1 = 7)$ are towns and the remainder $(N_2 = 25)$ are rural districts. A stratified simple random sample of eight $(n = 8)$ municipalities is drawn, and the appropriate stratum sample

Table 3.1   Auxiliary information from the *Province'91* population.

Parameter	Variable	Stratum 1	Stratum 2	All
Mean	UE91	1146	283	472
Total	UE91	8022	7076	15 098
Standard deviation	UE91	1318	331	743
Coefficient of variation	UE91	1.150	1.170	1.572
Total	HOU85	49 842	41 911	91 753

sizes are calculated under (a) proportional, (b) optimal and (c) power allocation schemes. For the power allocation scheme, the HOU85 is the auxiliary variable z. Certain background information for the strata is displayed in Table 3.1.

From Table 3.1, n_h for each stratum under various allocation schemes can be calculated.

(1) Proportional allocation:

$$n_{h,pro} = n\frac{N_h}{N} \longrightarrow \left\{ \begin{array}{l} n_1 = (8)\dfrac{7}{32} = 1.75 \\[2mm] n_2 = (8)\dfrac{25}{32} = 6.25 \end{array} \right\}$$

(2) Optimal allocation:

$$n_{h,opt} = n\frac{N_h S_h}{\sum_{h=1}^{H} N_h S_h} \longrightarrow \left\{ \begin{array}{l} n_1 = (8)9226/(9226+8275) = 4.22 \\[2mm] n_2 = (8)8275/(9226+8275) = 3.78 \end{array} \right\}$$

(3) Power allocation with $a = 1$, $a = \frac{1}{2}$ and $a = 0$:

$$n_{h,a=1} = n\frac{T_{zh}\,\text{c.v.}\,(y_h)}{\sum_{h=1}^{H} T_{zh}\,\text{c.v.}\,(y_h)} \longrightarrow \left\{ \begin{array}{l} n_1 = (8)\dfrac{49842 \times 1.150}{57318+49036} = 4.31 \\[2mm] n_2 = (8)\dfrac{41911 \times 1.170}{57318+49036} = 3.69 \end{array} \right\}$$

$$n_{h,a=1/2} = n\frac{\sqrt{T_{zh}}\,\text{c.v.}\,(y_h)}{\sum_{h=1}^{H} \sqrt{T_{zh}}\,\text{c.v.}\,(y_h)} \longrightarrow \left\{ \begin{array}{l} n_1 = (8)\dfrac{\sqrt{49842} \times 1.150}{257+240} = 4.13 \\[2mm] n_2 = (8)\dfrac{\sqrt{41991} \times 1.170}{257+240} = 3.86 \end{array} \right\}$$

$$n_{h,a=0} = n\frac{\text{c.v.}\,(y_h)}{\sum_{h=1}^{H} \text{c.v.}\,(y_h)} \longrightarrow \left\{ \begin{array}{l} n_1 = (8)\dfrac{1.150}{1.150+1.170} = 3.97 \\[2mm] n_2 = (8)\dfrac{1.170}{1.150+1.170} = 4.03 \end{array} \right\}$$

These calculations lead to the following results. With proportional allocation the individual stratum sample sizes are $n_1 = 2$ and $n_2 = 6$, whilst with the optimal and power allocations $n_1 = n_2 = 4$. Note that the so-called *equal allocation*, in which the sample sizes in each stratum are equal ($n_h = n/H$), also gives $n_1 = n_2 = 4$. The efficiency of optimal allocation and power allocation over proportional allocation can be inferred from the corresponding DEFF-values, which are 0.44 (for optimal and power allocation) and 0.74 (for proportional allocation).

Sample Selection

Sample selection is carried out independently in each stratum, which provides an opportunity to use different selection schemes in different strata. However, for convenience the same selection scheme often is used. Many standard computer packages such as SPSS, SAS and BMDP include subroutines for simple random sampling, or else relevant programming can be easily written for such samples, as was illustrated in Section 2.3. In STR sampling, the total population should be allocated to the appropriate strata and then an SRS sample is selected in each stratum. Or, alternatively, systematic sampling or PPS sampling can be applied to individual strata.

Stratified sampling is often an essential part of a sampling design in a large-scale survey. In complex surveys, the population of clusters is usually stratified. In the Mini-Finland Health Survey (see Section 5.1), for example, the regional clusters were stratified by whether they were urban or rural and the shares of the population in manufacturing industry and agriculture, and PPS sampling was used in sampling of the clusters. In the Finnish Health Security Survey (see Section 9.2), regional stratification of the population households was used, and in the Second National Assessment of the Comprehensive School (see Section 9.3) the population of schools was stratified into regional strata, whilst in the Occupational Health Care Survey (see Section 6.1) there were two types of stratification criteria: size and type of industry of establishment. In these surveys, systematic sampling of the clusters was used and moreover, in the Mini-Finland Health Survey and Occupational Health Care Survey, the sampling at the element-level also was based on systematic sampling. Let us next consider stratified sampling under a more restricted sampling situation from the *Province'91* population.

Example 3.2

Stratified simple random sampling from the *Province'91* population using optimal allocation. The demonstration population is divided into two strata, rural and urban municipalities. The allocation scheme is the optimal method which leads to equal stratum sample sizes $n_1 = n_2 = 4$ when the population total T is estimated, as previously shown. Under this allocation, a stratified simple random sample is selected (Table 3.2). Once the sample is drawn, the relevant design identifiers should be added to the data set as new variables (STR, CLU and WGHT) and used in the estimation procedure provided by PC CARP. The three estimation problems are considered as before. The estimator $\hat{t}_{str}$ of the total number of unemployed persons UE91 demonstrates clearly how stratification decreases the standard error in this case (Table 3.3). A similar effect is noted also for the ratio estimator $\hat{r}$ of the unemployment rate UE91/LAB91. For the third estimator $\hat{m}$, the median of the population distribution of UE91, no gain is achieved by using the stratification and optimal allocation.

Table 3.2 An optimally allocated stratified simple random sample from the *Province '91* population.

Sample design identifiers			Element	Study variables	
STR	CLU	WGHT	LABEL	UE91	LAB91
1	1	1.75	Jyväskylä	4 123	3 3786
1	2	1.75	Jämsä	666	6016
1	4	1.75	Keuruu	760	5919
1	6	1.75	Suolahti	457	3022
2	21	6.25	Leivonmäki	61	573
2	25	6.25	Petäjävesi	262	1737
2	26	6.25	Pihtipudas	331	2543
2	27	6.25	Pylkönmäki	98	545

Sampling rates: Stratum 1 = 4/7 = 0.57
                    Stratum 2 = 4/25 = 0.16

The stratum identifier has the value $STR = 1$ for a town and $STR = 2$ for a rural municipality. Cluster identifier CLU refers to the groups of elements; here each cluster contains a single element and the ID number of each municipality is chosen as the cluster identifier. The weight variable has to be calculated for each stratum separately from the stratum size and stratum sample-size figures. The weight WGHT for the first stratum is $w_{1k} = N_1/n_1 = 7/4 = 1.75$ are for the second $w_{2k} = N_2/n_2 = 25/4 = 6.25$. In addition, for simple random sampling without replacement, PC CARP requires sampling rates for each stratum, given by $4/7 = 0.57$ for the first stratum and $4/25 = 0.16$ for the second.

The estimation results with the values of the corresponding population parameters are shown in Table 3.3. The point estimates $\hat{t}_{str}$ and $\hat{r}_{str}$ for the total and the ratio are close to the values of the population parameters T and R. However, for the median the estimate $\hat{m}_{str} = 177$ deviates considerably from the true median $M = 229$. The optimally allocated stratified SRS design seems to be very efficient for the estimation of the total and the ratio in this case, with design-effect estimates $\text{deff}(\hat{t}_{str}) = 0.21$ and $\text{deff}(\hat{r}_{str}) = 0.38$. However, the estimation of the median is less efficient than under the unstratified SRS design, because $\text{deff}(\hat{m}_{str}) = 1.68$ is considerably greater than one.

Table 3.3 Estimates from an optimally allocated stratified simple random sample (n=8); the *Province'91* population.

Variable	Parameter	Estimate	s.e.	c.v.	deff
Total UE91	15 098	15 211	4 286	0.28	0.21
Ratio (%) UE91/LAB91	12.65	12.78	0.3	0.02	0.38
Median UE91	229	177	64	. . .	1.68

Summary

A small population split into two strata was considered with various allocation schemes. Stratification with proportional allocation yielded the sample size $n_1 = 1.75$ in the first stratum (towns), which is equal to the expected number of elements under unstratified SRS. Note that at least two elements per stratum are required to estimate the within-stratum variance. Stratification appeared useful in this case, since by using other allocation schemes more elements could be drawn from the first stratum. Stratification can be used to avoid under-representation of a smaller stratum, as shown from the example when optimal, power and equal allocation schemes were used.

In the estimating of the total, ratio and median, stratified sampling with optimal allocation produced deff estimates that indicated gain in efficiency for the total and ratio estimators; however, the estimated median had a deff estimate greater than one. Generally, the gain of precision attained in stratified sampling depends on the stratification scheme and on the allocation of the sample between the strata. It was demonstrated that under proportional allocation the design effect will be always less than one when a population total is estimated.

Stratification provides a powerful tool for improving the efficiency and, being suitable for various sampling situations, it is commonly used in practice. In addition to element sampling, stratified sampling is often present in sampling designs for complex surveys where the population of clusters is stratified.

3.2 CLUSTER SAMPLING

In complex surveys naturally formed groups of population elements such as households, villages, city blocks, schools, establishments or other are often used for sampling and data collection. For example, a household can be chosen as the unit of data collection in an interview survey. In addition to the original person-level population there is the additional population of households. Assuming that a suitable frame is available, a sample of households is drawn for the interviewing of the sample household members. This is an example of *one-stage cluster sampling*. If a household population frame is not available but a block-level frame is, a sample from the register of blocks can be drawn, and a sample of households can then be drawn from the sampled blocks by using lists of dwelling-units prepared from only the selected blocks. This is an example of *two-stage cluster sampling*.

Cluster sampling in social, health, and other surveys is motivated by the need for practical, economic, and sometimes also administrative efficiency. An important advantage of cluster sampling is that a sampling frame at the element level is not needed. The only requirements are for cluster-level sampling frames and frames for subsampling elements from the sampled clusters. Cluster-level frames are often easily accessible, for example, for establishments, schools, blocks or block-like units etc. Moreover, these existing structures provide the opportunity to include important structural information as a part of the analysis. For instance, in an educational survey it is practical to use the information that pupils are clustered within schools and further clustered as classes or teaching groups within schools. Schools can be taken as the population of clusters from which a sample of schools is first drawn and then a further sample of teaching groups can be drawn from those schools which have been sampled. If all the pupils in the sampled teaching groups are measured then the design belongs to the class of two-stage cluster-sampling designs. And in addition to sample selection and data collection, the multi-level structure can be used in the analysis, e.g. for examining differences between schools.

Thus, in *multi-stage sampling*, a subsample is drawn from the sampled clusters at each stage except the last. At this stage, all the elements from the sampled clusters can be taken in an element-level sample, or a subsample of the elements can be drawn. One-stage and two-stage cluster sampling are discussed in this chapter and demonstrated using the *Province'91* population. A more general setting for cluster sampling, also covering stratification of populations of clusters, will be demonstrated by various real surveys in Chapters 5 to 9.

The economic motivation for cluster sampling is the low cost of data collection per sample element. This is especially true for populations that have a large regional spread. Using cluster sampling, the travelling costs of interviewers can be substantially reduced as the workload for an interviewer can be regionally planned. The *cost efficiency* of cluster sampling can therefore be high. But there are also certain drawbacks in cluster sampling which concern statistical efficiency. If each cluster closely mirror the population structure we would attain efficient sampling such that standard errors of estimates would not exceed those of simple random sampling. However, in practice, clusters tend to be internally homogeneous, and this homogeneity increases standard errors and thus decreases *statistical efficiency*. We shall consider this more closely by considering the intra-cluster correlation. This concept will be used extensively in later chapters when analysing real data sets from cluster sampling using two approaches: by taking the intra-cluster correlation as a nuisance effect and by multi-level modelling methods.

Cost Efficiency in Cluster Sampling

Let us first use a simple case to illustrate the cost efficiency of cluster sampling relative to simple random sampling with replacement (SRSWR). The cost efficiency of cluster sampling can be assessed by a simple cost function C

$$C_{clu} = c_1(a) + c_2(a \times B),$$

where

C	is total sampling costs,
c_1	is sampling costs for a cluster,
c_2	is sampling costs for an element in a cluster,
B	are elements in a cluster (equal-sized clusters),
a	is number of sample clusters,
$n = a \times B$	is element sample size.

Under SRSWR the cost function is

$$C_{srswr} = c_1 n + c_2 n,$$

where n is the element sample size.

The constraint of equal total sampling costs $C = C_{clu} = C_{srswr}$ requires the following sample sizes for SRSWR and CLU sampling:

$$n_{srswr} = \frac{C}{c_1 + c_2}$$

$$n_{clu} = \frac{C}{(1/B)c_1 + c_2},$$

indicating that with a fixed sampling cost more population elements can be measured using cluster sampling than under SRSWR. Moreover, standard errors decrease inversely with square root of sample size, which in part compensates for the counter-effect of intra-cluster homogeneity upon standard errors. This implies that the design effect cannot serve as a single measure of the total efficiency of cluster sampling, so cost efficiency should also be taken into account.

Example 3.3

Cost efficiency under cluster sampling. The budget of a nation-wide survey based on personal interviews includes a grant of Fmk 80 000 to cover sampling and data-collection costs. Costs per interview are FIM 175, and average travelling expenses per interview are FIM 200. By firstly assuming that the sample is drawn by SRS, the sample size under fixed total costs is

$$n_{srs} = \frac{80\,000}{200 + 150} = 229.$$

Next, assuming that the population can be split into clusters each consisting of five people ($B = 5$), the sample size is

$$n_{clu} = \frac{80\,000}{200/5 + 150} = 421.$$

Cluster sampling nearly doubles the available sample size relative to SRS, since the costs of a single journey will cover five interviews instead of one.

One-stage Cluster Sampling

Let us introduce the principles of cluster sampling under the simplest design of this sort, namely one-stage cluster sampling. In one-stage cluster sampling, it is assumed that the N population elements are clustered into A groups, i.e. clusters. Making the somewhat unrealistic assumption of equal-sized clusters, each cluster is taken to consist of B elements. In a more general case, it is assumed that the population is clustered such that the size of cluster α is B_α elements. In both cases, a sample of a clusters is drawn from the population of A clusters, and all the elements of the sampled clusters are taken into the element-level sample. Remember, there is only a single sampling stage, namely that of the clusters, and therefore this design is known as one-stage cluster sampling.

The sample of a clusters is drawn from the population of clusters by using a specific element sampling technique such as simple random sampling, systematic sampling or PPS sampling. Because standard element sampling schemes can be used in one-stage cluster sampling, the selection algorithms previously described are readily available. The only difference is that a cluster, i.e. a group of population elements, constitutes the sampling unit instead of a single element of the population. Moreover, if the selection of the clusters is with equal inclusion probabilities, for example by using SRS or SYS, then the inclusion probabilities for the population elements are also equal, and this is independent of cluster sizes being equal.

In the simple case of equal-sized clusters, the element sample size is fixed and is $n = a \times B$. If the cluster sizes can vary, as is often the case in practice, the sample size $n = \sum_{\alpha=1}^{a} B_\alpha$ cannot be fixed in advance and it depends upon which clusters happen to be drawn in the sample. The expected sample size $\frac{a}{A} \times N$ and the actual sample size n can differ considerably if the variation in cluster sizes is large. This inconvenience can usually be controlled by using an appropriate sampling scheme. For example, if the sizes of the population clusters are (even roughly) known as auxiliary information, the clusters can be stratified by size, making it possible to approximately control the element sample size, n.

We introduce the basics of the estimation under one-stage cluster sampling in the case in which A unequal-sized clusters are present with cluster sizes B_α, and SRS without replacement is used to sample the a clusters; we call this the one-stage CLU design. Equally sized clusters where $B_\alpha = B$ is a special case of this. The element-level population size is thus given by $N = \sum_{\alpha=1}^{A} B_\alpha$ elements. Our aim is to estimate the population total T. For this, formulae from

SRS in Section 2.3 can be used and applied to the cluster totals. Certain alternative estimators are also given.

Let the value of the study variable be denoted $Y_{\alpha\beta}$, $\alpha = 1, \ldots, A$ in the population, and in the sample $y_{\alpha\beta}$, $\alpha = 1, \ldots, a$, and in both instances $\beta = 1, \ldots, B_\alpha$. The cluster-wise totals Y_α in the population are

$$Y_\alpha = \sum_{\beta=1}^{B_\alpha} Y_{\alpha\beta} = B_\alpha \bar{Y}_\alpha, \qquad \alpha = 1, \ldots, A,$$

where $\bar{Y}_\alpha$ is the mean per element in cluster α. The corresponding sample estimators $\bar{y}_\alpha$ are

$$y_\alpha = \sum_{\beta=1}^{B_\alpha} y_{\alpha\beta} = B_\alpha \bar{y}_\alpha, \qquad \alpha = 1, \ldots, a,$$

where $\bar{y}_\alpha$ is the mean per element in sample cluster α. An unbiased estimator of the population total $T = \sum_{\alpha=1}^{A} Y_\alpha$ is given by

$$\hat{t} = (A/a) \sum_{\alpha=1}^{a} y_\alpha = (A/a) \sum_{\alpha=1}^{a} B_\alpha \bar{y}_\alpha. \qquad (3.5)$$

The design variance $V_{\text{clu}}(\hat{t})$ of $\hat{t}$ and its unbiased estimator $\hat{v}_{\text{clu}}(\hat{t})$ can be derived from the corresponding SRS equations, because the only source of variation is that of the cluster totals Y_α around the overall mean per cluster $\bar{Y}$. The design variance of $\hat{t}$ is given by

$$V_{\text{clu}}(\hat{t}) = A^2 (1 - a/A) \sum_{\alpha=1}^{A} (Y_\alpha - \bar{Y})^2 / a(A - 1), \qquad (3.6)$$

where $\bar{Y} = T/A$ is the population mean per cluster. An unbiased estimator of the design variance is given by

$$\hat{v}_{\text{clu}}(\hat{t}) = A^2 (1 - a/A) \sum_{\alpha=1}^{a} (y_\alpha - \bar{y})^2 / a(a - 1), \qquad (3.7)$$

where $\bar{y} = \sum_{\alpha=1}^{a} y_\alpha / a$ is an estimator of the mean $\bar{Y}$. It can be inferred from (3.6) that if the cluster sizes B_α are equal or nearly so and if the cluster means $\bar{Y}_\alpha$ vary little, then the cluster totals $B_\alpha \bar{Y}_\alpha$ will also vary little and so a small design variance will be obtained. On the other hand, if the variation in the cluster sizes is large, the cluster totals will vary greatly and the design variance becomes large, showing inefficient estimation. However, the efficiency can be improved by using a ratio estimator where the cluster sizes B_α are used as an auxiliary size measure z. We can then have an estimator for the total given by

$$\hat{t}_{rat} = N\frac{\sum_{\alpha=1}^{a} y_\alpha}{\sum_{\alpha=1}^{a} B_\alpha} = N \times \bar{\bar{y}},$$

where $\bar{\bar{y}} = \sum_{\alpha=1}^{a} y_\alpha / \sum_{\alpha=1}^{a} B_\alpha$ is the sample mean per element which is an estimator of the population mean per element $\bar{\bar{Y}} = T/(A \times B)$. This ratio estimator is a special case of the ratio estimator considered later in Section 3.3. Assuming a large number of sample clusters an approximative design variance of $\hat{t}_{rat}$ is given by

$$V_{clu}(\hat{t}_{rat}) = A^2(1 - a/A)\sum_{\alpha=1}^{A} B_\alpha^2(\bar{Y}_\alpha - \bar{\bar{Y}})^2/a(A - 1).$$

The variation in the cluster means per element $\bar{Y}_\alpha$ around the population mean per element $\bar{\bar{Y}}$ can usually be expected to be smaller than that of the cluster totals Y_α around the grand mean $\bar{Y}$. If so, the estimation will be more efficient. Hence an estimator of the design variance is

$$\hat{v}_{clu}(\hat{t}_{rat}) = A^2(1 - a/A)\sum_{\alpha=1}^{a} B_\alpha^2(\bar{y}_\alpha - \bar{\bar{y}})^2/a(a - 1).$$

A similar effect on the efficiency can be expected when using PPS sampling for the clusters if their sizes B_α are known in advance. Then, the corresponding PPS estimators from Section 2.5 can be used.

It is also possible to base the estimation of the total T on the mean $\bar{y}_c$ of the cluster means $\bar{y}_\alpha$ given by

$$\bar{y}_c = \sum_{\alpha=1}^{a} \bar{y}_\alpha/a,$$

which is an estimator of the population mean $\bar{Y}_c = \sum_{\alpha=1}^{A} \bar{Y}_\alpha/A$ of the cluster means. If the clusters are equal-sized, i.e. if $B_\alpha = B$, then the resulting estimator

$$\hat{t}_c = N\bar{y}_c = N\sum_{\alpha=1}^{a} \bar{y}_\alpha/a$$

is unbiased and equal to $\hat{t}$ given in (3.5), and $\hat{t}_{rat}$. But $\hat{t}_c$ can be biased and even unconsistent under unequal-sized clusters. This can be seen by looking more closely at the bias. The bias is given by

$$\text{BIAS}(\hat{t}_c) = -\sum_{\alpha=1}^{A}(B_\alpha - \bar{B})(\bar{Y}_\alpha - \bar{Y}),$$

where $\bar{B}$ is the average cluster size. The equation for the bias indicates that the estimator $\hat{t}_c$ is unbiased if the cluster sizes B_α do not correlate with the cluster means $\bar{Y}_\alpha$, which is the case when the cluster sizes are equal. Therefore, if $\hat{t}_c$ is intended to be used, the relation of the cluster sizes and cluster means should be examined carefully.

Under equal-sized clusters the design variance of $\hat{t}_c$ can be also written as

$$V_{clu}(\hat{t}_c) = (A \times B)^2 (1 - a/A) S_b^2 / a, \qquad (3.8)$$

where the between-cluster variance S_b^2 can be derived from the cluster means $\bar{Y}_\alpha$ and their mean $\bar{Y}_c$ by

$$S_b^2 = \sum_{\alpha=1}^{A} (\bar{Y}_\alpha - \bar{Y}_c)^2 / (A - 1).$$

Because of equality of cluster sizes, $\hat{t}$ and $\bar{\bar{Y}}$ can be used in place of $\hat{t}_c$ and $\bar{Y}_c$ in (3.8) and in S_b^2.

We shall next study the efficiency of one-stage cluster sampling by inspecting the design effect DEFF of a total estimator under the one-stage CLU design in the simple case where the clusters are assumed equal-sized.

Example 3.4

Efficiency of one-stage cluster sampling from the *Province'91* population. We consider the efficiency of one-stage cluster sampling in the estimation of the total number T of unemployed persons (UE91) by calculating the DEFF of an estimator of T. Clusters are formed by combining groups of four neighbouring municipalities into eight clusters. The $N = 32$ municipalities of the province are divided into $A = 8$ equal-sized clusters so that $B_\alpha = B = 4$. It should be noticed that in real surveys the cluster sizes are usually unequal and, moreover, the number of population clusters is noticeably larger than here; therefore, the calculations should be taken hypothetically with the aim of illustrating the principles of the estimation. The cluster means $\bar{Y}_\alpha$ and totals Y_α of UE91 in all the population clusters are displayed in Table 3.4.

The total $T = 15\,098$, and the mean per cluster is $\bar{Y} = 1887$. The mean per element $\bar{\bar{Y}}$ and the mean of the cluster means $\bar{Y}_c$ are both 472 because of the equality of the cluster sizes. Let the sample size be $a = 2$ clusters, then the element sample size is $n = a \times B = 2 \times 4 = 8$. Because the cluster sizes are equal the total estimators $\hat{t}$, $\hat{t}_{rat}$ and $\hat{t}_c$ would provide the same estimates, and any of the corresponding design variances could be used. To evaluate the efficiency we calculate the design variance of $\hat{t}$ by using equation (3.8).

First, the between-cluster variance is obtained as

$$S_b^2 = \frac{1}{(8-1)} \sum_{\alpha=1}^{8} (\bar{Y}_\alpha - 472)^2 = 340^2,$$

Table 3.4 Cluster means and totals in the *Province'91* population, where each areal cluster includes four neighbouring municipalities.

Cluster identifier	UE91 Mean	UE91 Total
1	1 206	4 824
2	535	2 141
3	427	1 709
4	172	686
5	481	1 923
6	109	436
7	556	2 223
8	289	1 156

Sum of cluster totals	15 098
Mean per cluster	1 887
Mean per element	472
Mean of cluster means	472

giving the design variance

$$V_{clu}(\hat{t}) = (8 \times 4)^2 (1 - 2/8) S_b^2 / 2 = 32^2 \times 3/4 \times 340^2 / 2 = 6663^2.$$

The between-cluster variance $S_b^2 = 340^2$ will also be used in two-stage cluster sampling. Hence the design effect of the total estimator $\hat{t}$ is

$$\text{DEFF}_{clu}(\hat{t}) = \frac{V_{clu}(\hat{t})}{V_{srs}(\hat{t})} = \frac{6663^2}{7283^2} = 0.84.$$

The one-stage cluster sampling design appears to be slightly more efficient relative to the SRS design in this case. However, under complex surveys, due to positive intra-cluster correlation, cluster sampling usually tends to be less efficient than SRS when measured by the estimated design effects, as shown in later chapters. The unexpected result here can be partly explained by the method of forming the clusters on an administrative basis which produces relatively internally heterogeneous clusters with respect to the variation of UE91. If the clusters were formed by some other criteria, e.g. on a travel-to-work area basis, different results might be obtained because unemployment may be more homogeneous in such areas than in the regionally neighbouring municipalities.

In the next example, where a one-stage cluster sample is drawn from the *Province'91* population, it appears that, based on the estimated variances, the efficiency can be worse than that of SRS. This result, however, is crucially dependent on the composition of the sample in this case because only two clusters will be drawn from the small population of clusters.

Example 3.5

Analysing a one-stage CLU sample drawn from the *Province'91* population. The *Province'91* population is divided by a regional basis into eight $(A = 8)$ clusters, each comprising four $(B = 4)$ neighbouring municipalities. Eight municipalities are desired in the sample, so the element sample size is $n = 8$. Because the clusters are equal-sized the cluster-level sample size is $a = 2$. The sample of clusters is drawn by simple random sampling without replacement. As a result, the clusters 2 and 8 were drawn and we obtained the sample of eight municipalities as shown in Table 3.5.

The sample identifiers required for the analysis of the data set by PC CARP are the following three variables: STR is the stratum identifier which in this case is a constant because the population of clusters is not stratified, i.e. there is only one stratum. The cluster identification (2 or 8) is given by the variable CLU; and the weight variable is a constant WGHT = 4, i.e. the cluster size. The finite-population correction at the cluster level is $(1 - 0.25) = 0.75$, and so the sampling rate is 0.25.

Estimation results for the total $\hat{t}$, ratio $\hat{r}$ and median $\hat{m}$, and the values of the corresponding parameters T, R and M are displayed in Table 3.6. From there it can be seen that one-stage cluster sampling appears to be inefficient for all three estimators. The deff estimates are noticeably greater than one $(1.44 \leq \text{deff} \leq 1.92)$. Moreover, for this actual sample, the estimated $\text{deff}(\hat{t}) = 1.92$ differs noticeably from the corresponding parameter $\text{DEFF}(\hat{t}) = 0.84$. This is due to the small number of sample clusters which causes instability in the estimated design variances. The variance estimates depend

Table 3.5 A one-stage CLU sample of two clusters from the *Province'91* population.

| Sample design identifiers | | | Element | Study variables | |
STR	CLU	WGHT	LABEL	UE91	LAB91
1	2	4	Jämsä	666	6 016
1	2	4	Jämsänkoski	528	3 818
1	2	4	Keuruu	760	5 919
1	2	4	Kuhmoinen	187	1 448
1	8	4	Kinnula	129	927
1	8	4	Kivijärvi	128	819
1	8	4	Pihtipudas	331	2 543
1	8	4	Viitasaari	568	4 011

Sampling rate: $a/A = 2/8 = 0.25$

Table 3.6   Estimates from a one-stage CLU sample ($n=8$); the *Province'91* population.

Variable	Parameter	Estimate	s.e.	c.v.	deff
Total UE91	15 098	13 188	3 412	0.26	1.92
Ratio (%) UE91/LAB91	12.65	12.93	0.6	0.04	1.44
Median UE91	229	337	132	. . .	1.75

heavily on which clusters happen to be drawn; thus if selecting two clusters other than those just drawn, deff estimates noticeably less than one could be obtained. The problem of instability will be discussed in more detail in Chapter 6.

Two-stage Cluster Sampling

Subsampling from the sampled clusters is common when working with large clusters. This offers better possibilities for instance for the control of the element-level sample size n, when the cluster sizes vary. Moreover, with subsampling, the number of sample clusters can be increased when compared to one-stage cluster sampling for a fixed-element sample size, which can increase efficiency. A practical motivation is the availability of sampling frames which are only required for subsampling from the sampled clusters.

In *two-stage cluster sampling*, a sample of clusters is drawn from the population of clusters, i.e. primary sampling units (PSUs) at the first stage of sampling, by using the standard element-sampling techniques such as SRS, SYS or PPS. Moreover, the population of clusters can be stratified by using available auxiliary information. The simplest stratified two-stage cluster-sampling design, where exactly two clusters are drawn from each stratum, is often used in practice, offering the possibility to use a large number of strata and therefore increase the efficiency. At the second stage an element-level sample is drawn from the sampled clusters by again using standard element-sampling techniques. In practice, the cluster sizes in the population, and the cluster sample sizes, usually vary. Moreover, the inclusion probabilities can vary at each stage of sampling. But a sample with a constant overall sampling fraction can be obtained by an appropriate choice of the sampling fractions and selection techniques at each stage of sampling. This kind of multi-stage design is called an *epsem design* (equal probability of selection method).

In one-stage cluster sampling, all the elements in the sampled clusters make up the element-level sample and, thus, the only variation due to sampling was the between-cluster variation. But in two-stage cluster sampling an additional source of variation arises due to subsampling, namely the variation within the clusters, and this also contributes to the total variation.

For illustrating the basics of two-stage cluster sampling, we assume SRS without replacement at both stages of sampling and equality of the cluster sizes in the A population clusters, i.e. $B_\alpha = B$ for all α. The element-level population size is thus $N = A \times B$. Moreover, let us further assume that the element-level sample sizes also are equal for simplicity, i.e. $b_\alpha = b$ in all the a sample clusters; the sample size is thus $n = a \times b$. Cluster sampling under these assumptions results in equal inclusion probabilities for the population clusters, and they are equal also for the population elements, which provides an epsem sample. This can be seen by writing the sampling fractions a/A for the first stage and b/B for the second stage, giving a constant overall sampling fraction $a/A \times b/B = n/N$.

The main interest in the estimation usually concentrates at the second stage, i.e. element-level parameters. Let us consider the estimation of the population total $T = \sum_{\alpha=1}^{A} Y_\alpha$, where $Y_\alpha = B \times \bar{Y}_\alpha$ is the population total in cluster α, and $\bar{Y}_\alpha = \sum_{\beta=1}^{B} Y_{\alpha\beta}/B$ is the mean per element in cluster α as previously. An unbiased estimator of the total T is given by

$$\hat{t} = (A \times B) \sum_{\alpha=1}^{a} \bar{y}_\alpha/a,$$

where $\bar{y}_\alpha = \sum_{\beta=1}^{b} y_{\alpha\beta}/b$ is the mean per element in sample cluster α. In the derivation of the design variance for $\hat{t}$ a decomposition of the total variance into the between-cluster variance and within-cluster variance components can be used. The design variance for the estimator $\hat{t}$ is the weighted sum of the *between-cluster* variance S_b^2 and *within-cluster* variance S_w^2:

$$V_{clu}(\hat{t}) = (A \times B)^2 \left[\left(1 - \frac{a}{A} \right) \frac{S_b^2}{a} + \left(1 - \frac{b}{B} \right) \frac{S_w^2}{ab} \right], \qquad (3.9)$$

where

$$S_b^2 = \frac{1}{(A-1)} \sum_{\alpha=1}^{A} (\bar{Y}_\alpha - \bar{\bar{Y}})^2,$$

$$S_w^2 = \frac{1}{A(B-1)} \sum_{\alpha=1}^{A} \sum_{\beta=1}^{B} (Y_{\alpha\beta} - \bar{Y}_\alpha)^2,$$

and $\bar{\bar{Y}} = T/(A \times B)$ is the overall population mean per element. The between-cluster variance term is due to the first-stage sampling of the clusters and is similar to one-stage cluster sampling, and the additional within-cluster variation is due to the subsampling. In one-stage cluster sampling the within-cluster variance component is zero because all the B elements were taken from the sampled clusters, i.e. $b = B$.

Estimators of the variance terms S_b^2 and S_w^2 are obtained by inserting the sample counterparts in place of the population values. We hence obtain

$$\hat{s}_b^2 = \frac{1}{(a-1)} \sum_{\alpha=1}^{a} (\bar{y}_\alpha - \bar{\bar{y}})^2,$$

$$\hat{s}_w^2 = \frac{1}{a(b-1)} \sum_{\alpha=1}^{a} \sum_{\beta=1}^{b} (y_{\alpha\beta} - \bar{y}_\alpha)^2,$$

where $\bar{\bar{y}} = \sum_{\alpha=1}^{a} \bar{y}_\alpha / a$ is the sample mean per element. The estimator of the design variance of $\hat{t}$ is then given by

$$\hat{v}_{clu}(\hat{t}) = (A \times B)^2 \left[\left(1 - \frac{a}{A} \right) \frac{\hat{s}_b^2}{a} + \left(1 - \frac{b}{B} \right) \frac{a}{A} \frac{\hat{s}_w^2}{ab} \right]. \tag{3.10}$$

From (3.10) it can be inferred that if the first-stage sampling fraction a/A is small then the second component in the variance estimator becomes negligible. Then, a variance estimator based on only the between-cluster variation can be used as a slightly negatively biased approximation of the design variance of $\hat{t}$, which has the convenient property that it is only computed from cluster-level quantities. Further, if a/A is small, the first-stage finite-population correction would be close to one and thus can be omitted, leading to a with-replacement-type variance estimator. This kind of variance approximation will be extensively used when discussing on survey analysis in later chapters. Alternatively, if the fraction a/A is not negligible, the within-variance component can contribute substantially to the variance estimate.

In practice the cluster sizes B_α and the sample sizes from the sampled clusters b_α usually vary, and moreover the population of clusters can be stratified. Appropriate estimators for the total and the design variance of the total estimator should be used to properly account for the stratification and the variation in the cluster sample sizes. For the total, a ratio-type estimator or an estimator based on PPS sampling of the clusters can be used with the cluster sizes as the auxiliary size measure. The estimation of the design variance of a ratio-type estimator under two-stage stratified cluster sampling will be discussed in Chapter 5. There, various approximate variance estimators are introduced.

The inconvenient effect of the variation in the cluster sizes which produces a random element sample size, which is present when sampling under the two-stage CLU design, can be controlled by using PPS sampling of the clusters. Let us suppose that an epsem sample is desired with a fixed size of n elements. This can be attained by drawing a constant number $b_\alpha = b$ of elements from each of the a unequally sized sample clusters when the clusters are selected with PPS with inclusion probabilities proportional to the cluster sizes B_α, as can be inferred from the following formula:

$$\frac{n}{N} = \frac{a \times B_\alpha}{\sum_{\alpha=1}^{A} B_\alpha} \times \frac{b}{B_\alpha},$$

where a is the desired number of sample clusters and $b = n/N \times \sum_{\alpha=1}^{A} B_\alpha/a$.

In the next example, we evaluate the efficiency of the two-stage CLU design in the simple situation of equal-sized clusters, based on the calculation of the DEFF. Comparison is made with the one-stage CLU design.

Example 3.6

Efficiency of two-stage cluster sampling from the *Province'91* population. The number of clusters consisting of neighbouring municipalities is eight, so that $A=8$, and each cluster comprises $B=4$ municipalities. We compare the efficiency of one-stage and two-stage CLU designs in the estimation of the total T. Both designs involve equal clustering at the first stage. In one-stage cluster sampling two clusters $(a=2)$ were drawn, and all the four municipalities from the sampled clusters were taken into the element-level sample. The sample size was thus $n = a \times B = 2 \times 4 = 8$ municipalities. In two-stage cluster sampling we take $a = 4$ clusters in the first-stage sample, and we draw $b = 2$ municipalities from each sampled cluster at the second stage. The element-level sample size is then also $a \times b = 4 \times 2 = 8$ municipalities.

Under the one-stage CLU design the design variance was calculated as $V_{clu}(\hat{t}) = 6663^2$, and the design effect was $\mathrm{DEFF}(\hat{t}) = 0.84$. Under the two-stage CLU design we must first calculate the between-cluster and within-cluster variance components. The between-cluster variance in Example 3.4 was calculated as $S_b^2 = 340^2$. The within-cluster variance is

$$S_w^2 = \frac{1}{8(4-1)} \sum_{\alpha=1}^{8} \sum_{\beta=1}^{4} (Y_{\alpha\beta} - \bar{Y}_\alpha)^2 = 660^2.$$

The design variance of $\hat{t}$ is thus

$$V_{clu}(\hat{t}) = (8 \times 4)^2 \left[\left(1 - \frac{4}{8}\right) \frac{340^2}{4} + \left(1 - \frac{2}{4}\right) \frac{660^2}{4 \times 2} \right] = 6532^2$$

and the design effect of $\hat{t}$ for the two-stage design is

$$\mathrm{DEFF}_{clu}(\hat{t}) = 6532^2/7283^2 = 0.80.$$

When compared to the one-stage CLU design, the two-stage design is slightly more efficient. This is in part due to the property of the two-stage design that, with a given n, more first-stage units can be drawn than for the one-stage design. In this case, the number of sample clusters is doubled, which decreases the first-stage variance component. Of the total variance, 35% is contributed by the first stage (between-cluster) and 65% by the second stage (within-cluster). Thus, the within-cluster contribution dominates, which is in part due to the relative heterogeneity of the clusters. It should, however, be noticed that in the *Province '91* population the population of clusters is small and so are the cluster sample size and the sample size in subsampling. Therefore, these calculations should be taken as a hypothetical example, because in a real survey the corresponding figures are larger, the clusters tend to be relatively homogeneous, and a major share of the design variance is often due to the between-cluster variation.

The next example demonstrates computational results based on a sample drawn from the *Province'91* population using the two-stage CLU design. The efficiency is studied based on estimated design variances. The efficiency is also compared with that of the one-stage CLU sample from Example 3.5.

Example 3.7

Analysing a two-stage CLU sample drawn from the *Province '91* population. In the first stage, the clusters numbered 2, 3, 4 and 7 were drawn. In the second stage, two municipalities were drawn from each sample cluster. The data set of eight sample municipalities is displayed in Table 3.7.

In the analysis of the data from the two-stage CLU design the following design identifiers are required: the stratum identifier STR which is a constant 1 for all the sample elements, the cluster identifier CLU which has the values 2,

Table 3.7   A two-stage CLU sample of four clusters from the *Province '91* population.

Sample design identifiers			Element	Study variables	
STR	CLU	WGHT	LABEL	UE91	LAB91
1	2	4	Keuruu	760	5919
1	2	4	Kuhmoinen	187	1448
1	3	4	Äänekoski	767	5823
1	3	4	Konginkangas	142	675
1	4	4	Kyyjärvi	94	831
1	4	4	Pylkönmäki	98	545
1	7	4	Petäjävesi	262	1737
1	7	4	Uurainen	219	1330

Sampling rates: First stage $4/8 = 0.50$
Second stage $2/4 = 0.50$

Table 3.8   Estimates from a two-stage CLU sample ($n=8$); the *Province '91* population.

Variable	Parameter	Estimate	s.e.	c.v.	deff
Total UE91	15 098	10 116	2 659	0.26	0.93
Ratio (%) UE91/LAB91	1 2.65	13.81	0.5	0.04	0.99
Median UE91	229	192	48	. . .	1.07

3, 4 and 7 corresponding to the sampled clusters, and the weight variable WGHT which is a constant (4) for all the sample elements. It should be noted that the weight would vary between the clusters if the cluster sizes varied and the selection rates in the clusters were not equal. Using software-specific statements and options, this sampling design information is then supplied for the analysis. PC CARP is used with the analysis option TWO-STAGE. Because simple random sampling without replacement was used at both stages, the first-stage sampling rate $4/8$ and the second-stage sampling rate $2/4$ are also supplied, giving the weights $w_{\alpha\beta} = (A \times B)/(a \times b) = (8 \times 4)/(4 \times 2) = 4$ for all the sample elements. Estimation results on the total number of unemployed $\hat{t}$, the unemployment rate $\hat{r}$ and the median unemployment $\hat{m}$, as well as the values of the corresponding parameters T, R and M are displayed in Table 3.8.

The estimated design effects deff for the total, ratio and median estimators are close to one, indicating that the two-stage CLU sample does not differ greatly from SRS in efficiency. But the efficiency differs considerably from that of the one-stage counterpart where design-effect estimates noticeably larger than one were obtained for all the estimators. In the one-stage design, the number of sample clusters was very small, thus resulting in serious instability in the variance estimates. In the two-stage design, on the other hand, one half of all the population clusters were drawn and, therefore, the design is not as sensitive to instability and, in addition, the population clusters were relatively heterogeneous. It should be noticed, however, that in this example the clustering was an illustration of the estimation under two-stage cluster sampling, not an example of cluster sampling in real surveys. These will be considered in later chapters.

Intra-cluster Correlation and Efficiency

Efficiency of cluster sampling depends strongly on the internal composition of the clusters. Cluster sampling would be as efficient as simple random sampling if the clusters were internally heterogeneous so that each of them closely mirrored the overall composition of the element population. Efficiency decreases if the clusters are internally homogeneous and if the between-cluster variation is large. In practice, many naturally formed population subgroups are of this latter type.

The efficiency can be studied by the *intra-cluster correlation* which is a measure of the internal homogeneity of the clusters. This correlation can be included in the design variance equations of estimators from cluster sampling. Recall that in systematic sampling, a similar coefficient (the intra-class correlation) also played a crucial role; systematic sampling can indeed be taken as a special case of one-stage cluster sampling where only one cluster is drawn.

Let us assume equal-sized clusters $B_\alpha = B$ in all the A population clusters. We first study the decomposition $\sigma^2 = \sigma_b^2 + \sigma_w^2$ of the total variation σ^2 of the study variable y into the variation between the clusters σ_b^2 and within the clusters σ_w^2. The total variation σ^2 in the population can be written as

$$\sigma^2 = \sum_{\alpha=1}^{A}\sum_{\beta=1}^{B}(Y_{\alpha\beta} - \bar{\bar{Y}})^2/(A \times B),$$

where $Y_{\alpha\beta}$ is the population value of the study variable for an element β from cluster α and $\bar{\bar{Y}}$ is the overall mean per element as previously given. The total variation σ^2 decomposes into the between-cluster variation

$$\sigma_b^2 = \sum_{\alpha=1}^{A}(\bar{Y}_\alpha - \bar{\bar{Y}})^2/A$$

and into the within-cluster variation

$$\sigma_w^2 = \sum_{\alpha=1}^{A}\sum_{\beta=1}^{B}(Y_{\alpha\beta} - \bar{Y}_\alpha)^2/(A \times B),$$

where $\bar{Y}_\alpha$ is the mean per element in cluster α as before.

By using the formulae for intra-class correlation ω derived in Section 2.4 under systematic sampling for cluster sampling, we get

$$\omega = \frac{\sigma_b^2 - \sigma_w^2/(B-1)}{\sigma^2} = \frac{B}{B-1} \times \frac{\sigma_b^2}{\sigma^2} - \frac{1}{B-1}, \qquad (3.11)$$

so that the intra-cluster correlation can be expressed by using the between-cluster variance component σ_b^2 and the total variance σ^2.

The interpretation of the intra-cluster correlation depends on the share of the total variation σ^2 between the two variance components σ_b^2 and σ_w^2. First, if all the variation is within the clusters and there is no between-cluster variation, then the cluster means are equal and thus, $\sigma_b^2 = 0$. In this case $\sigma_w^2 = \sigma^2$ and the intra-cluster correlation coefficient is at minimum $\omega = -1/(B-1)$. If, on the other hand, all the variation is between the clusters, in which case the clusters are internally completely homogeneous and $\sigma_w^2 = 0$, then $\sigma_b^2 = \sigma^2$ and the coefficient has its maximum $\omega = 1$. And with the value $\omega = 0$ the between-cluster variation has the value σ^2/B. This corresponds to the case where elements are assigned to clusters at random.

Let us consider the efficiency of one-stage CLU sampling with respect to SRS of the same size n. The design variance of an estimator $\hat{t}$ of the total T was in equation (3.8) under the CLU design given as

$$V_{clu}(\hat{t}) = (A \times B)^2 \left(1 - \frac{a}{A}\right) \frac{S_b^2}{a},$$

where S_b^2 is the between-cluster variance component. From equation (3.11) it can be noted that the between-cluster variance component σ_b^2 can be written as

$$\sigma_b^2 = \frac{\sigma^2}{B}[1 + (B - 1)\omega],$$

and noting that $\sigma_b^2 = (A - 1)S_b^2/A$ and $\sigma^2 = (N - 1)S^2/N$ we obtain

$$V_{clu}(\hat{t}) = (A \times B)^2 \left(1 - \frac{a}{A}\right) \frac{S^2}{a} \left[\frac{1}{B}(1 + (B - 1)\omega)\right] \times \frac{N - 1}{N} \times \frac{A}{A - 1}.$$

Assuming large N and A, the last two terms become close to one and can thus be dropped. We hence obtain for the design variance of $\hat{t}$ an expression based on the total variance S^2 and the intra-cluster correlation ω :

$$V_{clu}(\hat{t}) \doteq (A \times B)^2 \left(1 - \frac{a}{A}\right) \frac{S^2}{n}[1 + (B - 1)\omega],$$

because $aB = n$. But the corresponding SRS design variance of $\hat{t}$ can be written as

$$V_{srs}(\hat{t}) = (A \times B)^2 \left(1 - \frac{n}{N}\right) \frac{S^2}{n},$$

which leads to the DEFF of $\hat{t}$ being given by

$$\text{DEFF}(\hat{t}) = \frac{V_{clu}(\hat{t})}{V_{srs}(\hat{t})} = 1 + (B - 1)\omega,$$

because $a/A = n/N$ in the finite population correction term of $V_{clu}(\hat{t})$.

The equation of DEFF indicates that if ω is positive, which is usually the case in practice, then cluster sampling is less effective than simple random sampling. And for a given ω, the DEFF increases with increasing the cluster size B. In the final example of this section the efficiency is further discussed as a function of the cluster size and the intra-cluster correlation.

Example 3.8

Cluster size and intra-cluster correlation in the *Province '91* population. Intra-cluster homogeneity appeared crucially to determine the efficiency of cluster sampling. When measured by the design effect DEFF, intra-cluster homogeneity depends on the value of the intra-cluster correlation coefficient ω and the size B of the clusters. We consider this more closely in the *Province'91* population. The clusters were formed by collecting regionally neighbouring municipalities into four clusters whose sizes were 2, 4, 8 and 16 municipalities. For example, if a cluster size of $B = 8$ municipalities is desired the province will be divided into $A = 4$ clusters. The results for the intra-cluster correlation ω and the corresponding design effects DEFF of an estimator $\hat{t}$ of the total under one-stage CLU design with various cluster sizes B are displayed in Table 3.9.

Table 3.9   Cluster size, intra-cluster correlation and DEFF in the *Province '91* population.

Cluster size B	Number of clusters A	Intra-cluster correlation	DEFF
2	16	0.167	1.167
4	8	−0.082	0.754
8	4	−0.060	0.580
16	2	−0.027	0.595

The design effects indicate that the cluster size of $B = 8$ municipalities leads to the most efficient estimation. Cluster size is thus an important determinant of the efficiency. It would thus be helpful in actual clustering procedures to at least roughly evaluate the intra-cluster correlation.

Summary

Cluster sampling is commonly used in practice because many populations are readily clustered into natural subgroups. Typical clusters met with in real surveys are regional administrative units, city blocks or block-like units, households, business firms or establishments, schools or school classes. Often for practical and economical reasons, these kinds of clusters are used in sampling and in data-collection procedures. A practical motivation is that sampling frames for subsampling are needed only for the sampled clusters. And an economical motivation is that the cost efficiency of cluster sampling can be fairly high. Good examples of various cluster sampling designs are to be found in Chapters 5, 6 and 9. A drawback in cluster sampling, however, is that due to the relative homogeneity of the clusters, as is often the case in practice, the statistical efficiency can be less than that of simple random sampling. But high cost efficiency can successfully compensate for this inconvenience

Our demonstration data, the *Province'91* population, appeared restrictive for thorough demonstration of cluster sampling and was thus used for illustrating the basic principles of sampling and estimation under one-stage and two-stage designs. In large-scale surveys there are usually a large number of clusters both in the population and in the sample. Moreover, the population of clusters can be stratified, and sampling can be achieved using several stages. In the analysis of such data, ratio-type estimators with approximative variance estimators are usually used in the estimation. These topics will be considered in detail in Chapter 5.

Cluster sampling is discussed in most textbooks on survey sampling. As further reading, Kish (1965), Frankel (1983), Kalton (1983) and Levy and Lemeshow (1991) can be recommended, covering introductory, advanced and more theoretical topics on cluster sampling.

3.3 MODEL-ASSISTED ESTIMATION

Introduction

In the techniques discussed so far, auxiliary information of the population elements is used in the sampling phase to attain an efficient sampling design. We now turn to a different way of utilizing auxiliary information. Our aim is to introduce estimators that can be used for the selected sample to obtain better estimates of the parameters of interest, relative to the estimates calculated with estimators based on the sampling design used.

Let us assume that appropriate auxiliary data are available from the population as a set of auxiliary variables. Of these variables, some might be categorical and some continuous. Some auxiliary data are perhaps used for the sampling procedure. Others can be used for improving the efficiency; a way to do this is, for example, to use an auxiliary variable z which is related to our study variable y for a reduction of the design variance of the original estimator of the population total of y. Recently, techniques using auxiliary information in this way have developed rapidly. In Särndal *et al.* (1992), these techniques are discussed in the context of *model-assisted design-based estimation*. *Model-assisted* estimation refers to the property of the estimators that models such as linear regression are used in the estimation of the finite-population parameters of interest. Model-assisted estimation should be distinguished from the multivariate survey analysis methods to be discussed in Chapter 8. There, models are also used but auxiliary information from the population is not assumed.

In the following, a brief review is given on model-assisted estimation. More specifically, *poststratification, ratio estimation* and *regression estimation* are considered. All these methods share the property that they are aimed at improving the estimation from a given sample by using available auxiliary information from the population. This can result in estimates closer to the true population value and reduction in the design variance of an estimator calculated from the sampled data.

In model-assisted estimation an auxiliary variable z, which is related to the study variable y, is required. If this variable is categorical, the target population U can be partitioned into subpopulations $U_1, \ldots, U_l, \ldots, U_L$ according to some classification principle. In poststratification these subpopulations are called *poststrata*. If the poststrata are internally homogeneous, this partitioning can capture a great deal of the total variance of the study variable y, resulting in a decrease of the design-based variance of an estimator. Moreover, poststratification can be used to obtain more accurate point estimates and reduce the bias of sample estimates caused by nonresponse.

The auxiliary variable z is often continuous. If it correlates strongly with the study variable y, a linear regression model can be assumed with y as the dependent variable and z as the predictor. This regression can be estimated from the observed sample and used in the estimation of the original target parameter. For this, ratio estimation and regression estimation can be used. By these methods, substantial gains in efficiency and increased accuracy are often reached.

Because model-assisted estimators are not solely design-based but also use model assumptions, the concept of *estimation strategy* will be used when referring to a combination of the sampling design and the appropriate estimator (see Section 2.1). To construct a model-assisted estimator under a given strategy, two kinds of weights are considered. The preliminary weights are the usual sampling design weights w_k, which generally are the inverses of the inclusion probabilities π_k; these weights are extensively used in this book. The other type of weights are called g *weights* and their values g_k depend both on the selected sample and on the chosen estimator. The product $w_k^* = w_k \times g_k$ gives a new adjusted weight variable w^*, which is used in the model-assisted estimators. The g weights will be explicitly given for poststratification and ratio estimation.

The basic principles of model-assisted estimation are most conveniently introduced for simple random sampling without replacement, although natural applications are often under more complex designs. A further simplification is that only one auxiliary variable is assumed. Also this assumption can be relaxed if multiple auxiliary variables are available. In the following, the pure design-based estimator under simple random sampling without replacement is considered as the reference strategy. The estimators and the corresponding strategies are the following:

Estimator	Strategy
Design-based estimator	SRS
Poststratified estimator	SRS*pos
Ratio estimator	SRS*rat
Regression estimator	SRS*reg

Poststratification

Poststratification can be used for improvement of efficiency of an estimator if a discrete auxiliary variable is available. This variable is used to stratify the sample data set after the sample has been selected. Recall from Section 3.1 that stratification of the element population as a part of the sampling design often gave a gain in efficiency. This was achieved by an appropriate choice of the stratification variables so that the variation in the study variable y within the strata would be small. Poststratification has a similar aim. To avoid confusion with the usual (pre)stratification, the population is partitioned into $1, \ldots, l, \ldots, L$ groups which are called *poststrata*.

To carry out poststratification, the sample data are first combined with the appropriate auxiliary data obtained perhaps from administrative registers or official statistics. Combining the sampled data with poststratum information and the corresponding selection probabilities, we can proceed with the estimation in basically the same way as if it were being done by ordinary (pre)stratification. Certain differences exist, however. Because we are stratifying after the sample selection or, more usually, after the data collection, we cannot assume any specific allocation scheme. The sample size n is fixed but how it is allocated to the different strata is not known until the sample is drawn. This property causes no harm for the estimation of, for example, the total, but the estimation of the variance of the total estimator requires more attention.

The *poststratified estimator* for the total T of y is given by

$$\hat{t}_{pos} = \sum_{l=1}^{L} \hat{t}_l = \sum_{l=1}^{L} \sum_{k=1}^{n_l} w_{lk}^* y_{lk}, \tag{3.12}$$

where $\hat{t}_l = N_l \bar{y}_l$ is an estimator of the poststratum total T_l and N_l is the size of the poststratum l. The poststratum weights are $w_{lk}^* = g_{lk} w_{lk}$, where the g weights are $g_{lk} = N_l / \hat{N}_l$ with the *estimated poststratum sizes* in the denominator, and w_{lk} are the original sampling weights. The calculation of w_{lk}^* will be illustrated in Example 3.9. The variance of $\hat{t}_{pos}$ can be determined in various ways, depending on how one uses the configuration of the observed sample. The configuration refers to how the actual poststratum sample sizes n_l are distributed, and if this is taken as given, the *conditional variance* is simply the same as the usual variance for stratified samples:

$$V_{srs,con}\left(\hat{t}_{pos} \mid n_1, \ldots, n_l, \ldots, n_L\right) = \sum_{l=1}^{L} N_l^2 \left(1 - \frac{n_l}{N_l}\right)\frac{S_l^2}{n_l}, \qquad (3.13)$$

where the poststratum variances are given by $S_l^2 = \sum_{k=1}^{N_l}(Y_{lk} - \bar{Y}_l)^2/(N_l - 1)$. By averaging (3.13) over all possible configurations of n, the *unconditional variance* is obtained. This gives an alternative variance formula,

$$V_{srs,unc}\left(\hat{t}_{pos}\right) = \sum_{l=1}^{L} N_l^2 \left(1 - \frac{E(n_l)}{N_l}\right)\frac{S_l^2}{E(n_l)}, \qquad (3.14)$$

where $E(n_l)$ is the expected poststratum sample size. This variance can be approximated in various ways. One of the approximations is

$$V_{srs,unc}\left(\hat{t}_{pos}\right) \doteq N^2\left(1 - \frac{n}{N}\right)\left(\frac{1}{n}\right)\left[\sum_{l=1}^{L}\left(\frac{N_l}{N}\right)S_l^2 + \left(\frac{1}{n}\right)\sum_{l=1}^{L}\left(1 - \frac{N_l}{N}\right)S_l^2\right]. \qquad (3.15)$$

The difference between the conditional and unconditional variances could be considerable if the sample size is small. The corresponding variance estimators $\hat{v}_{srs,con}(\hat{t}_{pos})$ and $\hat{v}_{srs,unc}(\hat{t}_{pos})$ are obtained by inserting $\hat{s}_l^2$ for S_l^2, where $\hat{s}_l^2 = \sum_{k=1}^{n_l}(y_{lk} - \bar{y}_l)^2/(n_l - 1)$. For illustrative purposes, both variances $V_{srs,con}$ and $V_{srs,unc}$ are estimated in the next example. There, the population ratio and median are also estimated in addition to the total, using PC CARP.

Example 3.9

Estimation with poststratification from a simple random sample drawn without replacement from the *Province'91* population. The sample used is drawn with SRS from the *Province'91* population in Section 2.3 (see Example 2.1). The sample is poststratified according to administrative division of the municipalities into urban and rural municipalities. The target population contains $N_1 = 7$ urban and $N_2 = 25$ rural municipalities. The two poststrata have the value 1 for urban and 2 for rural municipalities.

There are two analysis options in PC CARP for poststratification. Either a subprogram POST CARP can be used or the analysis of the poststratified SRS sample can be done by introducing poststratum identifiers in place of the original stratum identifiers. Both of the methods presuppose that the poststratum weights w_k^* replace original element weights. In Table 3.10, the sample information used for the estimation with poststratification is displayed.

Table 3.10 A simple random sample drawn without replacement from the *Province'91* population with poststratum weights.

| Sample design identifiers | | | | Study variables | | Poststratification | | |
STR	CLU	WGHT	Element LABEL	UE91	LAB91	POSTSTR	g WGHT	Post. WGHT
1	1	4	Jyväskylä	4 123	33 786	1	0.5833	2.3333
1	4	4	Keuruu	760	5 919	1	0.5833	2.3333
1	5	4	Saarijärvi	721	4 930	1	0.5833	2.3333
1	15	4	Konginkangas	142	675	2	1.2500	5.0000
1	18	4	Kuhmoinen	187	1 448	2	1.2500	5.0000
1	26	4	Pihtipudas	331	2 543	2	1.2500	5.0000
1	30	4	Toivakka	127	1 084	2	1.2500	5.0000
1	31	4	Uurainen	219	1 330	2	1.2500	5.0000

Sampling rate for calculation of *unconditional variance*: 8/32=0.25
Sampling rates for calculation of *conditional variance*:
Stratum 1 (Urban)= 3/7=0.43
Stratum 2 (Rural)= 5/25=0.20

Let us consider more closely the estimation of the total T. The poststratum totals of UE91 estimated from the table are $\hat{t}_1 = N_1\bar{y}_1 = 7\times 1868 = 13\,076$ and $\hat{t}_2 = N_2\bar{y}_2 = 25 \times 201.2 = 5030$. By using these estimates, the poststratified estimate for T is $\hat{t}_{pos} = \hat{t}_1 + \hat{t}_2 = 18\,106$.

Alternatively, the total estimate $\hat{t}_{pos}$ can be calculated using the poststratum weights w_k^*. To calculate w_k^*, the original sampling weights w_k should be corrected by the sample dependent g_k weights. For this, first the estimate of the poststratum size is determined. Denoting by w_{lk} the original element weight of a sample element which belongs to the poststratum l, an estimate for poststratum size $\hat{N}_l$ is given by summing up these original weights. Then, the corresponding g weight for an element k in poststratum l is simply $g_{lk} = N_l/\hat{N}_l$, where N_l is the exact size of poststratum l. For example, in Table 3.10 the original sampling weight under SRS is $w_k = 4$, or a constant for each population element. In the first poststratum, the poststratum size is $N_1 = 7$ and its estimated size is $\hat{N}_1 = 4 + 4 + 4 = 12$, because there are three sampled elements in the first poststratum. Thus the corresponding g weight is $g_{1k} = N_1/\hat{N}_1 = 7/12 = 0.5833$. Finally, the postratum weights are given for the first poststratum by $w_{1k}^* = g_{1k} \times w_{1k} = 0.5833 \times 4 = 2.3333$. This value turns out to be the same for all the sampled elements for the first poststratum (urban municipalities). Using the poststratum weights, the estimate $\hat{t}_{pos}$ will be equal to that previously calculated.

Estimation results for all the estimators are displayed in Table 3.11. The estimation of the conditional variance of the poststratified estimators $\hat{t}_{pos}$, $\hat{r}_{pos}$ (ratio) and $\hat{m}_{pos}$ (median) is performed using the option POST CARP. The original setting of sample identifiers remains, say STR = 1 and CLU = ID, but

Table 3.11 Poststratified estimates from a simple random sample drawn without replacement from the *Province '91* population.

(1) Poststratified estimates (conditional)

Variable	Estimate	s.e.	c.v.	deff
Total UE91	18 106	6014	0.33	0.33
Ratio (%) UE91/LAB91	12.97	0.45	0.03	0.59
Median UE91	194	36	. . .	1.09

(2) Poststratified estimates (unconditional)

Variable	Estimate	s.e.	c.v.	deff
Total UE91	18 106	7364	0.41	0.50
Ratio (%) UE91/LAB91	12.97	0.49	0.03	0.70
Median UE91	194	50	. . .	1.12

(3) Pure design-based estimates under SRS

Variable	Estimate	s.e.	c.v.	deff
Total UE91	26 440	13 282	0.50	1.00
Ratio (%) UE91/LAB91	12.78	0.41	0.03	1.00
Median UE91	226	149	. . .	1.00

the element weights are to be replaced by the poststratum weights, and the sampling rate is 0.43 for the first poststratum and 0.20 for the second poststratum. The unconditional variance is estimated by changing the stratum identifier and using PC CARP directly without the poststratification option. Original sampling weights are used and the sampling rate is 0.25 for both poststrata. Note that this procedure roughly approximates the formula given in (3.15). For comparison, the pure design-based estimates $\hat{t}$, $\hat{r}$ and $\hat{m}$ obtained under SRS are included.

The comparison shows how poststratification affects point estimates. The biggest gain is obtained when estimating the population total. The estimate of the number of unemployed is $\hat{t}_{pos} = 18\,106$, which is closer to the true value $T = 15\,098$ than the design-based estimate $\hat{t} = 26\,440$. The ratio estimate changes only slightly. The median behaves somewhat peculiarly, as has been seen previously.

The reason for a more accurate estimate for the total is obvious. Under SRS, one should have drawn urban and rural municipalities approximately by their respective proportions: $(8/32) \times (7) \approx 2$ towns and $(8/32) \times (25) \approx 6$ rural municipalities. The urban municipalities have larger population and unemployment figures. If by chance they are over-represented in the sample, then the pure design-based estimator will overestimate the population total. But poststratification can correct (at least partially) skewnesses. Therefore, we could also get a point estimate closer to its true value.

Poststratification can also improve efficiency. Again, this is true especially for the total. The estimated variance of $\hat{t}_{pos}$ under the conditional assumption is reduced to one-third when compared with the pure design-based estimate $\hat{t}$, which is indicated by deff $= 0.33$. If the unconditional variance is used as a basis, then deff $= 0.50$. The unconditional variance estimate is greater than the conditional variance estimate, because the poststratum sample sizes n_l are by definition random variables whose variance contribution increases the total variance.

Ratio Estimation of Population Total

The estimation of the population total T of a study variable y was considered previously under poststratification using the sample data and a discrete auxiliary variable. *Ratio estimation* can also be used to improve the efficiency of the estimation of T, if a continuous auxiliary variable z is available. The population total T_z and the n sample values z_k of z are required for this method. Such information can often be obtained from administrative registers or official statistics. This information can be used to improve the estimation of T by first calculating the sample estimator $\hat{r} = \hat{t}/\hat{t}_z$ of the ratio $R = T/T_z$ and multiplying $\hat{r}$ by the known total T_z. Ratio estimation of the total can be very efficient if the ratio Y_k/Z_k of the values of the study and auxiliary variables is nearly constant across the population.

Ratio estimators are usually effective but slightly biased. Because of bias, the mean square error (MSE) could be used instead of the variance when examining the sampling error. It has been shown that the proportional bias of a ratio estimator is $1/n$ and so becomes small when the sample size increases. Thus, the variance serves as an approximation to the MSE in large samples. The properties of ratio estimators have been studied widely in classical sampling theory.

Let us consider ratio estimation of the total T of y under simple random sampling without replacement. We are interested in a *ratio-estimated total* given by

$$\hat{t}_{rat} = \hat{r} \times T_z = \sum_{k=1}^{n} w_k^* y_k \,, \tag{3.16}$$

where $\hat{r} = \hat{t}/\hat{t}_z = N\bar{y}/N\bar{z} = \sum_{k=1}^{n} y_k / \sum_{k=1}^{n} z_k$ and T_z is the population total of the auxiliary variable z. The adjusted weights are $w_k^* = g_k w_k = (T_z/\hat{t}_z) w_k$.

In the estimator (3.16), $\hat{r}$ is a random variable and the total T_z is a constant. Thus the variance of $\hat{t}_{rat}$ can be written simply as $V_{srs}(\hat{t}_{rat}) = T_z^2 \times V_{srs}(\hat{r})$. If the SRS design variance of the estimator $\hat{r}$ of a ratio (equation (2.8)) is introduced here, an approximative variance of the ratio-estimated total is given by

$$V_{srs}(\hat{t}_{rat}) \doteq N^2 \left(1 - \frac{n}{N}\right) \left(\frac{1}{n}\right) \sum_{k=1}^{N} \frac{(Y_k - R \times Z_k)^2}{(N-1)} \,, \tag{3.17}$$

whose estimator is given by

$$\hat{v}_{srs}(\hat{t}_{rat}) = N^2 \left(1 - \frac{n}{N}\right)\left(\frac{1}{n}\right) \sum_{k=1}^{n} \frac{(y_k - \hat{r}z_k)^2}{(n-1)}. \tag{3.18}$$

By studying the sum of squares in the variance equation (3.17) it is possible to find the condition under which ratio estimation results in an improved estimate of a total. The total sum of squares can be decomposed as follows:

$$\sum_{k=1}^{N}(Y_k - R \times Z_k)^2/(N-1) = \sum_{k=1}^{N}[(Y_k - \bar{Y}) - R(Z_k - \bar{Z})]^2/(N-1)$$

$$= \sum_{k=1}^{N}[(Y_k - \bar{Y})^2 - R^2(Z_k - \bar{Z})^2$$

$$- 2R(Y_k - \bar{Y})(Z_k - \bar{Z})]/(N-1)$$

$$= S_y^2 + R^2 S_z^2 - 2R\rho_{yz}S_yS_z,$$

where ρ_{yz} is the finite-population correlation coefficient of the variables y and z. Consider the difference

$$V_{srs}(\hat{t}) - V_{srs}(\hat{t}_{rat}) = N^2 \left(1 - \frac{n}{N}\right)\left(\frac{1}{n}\right) S_y^2 - [S_y^2 + R^2 S_z^2 - 2R\rho_{yz}S_yS_z].$$

The ratio estimator improves efficiency if $V_{srs}(\hat{t}) > V_{srs}(\hat{t}_{rat})$, which occurs when

$$R^2 S_z^2 < 2R\rho_{yz}S_zS_y$$

is valid or

$$2\rho_{yz} > \frac{RS_z}{S_y}.$$

It should be noted that $R = \bar{Y}/\bar{Z}$, and that the former condition expressed in terms of coefficients of variation (c.v.) of the variables z and y is given by

$$\rho_{yz} > \left(\frac{1}{2}\right)\frac{\text{c.v. }(y)}{\text{c.v. }(z)},$$

where c.v. $(y) = S_y/\bar{Y}$ and c.v. $(z) = S_z/\bar{Z}$ are the coefficients of variation of y and z, respectively. Therefore, improvement in efficiency depends on the correlation between the study and auxiliary variables y and z and the coefficient of variation of each variable.

Example 3.10

Efficiency of a ratio-estimated total in the *Province'91* population. The variable UE91 is the study variable y and HOU85 is chosen as the auxiliary variable z. The correlation coefficient between UE91 and HOU85 is $\rho_{yz} = 0.9967$, and the corresponding coefficients of variation are c.v. $(y) = S_y/\bar{Y} = 743/472 = 1.57$ and c.v. $(z) = S_z/\bar{Z} = 4772/2867 = 1.66$. Thus, the condition given above is valid since

$$\rho_{yz} = 0.9967 > 0.4729 = \frac{1}{2} \times \frac{1.57}{1.66}.$$

It can be seen that the ratio estimation improves the efficiency. The improvement can also be measured directly as a design effect. In addition to the parameters given, the ratio $R = \bar{Y}/\bar{Z} = 472/2867 = 0.1646$ is required. The value of the design effect of the ratio-estimated total $\hat{t}_{rat}$ in the *Province'91* population is given by

$$
\begin{aligned}
\mathrm{DEFF}_{srs}(\hat{t}_{rat}) &= \frac{S_y^2 + R^2 S_z^2 - 2R\rho_{yz}S_y S_z}{S_y^2} \\
&= \frac{743^2 + 0.1646^2 \times 4772^2 - 2 \times 0.1646 \times 0.9967 \times 743 \times 4772}{743^2} \\
&= 0.0102
\end{aligned}
$$

which is close to zero. This substantial improvement in efficiency is due to the favourable relationship between UE91 and HOU85 such that the ratio Y_k/Z_k is nearly constant across the population.

The ratio-estimated total is in practice calculated by using the available survey data under the actual sample design. If the design is, say, stratified simple random sampling, the corresponding parameters would be estimated by using appropriate stratum weights. The present example was evaluated under SRS, which will also be used in the following example. There, use of g weights will also be illustrated.

Example 3.11

Calculating a ratio-estimated total from a simple random sample drawn without replacement from the *Province'91* population. We use again UE91 as the study variable and HOU85 as the auxiliary variable. The estimated ratio is $\hat{r} = \bar{y}/\bar{z} = 0.1603$, which is calculated from the sample in Table 3.12. The sample identifiers are STR = 1, ID is the cluster identifier, and the weight is WGHT = 4.

To carry out ratio estimation of the total, the adjusted weights w_k^* are first calculated. The sampling weight w_k is a constant $w_k = N/n = 32/8 = 4$ as

Table 3.12   A simple random sample drawn without replacement from the *Province'91* population prepared for ratio estimation.

Sample design identifiers			Element	Study var.	Aux. var.	g	Adj.
STR	CLU	WGHT	LABEL	UE91	HOU85	WGHT	WGHT
1	1	4	Jyväskylä	4 123	26 881	0.5562	2.2248
1	4	4	Keuruu	760	4 896	0.5562	2.2248
1	5	4	Saarijärvi	721	3 730	0.5562	2.2248
1	15	4	Konginkangas	142	556	0.5562	2.2248
1	18	4	Kuhmoinen	187	1 463	0.5562	2.2248
1	26	4	Pihtipudas	331	1 946	0.5562	2.2248
1	30	4	Toivakka	127	834	0.5562	2.2248
1	31	4	Uurainen	219	932	0.5562	2.2248

Sampling rate : 8/32 = 0.25.

before. The values of the g weight are $g_k = T_z / \hat{t}_z$. The population total of the auxiliary variable is $T_z = 91\,753$ and its estimate calculated from the sample is $\hat{t}_z = 164\,952$. Thus the g weight is the constant $g_k = 91\,753/164\,952 = 0.5562$. Multiplying the weight w_k by the g weight gives the value for the adjusted weight $w_k^* = 4 \times 0.5562 = 2.2248$.

The ratio estimate for the total is calculated as

$$\hat{t}_{rat} = \sum_{k=1}^{n} w_k^* y_k = \hat{r} \times T_z = 0.1603 \times 91\,753 = 14\,707,$$

which is much closer to the population total $T = 15\,098$ than the SRS estimate $\hat{t} = 26\,440$ for the total number of unemployed. The variance estimate for the total estimator is

$$\hat{v}_{srs}(\hat{t}_{rat}) = 32^2 \frac{(1 - 0.25)}{8} \times 91^2 = 892^2.$$

The corresponding deff estimate is

$$\text{deff}_{srs}(\hat{t}_{rat}) = \frac{\hat{v}_{srs}(\hat{t}_{rat})}{\hat{v}_{srs}(\hat{t})} = 892^2/13282^2 = 0.0045,$$

which also shows that ratio estimation improves the efficiency. The minimal auxiliary information of the population total T_z and the sample values of z yield good results.

It is also possible to calculate the DEFF when using the ratio-estimated total since the variance of $V_{srs}(\hat{t}_{rat})$ is

$$V_{srs}(\hat{t}_{rat}) \doteq N^2 \left(1 - \frac{n}{N}\right)\left(\frac{1}{n}\right) \sum_{k=1}^{N} \frac{(Y_k - R \times Z_k)^2}{(N-1)}$$

$$= 32^2 \frac{(1 - 0.25)}{8} \times 75^2 = 736^2.$$

Division by the corresponding SRS design variance of $\hat{t}$ gives

$$\text{DEFF}_{srs}(\hat{t}_{rat}) = \frac{V_{srs}(\hat{t}_{rat})}{V_{srs}(N\bar{y})} = 736^2/7283^2 = 0.0102,$$

which is the same figure presented previously in Example 3.10.

For these data, ratio estimation considerably improves efficiency and brings the point estimate for the total close to its population value. The value of the ratio estimator is based on the fact that across the population, the ratio Y_k/Z_k remains nearly constant. It should be noted that even a high correlation between the variables does not guarantee this, because the ratio estimator assumes that the regression line of y and z goes near to the origin. Thus, an intercept term is not included in the corresponding regression equation. The ratio estimator may therefore be unfavourable if the population regression line intercepts the y-axis far from the origin, even if the correlation is not close to zero. For these situations, the method to be presented next would be more appropriate.

Regression Estimation of Totals

Regression estimation of the population total T of a study variable y is based on the linear regression between y and a continuous auxiliary variable z. The linear regression can, for example, be given by $E(\tilde{y}_k) = \alpha + \beta Z_k$ with a variance $V(\tilde{y}_k) = \sigma^2$, where $\tilde{y}_k$ are independent random variables with the population values Y_k as their assumed realizations, α, β and σ^2 are unknown parameters, Z_k are known population values of z, and E and V refer respectively to the expectation and variance under the model. The finite-population analogues of α and β, denoted respectively by A and B, are estimated from the sample using weighted least squares estimation so that the sampling design is properly taken into account. It is immediately obvious that multiple auxiliary variables can also be incorporated in the model. Note that the model assumption introduces a new type of randomness; in the estimation considered previously, the sample selection was the only source of random variation.

We consider the basic principles of regression estimation for simple random sampling without replacement using the above regression model with a single auxiliary variable. The finite-population parameters A and B are estimated by the ordinary least squares method giving $\hat{b} = \hat{s}_{yz}/\hat{s}_z^2$ as an estimator of the slope B and $\hat{a} = \bar{y} - \hat{b}\bar{z}$ as an estimator of the intercept A. Using the estimator $\hat{b}$, the *regression estimator* of the total T of y is given by

$$\hat{t}_{reg} = N(\bar{y} + \hat{b}(\bar{Z} - \bar{z})) = \hat{t} + \hat{b}(T_z - \hat{t}_z) \tag{3.19}$$

where $\hat{t} = N\bar{y}$ is the SRS estimator of T, $\hat{t}_z = N\bar{z}$ is the SRS estimator of T_z and

$\bar{Z} = T_z/N$. Alternatively, if transformed values $z_k^* = \bar{Z} - z_k$ are used in the regression instead of z_k, an estimated intercept for this model is $\hat{a}^* = \hat{a} + \hat{b}\bar{Z}$ giving $\hat{t}_{reg} = N\hat{a}^*$, because (3.19) can be written also as $\hat{t}_{reg} = N\hat{a} + \hat{b}T_z$. Note that the regression estimation of the total T presupposes only knowledge of the population total T_z and the sample values z_k of the auxiliary variable z.

Regression estimators constitute a wide class of estimators. For example, the previous ratio estimator $\hat{t}_{rat} = \hat{r}T_z$ is a special case of (3.19) such that the intercept parameter A is assumed zero and the slope B is estimated by $\hat{b} = \hat{r} = \hat{t}/\hat{t}_z$.

The improvement gained in regression estimation, as compared with the corresponding simple-random-sampling estimators, depends on the value of the finite-population correlation coefficient $\rho_{yz} = S_{yz}/S_y S_z$ between the variables y and z. This can be seen in the variance formula. The approximate variance of the regression estimator $\hat{t}_{reg}$ is given by

$$V_{srs}(\hat{t}_{reg}) \doteq N^2\left(1 - \frac{n}{N}\right)\left(\frac{1}{n}\right)S_y^2(1 - \rho_{yz}^2),\qquad(3.20)$$

which is exact for large samples. It will be noted that the value of the correlation coefficient has a decisive influence on the possible improvement of the regression estimation. If ρ_{yz} is zero, the variance of the regression estimator $\hat{t}_{reg}$ equals that of the SRS counterpart $\hat{t}$. But with a nonzero correlation coefficient the variance obviously decreases.

An estimator of the variance of $\hat{t}_{reg}$ under simple random sampling is given by

$$\hat{v}_{srs}(\hat{t}_{reg}) = N^2\left(1 - \frac{n}{N}\right)\left(\frac{1}{n}\right)\sum_{k=1}^{n}[(y_k - \bar{y}) - \hat{b}(z_k - \bar{Z})]^2/(n - 2),\qquad(3.21)$$

which is valid for large samples.

Under certain conditions, the regression estimator of a total is more efficient than the ratio estimator. This will be demonstrated below by considering the variances of the SRS estimator, the ratio estimator and the regression estimator. Simple random sampling without replacement is assumed, and the constant (c) given in the formulae represents $c = N^2(1 - \frac{n}{N})(\frac{1}{n})$. The variances are:

Pure design-based estimator	$V_{srs}(\hat{t}) = cS_y^2$
Ratio estimator	$V_{srs}(\hat{t}_{rat}) = c(S_y^2 + R^2 S_z^2 - 2R\rho_{yz}S_y S_z)$.
Regression estimator	$V_{srs}(\hat{t}_{reg}) = cS_y^2(1 - \rho_{yz}^2)$

Studying the relationship between the regression coefficient B and the ratio R will reveal the condition where the regression-estimated total is more efficient than the ratio-estimated total. To find this condition, the difference between the two variances is

$$V_{srs}(\hat{t}_{rat}) - V_{srs}(\hat{t}_{reg}) = c[(S_y^2 + R^2 S_z^2 - 2R\rho_{yz}S_yS_z) - S_y^2 + \rho_{yz}^2 S_y^2]$$
$$= c[(R^2 S_z^2 - 2R\rho_{yz}S_yS_z) + \rho_{yz}^2 S_y^2].$$

Regression estimation is more efficient if the difference is positive:

$$R^2 S_z^2 - 2R\rho_{yz}S_yS_z + \rho_{yz}^2 S_y^2 > 0.$$

The condition can be rewritten as

$$-\rho_{yz}^2 S_y^2 < R^2 S_z^2 - 2R\rho_{yz}S_yS_z.$$

By dividing the inequality above by S_z^2 and inserting $\rho_{yz} = S_{yz}/S_yS_z$ and $B = S_{yz}/S_z^2$, gives

$$-B^2 < R^2 - 2RB.$$

Regression estimation, then, is more efficient than ratio estimation if

$$(B - R)^2 > 0.$$

Thus the squared difference between the finite-population regression coefficient and the ratio determines when the regression estimation is more efficient.

Example 3.12

A comparison of ratio and regression estimation of the total in the *Province'91* population. Using UE91 as the study variable and HOU85 as the auxiliary variable, the population parameters for the regression coefficient and ratio are calculated as $B = 0.1552$ and $R = 0.1645$, respectively. The condition established above gives

$$(B - R)^2 = (0.1552 - 0.1645)^2 = 0.0001,$$

so that both estimation methods give approximately the same efficiency. Sample-wise variation might, however, lead to this not holding true in every case. But considerable improvement is gained with ratio and regression estimation when compared against the standard error from simple random sampling. This is clearly seen in the design effect, which is

$$\text{DEFF}_{srs}(\hat{t}_{reg}) = (1 - 0.9967^2) = 0.0066.$$

In the next example, we compute a regression-estimated total from a sample data set by using PC CARP.

Example 3.13

Regression estimation of the total in the *Province'91* population. The previously selected simple random sample is used. There, the study variable UE91 is regressed with the auxiliary variable HOU85 using the option for linear regression of PC CARP. We conduct regression estimation in two ways, resulting in equals estimates. The original HOU85 is first used as the predictor and an estimate $\hat{t}_{reg}$ is computed using the estimated slope $\hat{b}$ and equation (3.19). A transformed variable DIFFHOU85 is then constructed, as the difference between the population mean $\bar{Z} = 2867$ and the sample values z_k of HOU85 (see Table 3.13). The estimated intercept $\hat{a}^*$ of the resulting regression model is used in the estimator $\hat{t}_{reg}$. In the table, the sample identifiers correspond to the SRS case, and the sampling rate is, as previously, 0.25.

Using UE91 as the dependent variable and the original HOU85 as the predictor, the slope is estimated as $\hat{b} = 0.152$, giving

$$\hat{t}_{reg} = \hat{t} + \hat{b}(T_z - \hat{t}_z) = 26\,440 + 0.152(91\,753 - 164\,952) = 15\,312.$$

The same estimate is obtained when using the transformed variable DIFFHOU85 as the predictor. Under this model, the estimator $\hat{t}_{reg} = N\hat{a}^*$ is used with an estimated intercept $\hat{a}^* = 478.5$ and $N = 32$. Since the SRS standard error of $\hat{a}$ is 20.25, the standard error of $\hat{t}_{reg}$ is s.e $= 32 \times 20.25 = 648$.

The corresponding design-based total estimate obtained under simple random sampling without replacement was $\hat{t} = 26\,440$, whose standard error was 13 282. Therefore, the deff estimate is deff $= 648^2/13\,282^2 = 0.002$, which is

Table 3.13 A simple random sample drawn without replacement from the *Province'91* population prepared for regression estimation.

Sample design identifiers			Element	Study variable		Auxiliary information	
STR	CLU	WGHT	LABEL	UE91	$\bar{Z}$	HOU85	DIFFHOU85
1	1	4	Jyväskylä	4 123	2867	26 881	$-24\,014$
1	4	4	Keuruu	760	2867	4 896	$-2\,029$
1	5	4	Saarijärvi	721	2867	3 730	-863
1	15	4	Konginkangas	142	2867	556	2 311
1	18	4	Kuhmoinen	187	2867	1 463	1 404
1	26	4	Pihtipudas	331	2867	1 946	921
1	30	4	Toivakka	127	2867	834	2 033
1	31	4	Uurainen	219	2867	932	1 935

Sampling rate: 8/32 = 0.25.

almost zero and is persuasive evidence of the superiority of regression estimation over pure design-based estimation for the present estimation problem. Improved efficiency is due to the strong linear relationship between UE91 and HOU85.

Regression estimation was illustrated in the simplest case where one auxiliary variable was used and simple random sampling without replacement was assumed. The method can also be applied for more complex designs, and multiple auxiliary variables can be incorporated in the estimation. For this, the weighted least squares regression option of PC CARP (or SUDAAN) can also be used. Although use of multivariate regression models for regression estimation under a complex design is technically straightforward, there are certain theoretical complexities when compared to regression estimation under simple random sampling. Therefore, a sensible rule is that a simple model is often more appropriate than a complicated one.

Another generalization is also obvious since discrete covariates can also be incorporated into a linear model. Using this kind of auxiliary variables for regression estimation leads to analysis-of-variance-type models. This approach is connected with poststratification and the so-called *raking ratio estimator* (see Chapter 4).

Comparison of Model-assisted Estimates

We finally compare the model-assisted estimation results obtained previously from a sample drawn with simple random sampling without replacement from the *Province'91* population. More specifically, poststratification, ratio estimation and regression estimation results for the population total T of UE91 are compared. The pure design-based estimate using the standard SRS formula is also included. The known population total $T = 15\,098$ of UE91 is the reference figure (see Table 3.14).

Two obvious conclusions can be drawn. Firstly, point estimates calculated using auxiliary information are closer to the population total than the pure

Table 3.14 Model-assisted estimation results for the population total of UE91 from an SRS sample of eight elements drawn from the *Province'91* population.

Estimator	Strategy	Estimate	s.e.	deff
Design-based estimator	SRS	26 440	13 282	1.0000
Poststratified estimator	SRS*pos	18 106	6 014	0.3323
Ratio estimator	SRS*rat	14 707	892	0.0045
Regression estimator	SRS*reg	15 312	648	0.0020
Population total		15 098		

SRS-based estimate. Secondly, the model-assisted estimators are much more efficient than the SRS counterpart.

The poststratified estimator uses as discrete auxiliary information the administrative division of municipalities into urban and rural municipalities. Improved estimates result, since this division is in relation to the variation of the study variable in such a way that the variation of unemployment figures is smaller in the poststrata than in the whole population. But the relation is not as strong as between UE91 and the continuous auxiliary variable HOU85, the number of households. This can be seen from the ratio and regression estimation results. Because ratio estimation assumes that the regression line of UE91 and HOU85 goes through the origin, and this is not exactly the case, regression estimation performs slightly better than ratio estimation.

Summary

Using auxiliary information from the population in the estimation of a finite-population parameter of interest is a powerful tool to get more precise estimates if the variation of the study variable has some strong relationship with an auxiliary covariate. If so, efficient estimators can be obtained such that they produce estimates close to the true population value and have a small standard error. The auxiliary variable can be a discrete variable, in which case poststratification can be used. If the covariate is a continuous variable, ratio estimation or regression estimation is appropriate. These methods are also available in commercial software for survey analysis such as PC CARP and SUDAAN.

Model-assisted estimation is often used in descriptive surveys to improve the estimation of the population total of a study variable of interest, whereas in multi-purpose studies, where the number of study variables may be large, it may be difficult to find good auxiliary covariates for this purpose. In such surveys, however, poststratification is often used to adjust for nonresponse.

We have examined here the elementary principles of model-assisted estimation supplemented with computational illustrations. For more details, the reader is encouraged to consult Särndal *et al.* (1992); there, model-assisted survey sampling, covering poststratification, ratio estimation and regression estimation, is extensively discussed. These methods are considered as special cases of *generalized regression estimation* which is used in many statistical agencies in the production of official statistics (see, e.g., Estevao *et al.* 1995). A clear overview of poststratification is to be found in Holt and Smith (1979). Smith (1991) recommends use of conditional inference if possible, although Rao (1985) argues that the design-based conditional theory for complex surveys may be intractable. Then, a useful alternative may be to apply the predictive approach (see e.g., Valliant 1993). Further, as a generalization of poststratification, Deville and Särndal (1992) and Deville *et al.* (1993) consider a class of weights calibrated to known marginal totals.

3.4   EFFICIENCY COMPARISON USING DESIGN EFFECTS

The design effect provides a convenient tool for the comparison of efficiency of the estimation of the population parameter of interest under various sampling designs. In this section we summarize the findings on efficiency evaluations from the preceding sections.

Efficiency is derived by comparing the variance of an estimator with that obtained under simple random sampling without replacement, and is measured as the population design effect DEFF, or as an estimated design effect deff calculated from the selected sample. We previously evaluated the efficiency in three ways: (1) analytically, by deriving the corresponding design variance formulae, (2) population-based, by calculating from the small fixed population, the *Province'91* population, the true value of the design variances, and (3) sample-based, by estimating the design variances from one realization of a sampling design applied to the *Province'91* population. For the total, evaluation by these methods covered all the basic sampling techniques considered. In the sample-based evaluation of the design effect using an estimated deff, we considered the estimators of a total, a ratio and a median.

Let us consider first the evaluation of efficiency for the estimation of the total T of a study variable y. The design effect DEFF is defined as a ratio of two design variances: the actual variance $V_{p(s)}(\hat{t}^{\star})$ of an estimator $\hat{t}^{\star}$ of the total, reflecting properly the sampling design, and the variance $V_{srs}(N\bar{y})$ derived assuming simple random sampling without replacement, where $\hat{t}^{\star}$ is the design-based estimator of the total under the design $p(s)$ and $N\bar{y} = \hat{t}$ is the corresponding SRS estimator. Note that the two estimators of the total may be different, and the same sample size is assumed as for the actual sampling design. The design effect is thus

$$\text{DEFF}_{p(s)}(\hat{t}^{\star}) = V_{p(s)}(\hat{t}^{\star})/V_{srs}(N\bar{y}) \tag{3.22}$$

as defined in Section 2.1. The equation indicates that if DEFF>1 then actual design is less efficient than SRS; if DEFF is approximately one then the designs are equally efficient; and if DEFF<1 the efficiency of the actual design is superior to SRS.

Analytical Evaluation of Design Effect

The analytical evaluation of DEFF is possible if the population parameters in the variance equations, such as the population variance S^2, cancel out in the formula of the design effect. For example, the design effect under simple random sampling with replacement can be calculated for a given sample size n and population size N. Hence we have DEFF $= (N-1)/(N-n)$ with the result that the design effect for SRSWR is greater than or equal to one. It is also sometimes possible to identify conditions when the design effect will be less than one and the actual design will be more efficient than SRS.

Analytical evaluation of the design effect for an estimator of a total is illustrated for stratified simple random sampling STR, PPS sampling with probabilities proportional to a size measure, and cluster sampling CLU. Systematic sampling is excluded because it can be considered as a special case of cluster sampling.

1) *Stratified sampling with proportional allocation (STR)* Factors affecting efficiency under STR are the possible heterogeneity of separate strata and internal homogeneity within each stratum. The design effect for an estimator $\hat{t}^{\star} = \hat{t}$ of the total T of the study variable y is

$$\mathrm{DEFF}_{str}(\hat{t}) \doteq \frac{\sum_{h=1}^{H} W_h S_h^2}{S^2}, \tag{3.23}$$

where S_h^2 are intra-stratum variances and S^2 is the population variance of y (see Section 3.1). In stratified sampling the DEFF is usually less than one, which happens when the strata are internally homogeneous with respect to the variation of the study variable, i.e. if the intra-stratum variances are small.

2) *Sampling with probability proportional to a measure of size (PPS)* The value of an auxiliary variable z measuring the size of a population element is required from all the units in the population. Assuming that the population regression line of y and z intercepts the y axis near to the origin, an approximate equation of the design effect of an estimator $\hat{t} = \hat{t}_{ht}$ (the *Horvitz–Thompson* estimator) is given by

$$\mathrm{DEFF}_{pps}(\hat{t}_{ht}) \doteq (1 - \rho_{yz}^2), \tag{3.24}$$

where ρ_{yz} is the finite-population correlation coefficient between the study variable y and the size measure z (see Section 2.5). Given the above condition, if z is a good size measure correlating strongly with y, a DEFF smaller than one is obtained.

3) *Cluster sampling (CLU)* The design effect under cluster sampling depends on the value of the intra-cluster coefficient ω of the study variable y measuring internal homogeneity of the population clusters. Assuming equal-sized clusters, an approximative equation of the design effect of an estimator $\hat{t}$ is given by

$$\mathrm{DEFF}_{clu}(\hat{t}) \doteq 1 + (B - 1)\omega, \tag{3.25}$$

where B is the cluster size (see Section 3.2). Because in cluster sampling the clusters are usually internally homogeneous, resulting in a positive ω, the DEFFs tend to be greater than one.

To fully utilize the above formulae in planning a sampling design, it would be necessary to know the variation of the study variable in the population. In choosing a sampling design the planner would also need knowledge about the variation at stratum and cluster levels, and information on the correlation of the study variable and the size measure. In practice, however, this kind of information is rarely available, but in some cases approximations can be taken from auxiliary sources, or by carrying out a smaller pilot study.

Population Design Effects

We next perform a numerical evaluation of the population DEFFs for the total by calculating the design variances by the corresponding formulae for the six sampling designs considered for the *Province'91* population. The fixed sample size is eight municipalities $(n = 8)$ drawn from the population of 32 municipalities $(N = 32)$. The values of the population design effects are displayed in Table 3.15.

PPS sampling with probability proportional to a measure of size appears to be the most efficient sampling design for the estimation of a total. The population DEFF is 0.01, which is very small. Improved efficiency is due to the relationship between UE91 and HOU85 (which was used as the size measure) such that the ratio of these variables is nearly constant across the population. It should be noted that the shape of the population distribution of the study variable UE91 also affects efficiency. The distribution of UE91 in the *Province '91* population is very skewed. However, under PPS large selection probabilities are given for large clusters, such that the possible samples drawn from the population will vary to a rather small extent in their composition. Sample totals are thus expected not to vary much from sample to sample and this leads to efficient estimation. For improved efficiency it is also beneficial if the study variable and the size measure are strongly correlated. In the case considered, the correlation was close to one.

Table 3.15 Population DEFFs for a total estimator under various sampling designs for the *Province '91* population.

Sampling design		S.E.	DEFF
Sampling proportional to size (wr)	PPS	720	0.01
Stratified sampling (power alloc.)	STR	4852	0.44
Systematic sampling (random start)	SYS	5408	0.55
Cluster sampling (two-stage)	CLU2	6532	0.80
Cluster sampling (one-stage)	CLU1	6663	0.84
Simple random sampling (wor)	SRS	7283	1.00

Stratified sampling also appears to be quite efficient for the estimation of a total because the DEFF is 0.44, but the difference in favour of PPS is still noticeable. The stratification divided the municipalities into urban and rural ones, and it appeared that in urban municipalities there are more unemployed on average than in the rural municipalities. The strata were thus internally homogeneous, a property which increases efficiency. The efficiency of systematic sampling is close to the STR design. Since there is a monotonic trend in the sampling frame, intra-class correlation becomes quite close to zero, leading to improved efficiency. Efficiency of two-stage cluster sampling is somewhat less than that of systematic sampling, and one-stage cluster sampling is slightly less efficient than two-stage cluster sampling.

Sample Design Effects

The previous efficiency comparisons were theoretical in the sense that we considered the design variances at the population level. We next evaluate the efficiency from a selected sample of size $n = 8$ units drawn from the *Province'91* population. We thus obtain an estimated deff, calculated by the corresponding variance estimates $\hat{v}_{p(s)}(\hat{\theta}^{\star})$ and $\hat{v}_{srs}(\hat{\theta})$, which for an estimator of a population parameter θ is given by

$$\text{deff}_{p(s)}(\hat{\theta}^{\star}) = \frac{\hat{v}_{p(s)}(\hat{\theta}^{\star})}{\hat{v}_{srs}(\hat{\theta})}, \qquad (3.26)$$

where $\hat{\theta}^{\star}$ is a design-based estimator of θ and $\hat{\theta}$ is the SRS counterpart.

By using the sample deff, the efficiency of estimation under the given sample obtained with the various sampling designs $p(s)$ is compared for the estimators $\hat{t}^{\star}$ (total), $\hat{r}^{\star}$ (ratio) and $\hat{m}^{\star}$ (median). There is a natural interpretation for these estimators in the *Province'*91 population. The total measures the total number of unemployed (UE91) in the province, the ratio measures the unemployment

Table 3.16   The sample deff estimates of the estimators of the total, the ratio and the median under the six different sampling designs; the *Province'91* population.

Sampling design		deff($\hat{t}^{*}$)	deff($\hat{r}^{*}$)	deff($\hat{m}^{*}$)
Sampling proportional to size	PPS	0.0035	0.19	2.02
Stratified sampling (power alloc.)	STR	0.21	0.38	1.68
Systematic sampling (implicit str.)	SYS	0.76	1.29	0.70
Cluster sampling (two-stage)	CLU2	0.93	0.99	1.07
Cluster sampling (one-stage)	CLU1	1.92	1.44	1.75
Simple random sampling	SRS	1.00	1.00	1.00

rate, and the median gives an average number of unemployed per municipality. The figures displayed in Table 3.16 are extracted from the PC CARP outputs presented in earlier sections.

The deff estimates vary not only between the sampling designs but also between the estimators for a given design. PPS and STR are the most efficient designs for the total because the deff estimates are close to zero. For the ratio, PPS and STR are superior to the others but have larger deffs than those calculated for the total. For the median, the deff estimates vary irregularly from one to two except under systematic sampling with implicit stratification where the deff estimate is smaller than one.

Summary

The design effect provides a practical tool for the evaluation of the efficiency of an estimator under a given sampling design. Using design effects it is also possible to compare the efficiency of different sampling designs. The design effect clearly shows the effect of complex sampling relative to simple random sampling. Even for a scalar-type estimator the sampling design can affect the design effect in various ways depending on the type of the estimator being considered. An estimator of a total is a linear-type estimator, a ratio estimator is a nonlinear estimator and a median is a robust estimator of a mean. These represent the types of estimator commonly used in statistical analysis. It is important to note that if an optimal design were desired for a given estimator, say for the total, so as to minimize its standard error, i.e. to produce a deff estimate close to zero, the optimality criterion would not necessarily be fulfilled for another estimator. In our examples an estimator $\hat{m}$ for median seemed to be almost untouched by the design effect.

Design effects can be successfully utilized in the analysis of complex survey data. In the preceding sections, we used design effects mainly for descriptive purposes to solve estimation problems concerning a small fixed population. In the following chapters, we present several analytical situations and give further practical examples of the use of design effects. There, estimation and testing problems are considered for complex survey data from large populations. It will be shown, for example, that using design effects (or their generalizations) it is possible to estimate standard errors and calculate observed values of test statistics so that the complexities of a sampling design are properly accounted for. For both descriptive and analytical purposes, design effects can be obtained by using commercial software for survey analysis. Moreover, design effects are good indicators of the effects of complex sampling inherent in the computations. The classical paper of Kish and Frankel (1974) is recommended as further reading on this topic.

4

Handling Missing Data

In the survey estimation methodology discussed so far, the only source of variation has been the sampling error, which has been measured by the standard error of an estimator. In addition to the sampling error, there are also other sources of variation in surveys causing so-called *nonsampling errors.* In particular, these errors can be present in large-scale surveys. The important types of nonsampling errors are *nonresponse, measurement errors* and *outlying observations.* If nonsampling errors can be assumed random, they do not cause bias or other complications in the estimation. This kind of nonsampling error is called *ignorable.* But nonsampling errors can also be harmful when they are nonrandom and cause biased estimation. Various techniques are available for adjusting for the undesirable effects of this *nonignorable* nonsampling error. In this section, we discuss in greater detail methods for adjusting for a particular source of nonsampling error, namely that caused by nonresponse.

Nonresponse

Failure to obtain all the intended measurements or responses from all the selected sample members is called nonresponse. Nonresponse causes *missing data,* i.e. results in a data set whose size for the study variable y is smaller than planned. Two types of missing data are distinguished for a sample element. First, a selected sample element can be totally missing, e.g. due to a refusal to participate in a personal interview. The *unit nonresponse* has thus arisen, because the entire y vector is missing for that sample element. On the other hand, if an interviewed person does not respond to all of the questions, an *item nonresponse* has arisen, because at least one item of the y vector is missing for that element. Missing data of either type can give biased estimates and erroneous standard error estimates. Moreover, in a large-scale survey, both types of nonresponse can be present.

The type of the missing data, unit nonresponse or item nonresponse, guides the selection of an appropriate method for adjusting for the nonresponse in an estimation procedure.

Various *reweighting methods* are available for appropriate adjustment for the unit nonresponse. And for the item nonresponse, the missing values can be *imputed* by various imputation methods. Reweighting and imputation are discussed separately in the following two sections.

Some examples of the main causes leading to nonresponse are:

1. *Non-contacts*	In a mail survey, the questionnaire may not reach some of the sampled persons. This situation arises if, for example, a register file used as the sampling frame is not up-to-date. This results in unit nonresponse.

2. *Refusals*	There can be various causes for a refusal in a survey. For example, the time for contact with a sample person can be inconvenient, and the interviewee refuses. This also results in unit nonresponse.

3. *Inability to answer all the questions*	Some questions can be related to subject matters which are salient or too difficult for the interviewee. So, one or more questions do not receive a valid response or any response at all. This results in item nonresponse.

These three causes of nonresponse, and others, can result in noticeable chunks of missing data in large-scale surveys. For example, in the 1981 Household Survey carried out by Statistics Finland, the intended sample size was 9970 households. The final interviewed sample amounted to 7360 households due to unit nonresponse. Thus, the nonresponse rate was 26%. In the 1985 Household Survey, the nonresponse rate increased to 30%. The main reason for nonresponse was refusal; altogether 80% of the total group of nonrespondents. The next most common reason was where a sample household could not be contacted due to a wrong address. The proportion of these households was 10% of the total group of the nonrespodents. Laaksonen (1992) has studied the structure of the missing data and the corresponding adjustment methods when analysing the Household Survey data. As another example, in the Occupational Health Care Survey, described in Section 6.1, where the data were gathered from industrial establishments, overcoverage was one of the causes of unit nonresponse; this has been modelled by Lehtonen (1988).

Impact of Unit Nonresponse

Nonresponse results in a sample data set whose size $n^{(r)}$ is smaller than the intended sample size n, thus increasing the standard errors of the estimates. This can be seen by considering the variance of an estimator $\hat{t}$ of a population total T. Under simple random sampling without replacement, this variance is $V_{srs}(\hat{t}) = N^2(1 - n/N)S^2/n$ where the denominator is the original sample size n. If the number of respondents decreases due to unit nonresponse, the denominator decreases and thus the variance increases. But a more serious consequence   is that   the estimation can become biased due to missing observations. This is particularly true if unit nonresponse is systematic, i.e. nonignorable, such that it is more common in certain subgroups of the sample. For example, as an extreme case, let us suppose that in an interview survey, a certain subgroup of the sample refuses totally to participate. In this case, the total population can be divided into two strata, one for the response group and one for the nonresponse group, whose stratum sizes are N_1 and N_2. The corresponding stratum weights are $W_1 = N_1/N$ and $W_2 = N_2/N$, where the weight W_1 is for the response group and W_2 is for the nonresponse group. After the fieldwork, all the sample data available for the estimation come only from the first stratum covering the response cases. Let the estimator for the total T be $\hat{t} = N \times \bar{y}^{(r)}$, where $\bar{y}^{(r)}$ is the mean of the respondent data. Because all the respondents are from the stratum 1, the expectation the respondent mean $\bar{y}^{(r)}$ equals, say $\bar{Y}_1$, the population mean of that stratum. If the population stratum means are unequal, or $\bar{Y}_1 \neq \bar{Y}_2$, then the estimator $\hat{t}$ is a biased estimator for the population total T, since

$$E(\hat{t}) - T = N\bar{Y}_1 - (N_1\bar{Y}_1 + N_2\bar{Y}_2) = N_2(\bar{Y}_1 - \bar{Y}_2) = \text{BIAS}(\hat{t}) \qquad (4.1)$$

In practice, this bias is difficult to evaluate. Although the stratum size N_2 could be roughly estimated, the stratum mean $\bar{Y}_2$ remains totally unknown. Moreover, to cause further inconvenience in this case, the variance of the estimated total will be underestimated. The bias due to unit nonresponse is illustrated in the following example.

Example 4.1

Unit nonresponse bias in the *Province'91* population. Let us assume that the southern municipalities were not able to complete the records for the unemployed in time. These municipalities are, for example, Kuhmoinen, Joutsa, Luhanka, Leivonmäki and Toivakka. The population of municipalities can thus be divided into two strata, the stratum of the respondents ($N_1 = 27$) and the stratum of the nonrespondents ($N_2 = 5$), whose stratum totals, sizes and means are:

$T_1 = 14\,475$	$N_1 = 27$	(response)	$\bar{Y}_1 = 536$
$T_2 = 623$	$N_2 = 5$	(nonresponse)	$\bar{Y}_2 = 125$
$T = 15\,098$	$N = 32$	(whole province)	$\bar{Y} = 472$

When drawing the sample by SRS, the selected sample would include both the response and the nonresponse municipalities. Thus, the expected value of the total estimator, based on the response group sample total $\hat{t}$, will be

$E(\hat{t}) = N \times \bar{Y}_1 = 32 \times 536 = 17\,152$. If this estimator is taken as the estimator of the population total, a biased estimate results, where the bias due to the unit nonresponse is

$$\text{BIAS}\,(\hat{t}) = E(\hat{t}) - T = N_2(\bar{Y}_1 - \bar{Y}_2) = 5 \times (536 - 125) = 2055,$$

and is noticeably large.

Adjustment Methods for Nonresponse

In practice, nonresponse can often be systematic since it relates to only certain subgroups of the target population. This can be seen, for example, in the Occupational Health Care Survey (see Section 6.1), where unit nonresponse concentrated in small single-site business firms that operated in the construction industry. The southern municipalities constituted the group of the nonrespondents in the example case from the *Province'91* Population, resulting in seriously biased estimation. Therefore, the amount and structure of missing data should be carefully examined, and adjusted if appropriate, before further estimation. This generally holds for both unit and item nonresponse.

The two main methods for adjustment for nonresponse are reweighting and imputation. The adjustment for unit nonresponse can be done by *reweighting* the sampling weights $w_k = 1/\pi_k$ by the inverses $1/\theta_k$ of response probabilities θ_k, providing new adjusted weights $w_k^* = 1/(\pi_k \theta_k)$. Reweighting methods for unit nonresponse are commonly used, for example, by national statistical agencies. Section 4.1 is devoted to the use of these methods.

In particular for item nonresponse, a missing value can be replaced by a predicted value obtained from an appropriate model for the item nonresponse. One of the simplest techniques used in practice, is to replace a missing value for a certain variable by the overall mean of that variable, obtained from the group of respondents. But this is not a totally satisfactory method, as will be seen later in Section 4.2, where various alternative *imputation methods* for item nonresponse are examined.

4.1   REWEIGHTING

Unit nonresponse refers to the situation where data are not available within the survey data set for a number of sampling units. Reweighting can then be used, applied to the observations from the respondents, using auxiliary information available for both the respondents and the nonrespondents. As a simple example, consider the estimation of a population total: the values obtained from the respondents can be multiplied by the inverse of the response rate. For a simple example, if the overall response rate in a survey is 71%, a suitable raising factor would be $1/0.71 = 1.41$. In this nonresponse model, it is assumed that each population element has the same probability θ of responding if selected in the sample, i.e. $\theta_k = \theta$ for all the population elements $k = 1, \ldots, N$. Under this rather naive assumption of a nonresponse mechanism, a *Horvitz–Thompson-type* estimator for the population total would be $\hat{t}_{ht} = \sum_{k=1}^{n^{(r)}} y_k / (\theta \times \pi_k)$, where y_k is the sample value of the study variable y for a respondent, π_k is the inclusion probability, and the superscript '(r)' refers to the respondents; so, $n^{(r)}$ denotes the number of respondents in the sample. The impact of unit nonresponse can be seen more clearly if the previous estimator is written in the form $\hat{t}_{ht} = (\frac{1}{\theta})\hat{t}_o$, where $\hat{t}_o$ is an ordinary *Horvitz–Thompson* estimator of the total based on the respondent data, i.e. $\hat{t}_o = \sum_{k=1}^{n^{(r)}} y_k / \pi_k$.

Although these kind of adjustment methods are used in practice, better estimation can be attained by modelling the response probability. Commonly used models for this are the *adjustment-cell weighting model*, *raking-ratio weighting model* and *logistic models* for the θ. For these nonresponse models, the observed sample is first partitioned into homogeneous response groups according to auxiliary information available from both the respondents and the nonrespondents. These groups are called *adjustment-cells*, and they are denoted by $1, \ldots, c, \ldots, C$. The cell sample sizes and the numbers of respondents in each cell are denoted correspondingly by $n_1, \ldots, n_c, \ldots, n_C$ and $n_1^{(r)}, \ldots, n_c^{(r)}, \ldots, n_C^{(r)}$. The homogeneity of the adjustment cells means that all the elements in a cell c are assumed to have the same response probability θ_c which is estimated by the cell response rate $\hat{\theta}_c = n_c^{(r)}/n_c$. Between the adjustment cells, however, the response probabilities, can vary. And in the reweighting, the inverses of the estimated response probabilities, i.e. cell response rates $\hat{\theta}_c$, can be used. This adjustment for the unit nonresponse can be more powerful than the previous one, because the response probabilities are modelled by using the information about the structure of the nonresponse more efficiently.

Some commonly used nonresponse modelling techniques are summarized now.

(1) *Constant weighting*, where it is assumed that $\theta_k = \theta$ for all $k = 1, \ldots, N$. This model assumes that each population element has a common probability of responding if selected in the sample. However, it is often unrealistic in practice, and it should mainly be used for comparative purposes.

(2) *Adjustment-cell weighting*, where it is asssumed that $\theta_{ck} = \theta_c$, $c = 1, \ldots, C$. We assume that the population is divided into a number of adjustment cells so that within each cell, all the population elements share the same probability of responding if selected in the sample. The adjustment-cell model will be dis-

cussed in more detail by introducing three different estimators for a population total.

(3) *Raking-ratio weighting*, where it is assumed that $\theta_{cdk} = \theta_{cd}$ for $c = 1, \ldots, C$, and $d = 1, \ldots, D$. In this model, it is assumed that the population is cross-classified by two auxiliary variables into a total of $C \times D$ adjustment cells. This model also assumes that there is for each population element in cell (cd) a constant probability of responding if selected in the sample. In addition, it assumes that the response probability in cell (cd) has a particular form $\theta_{cd} = \alpha_c \beta_d$, where α_c is the response probability for class c and β_d is that for the class d.

(4) *Logistic or probit models for response probability* θ_k. In this method, the adjustment cells $c = 1, \ldots, C$ are first formed, and then, the cell response probabilities are modelled by using, for example, a logit model

$$\log\left[\frac{\theta_c}{1 - \theta_c}\right] = \sum_{s=1}^{S} \beta_s z_{cs}, \quad c = 1, \ldots, C,$$

where z_{cs} are the values of the S auxiliary variables z_s, and β_s are the S regression coefficients.

However, all the methods, except the first, rely upon the construction of adjustment cells of some sort. Let us consider in more detail the second method, namely the adjustment-cell modelling method (see Little 1986).

Adjustment-cell Weighting

The method of adjustment-cell weighting resembles that of poststratification, which was discussed in Section 3.3. In the adjustment-cell method, a stratification variable is first chosen in order to attain adjustment-cells as homogeneous as possible. Then, an appropriate estimator is chosen from the set of available estimators, for the estimation of the desired parameter. This method of adjustment also allow us to study the bias caused by the unit nonresponse. Moreover, the mean square error MSE should be examined instead of the variance, where the MSE for an estimator $\hat{t}$ of the total can be written as

$$\text{MSE}\,(\hat{t}) = V(\hat{t}) + \text{BIAS}^2\,(\hat{t}), \tag{4.2}$$

where $\text{BIAS}\,(\hat{t}) = E(\hat{t}) - T$.

In the following, three reweighted estimators are considered in the estimation of the total. Their bias, variance and mean square errors are also calculated.

(1) *General adjustment-cell estimator* is given by

$$\hat{t}_R = N \sum_{c=1}^{C} p_c^{(r)} \bar{y}_c^{(r)}, \qquad (4.3)$$

where $\bar{y}_c^{(r)} = \sum_{k=1}^{n_c^{(r)}} y_{ck} / n_c^{(r)}$ is the respondent mean in adjustment cell c and $p_c^{(r)} = n_c^{(r)} / n^{(r)}$ is the proportion of respondents in cell c calculated from the total number of respondents in the sample. Note the difference between $n_c^{(r)}$ and $n^{(r)}$; and that between $p_c^{(r)}$ and $\hat{\theta}_c = n_c^{(r)} / n_c$.

(2) *Sample-based adjustment-cell weighting estimator* is given by

$$\hat{t}_S = N \sum_{c=1}^{C} p_c \bar{y}_c^{(r)}, \qquad (4.4)$$

where $\bar{y}_c^{(r)}$ is the respondent mean in adjustment cell c and $p_c = n_c / n$ is the estimated relative size of adjustment-cell c.

(3) *Population-based adjustment-cell weighting estimator* is given by

$$\hat{t}_A = N \sum_{c=1}^{C} P_c \bar{y}_c^{(r)}, \qquad (4.5)$$

where $\bar{y}_c^{(r)}$ is the respondent mean in the adjustment cell c and $P_c = N_c / N$ is the relative size of the adjustment cell. This corresponds to the exact post-stratum weight (see Section 3.3).

However, all of the estimators defined remain biased, so their mean square errors MSE should be calculated, using the equations for the variances and biases. To evaluate the biases, the population mean $\bar{Y}$, the population mean of respondents $\bar{Y}^{(r)}$ and the corresponding adjustment-cell means $\bar{Y}_c$ and $\bar{Y}_c^{(r)}$ are needed. In the formulae below, the term $S_c^{(r)2}$ refers to the population variance of respondents in adjustment-cell c. The formulae are given neglecting the finite-population corrections, because in practice nonresponse adjustments are usually carried out in large-scale surveys where there are thousands of observations, and so the finite-population correction is of little value and is often not used.

The bias of the estimator $\hat{t}_R$ is given by

$$\text{BIAS}\,(\hat{t}_R) = N \sum_{c=1}^{C} (p_c^{(r)} - P_c^{(r)})\bar{Y}_c^{(r)} + N(\bar{Y}^{(r)} - \bar{Y}),\qquad(4.6)$$

where $P_c^{(r)} = N_c^{(r)}/N^{(r)}$ is the population proportion of respondents in the adjustment cell c calculated from the total number of respondents in the population.

The variance of $\hat{t}_R$ is given by

$$V(\hat{t}_R) = N^2 \sum_{c=1}^{C} p_c^{(r)2} S_c^{(r)2}/n^{(r)}.\qquad(4.7)$$

The bias of the estimator $\hat{t}_S$ is given by

$$\text{BIAS}\,(\hat{t}_S) = N \sum_{c=1}^{C} (p_c - P_c)\bar{Y}_c^{(r)} + N \sum_{c=1}^{C} P_c(\bar{Y}_c^{(r)} - \bar{Y}_c),\qquad(4.8)$$

The variance of this estimator is given by

$$V(\hat{t}_S) = N^2 \sum_{c=1}^{C} p_c^2 S_c^{(r)2}/n_c^{(r)}.\qquad(4.9)$$

The bias of the estimator $\hat{t}_A$ is given by

$$\text{BIAS}\,(\hat{t}_A) = N \sum_{c=1}^{C} P_c(\bar{Y}_c^{(r)} - \bar{Y}_c),\qquad(4.10)$$

and the variance of the estimator is given by

$$V(\hat{t}_A) = N^2 \sum_{c=1}^{C} P_c^2 S_c^{(r)2}/n_c^{(r)}.\qquad(4.11)$$

In practice, the first reweighted estimator for the total, the general adjustment-cell weighting estimator, can be calculated from the respondent data set without any auxiliary information from the nonrespondents. In the sample-based adjustment-cell weighting estimator, information of the amount of nonrespondents in class c is required, which implies that auxiliary information is also required from the nonrespondents. And in the population-based cell-weighting estimator, auxiliary information for all population elements is required.

The statistical properties of the estimators depend on the success of the division of the adjustment cells into cells with approximately equal homogeneity, since a common response probability is strived for within a cell. For bias reduction, such auxiliary variables should strongly correlate with the study variable. However, in large-scale surveys with many diverse study variables, it is often impossible to attain an overall feasible construction for the adjustment cells. But in descriptive surveys this goal is more often attained.

The three reweighted estimators are calculated from the *Province'91* population in the following example, together with their respective variances and mean square errors.

Example 4.2

Adjustment by reweighting the unit nonresponse in the case where the sample is selected with SRS from the *Province'91* population. The reweighted estimators for the total T are the following: the respondent mean estimator, the sample-based and the population-based adjustment cell weighting estimators. The adjustment cells are formed by the division of municipalities into towns and rural municipalities. The corresponding reweights are calculated as follows. The respondent mean estimator uses reweights that are the actual sampling weights $w_k = 4$ multiplied by the inverse of the response probability $\hat{\theta} = n^{(r)}/n$, which inverse is in this case $(6/8)^{-1} = 1.3333$. Thus the R-WGHT is $1.3333 \times 4 = 5.3333$ for all responded units and zero for nonrespondents.

Let us consider next the response rate in adjustment cells. All the towns responded, so the response rate in adjustment cell c_1 is $\hat{\theta}_1 = 1$. In the second adjustment cell c_2 for rural municipalities, $\hat{\theta}_2 = 3/5$ and thus the inverse of the response probability is $(3/5)^{-1} = 1.6667$. In the case of the S-WGHT the sizes of the corresponding population cells are estimated by the counts included in the sample. The estimated size of the cell c_1 including towns is $\hat{N}_1 = n_1/n \times N = (3/8) \times 32 = 12$ and the corresponding size of the cell c_2 is $\hat{N}_2 = N - \hat{N}_1 = 32 - 12 = 20$. Thus the S-WGHT in adjustment cell c_1 is $1 \times 12/3 = 4.0000$ and in adjustment cell c_2 the S-WGHT is $1.6667 \times 20/5 = 6.6667$ for the respondent municipalities and zero for nonrespondents. The A-WGHT is similar except that the estimated size $\hat{N}_i$ in the cell c_i is replaced by the corresponding true cell size in the population. Those figures are $N_1 = 7$ for towns and $N_2 = 25$ for rural municipalities. The corresponding A-WGHTS are, in the first adjustment cell, $1 \times (7/3) = 2.3333$ and $1.6667 \times (25/5) = 8.3333$ for respondents and zero for the nonrespondents in the second adjustment cell. The sample data with the corresponding reweights are displayed in Table 4.1. Note that all reweights sum up to 32.

The three previously derived reweighted estimates are next calculated from the data presented in Table 4.1 by using the SRS option of PC CARP. The sampling rates are defined as the number of respondents in the sample divided by

Table 4.1 A simple random sample from the *Province'91* population including two missing values, and adjustment reweights for the correction of unit nonresponse.

STR	CLU	WGHT	Element LABEL	Response UE91	Adj. CELL	Reweights by adjustment model R-WGHT	S-WGHT	A-WGHT	Full SAMPLE
1	1	4	Jyväskylä	4123	1	5.3333	4.0000	2.3333	4123
1	4	4	Keuruu	760	1	5.3333	4.0000	2.3333	760
1	5	4	Saarijärvi	721	1	5.3333	4.0000	2.3333	721
1	26	4	Pihtipudas	331	2	5.3333	6.6667	8.3333	331
1	18	4	Kuhmoinen	.*	2	0.0000	0.0000	0.0000	187
1	31	4	Uurainen	219	2	5.3333	6.6667	8.3333	219
1	30	4	Toivakka	.*	2	0.0000	0.0000	0.0000	127
1	15	4	Konginkangas	142	2	5.3333	6.6667	8.3333	142

* A missing value is denoted as '.'.

the estimated or actual size of the adjustment cell in the target population. For the estimator $\hat{t}_R$ the sampling rate is $(n_1^{(r)} + n_2^{(r)})/N = (3 + 3)/32 = 0.1875$. For the sample-based adjustement cell estimator $\hat{t}_S$ the sampling rate is, in the first adjustment cell, $n_1^{(r)}/\hat{N}_1 = 3/12 = 0.2500$ and in the second adjustment cell $n_2^{(r)}/\hat{N}_2 = 3/20 = 0.1500$. The third estimator, $\hat{t}_A$, uses the true population sizes of adjustment cells. Thus the corresponding sampling rates are $n_1^{(r)}/N_1 = 3/7 = 0.4286$ and $n_2^{(r)}/N_2 = 3/25 = 0.1200$. For a fair comparison, the basic design-based estimator $\hat{t}_{ht}$ for a total is calculated from figures presented in the column headed 'Full sample' in Table 4.1. Sampling rate is in this case $n/N = 8/32 = 0.2500$.

Of the reweighting methods, the estimator $\hat{t}_A$, based on the population-based estimator, gives an estimate closest to the population total ($T =15\,098$). This estimator is almost equal to the poststratified estimator. The mean square error of a total estimator includes a term due to the response bias, and the variance within the adjustment-cells. For calculating the mean square error MSE, the population mean, the population respondent mean and corresponding adjustment-cell means are needed. They are: $\bar{Y} = 472$, $\bar{Y}^{(r)} = 536$, $\bar{Y}_1^{(r)} = 1146$ and $\bar{Y}_2^{(r)} = 323$. The adjustment-cell population variances are $S_1^{(r)2} = 1318^2$ and $S_2^{(r)2} = 360^2$. The bias, variance and mean square error of each estimator are given below.

1. The bias of the estimator $\hat{t}_R$ (general adjustment-cell estimator) is

$$\text{BIAS}\,(\hat{t}_R) = 32 \times \left[\left(\frac{3}{6} - \frac{7}{27}\right) \times 1146 + \left(\frac{3}{6} - \frac{20}{27}\right) \times 323\right]$$
$$+ 32 \times (536 - 472) = 8388.$$

The variance and the mean square error are

$$V(\hat{t}_R) = 8925^2 \quad \text{and} \quad \text{MSE}\,(\hat{t}_R) = 12\,248^2.$$

2. The bias of the estimator $\hat{t}_S$ (sample-based adjustment-cell estimator) is

$$\text{BIAS}\,(\hat{t}_S) = 32 \times \left[\left(\frac{3}{8} - \frac{7}{32}\right) \times 1146 + \left(\frac{5}{8} - \frac{25}{32}\right) \times 323\right.$$
$$\left. +0 + 32\left[\frac{25}{32}(323 - 283)\right] = 5115.$$

The variance and the mean square error are

$$V(\hat{t}_S) = 10\,033^2 \quad \text{and} \quad \text{MSE}\,(\hat{t}_S) = 11\,262^2.$$

3. The bias of the estimator $\hat{t}_A$ (population-based adjustment-cell estimator) is

$$\text{BIAS}\,(\hat{t}_A) = 0 + 32\left[\frac{25}{32}(323 - 283)\right] = 1000.$$

The variance and the mean square error are

$$V(\hat{t}_A) = 7441^2 \quad \text{and} \quad \text{MSE}\,(\hat{t}_A) = 7508^2.$$

The estimation results are summarized in Table 4.2.

In the table, the second last row, including the SRS estimates from the original full sample, can be used as the reference when comparing the results to those obtained from data where unit nonresponse is not present. The sample-based adjustment works most reasonably, providing an estimate for the

Table 4.2 Estimates of the total and its mean square error MSE under various reweighting models; a simple random sample from the *Province'91* population.

Model	Estimate	s.e. $(\hat{t})$	MSE$(\hat{t})$
No reweighting ($n = 6$)	33 576	17 988	12 248
Respondent mean adjustment $\hat{t}_R$	33 576	18 756	12 248
Sample-based cell adjustment $\hat{t}_S$	27 029	11 882	11 262
Population-based cell adjustment $\hat{t}_A$	18 843	6 469	7 508
Full sample ($n = 8$)	26 440	13 282	7 280
True population value ($N = 32$)	(15 098)		

See text for sampling rates.

total that is close to that derived from the full sample. On the other hand, when the results are referred to the true value of the total (the last row in the table), the population-based method gives an estimate closest to T. These results are reasonable, considering the way that the auxiliary information is used in the methods. The population-based method thus tends to adjust for the nonresponse bias in a similar way to ordinary poststratification.

Three reweighted estimators for a total have been discussed. The population-based adjustment-cell estimator gave an estimate closest to the population total. In addition, its mean square error was the smallest. This method, however, requires auxiliary information from all the elements of the population, whereas in the other two methods, auxiliary information for the selected sample, or no auxiliary information at all, is required. Therefore, in practice, the choice of the reweighting method depends on the availability of auxiliary information.

The efficiency of the reweighting methods depends mainly on a successful choice of adjustment cells. To reach this goal, a versatile use of auxiliary information gathered from the selected sample and from the target population is beneficial.

4.2 IMPUTATION

Item nonresponse means that missing values are distributed more or less randomly among the observations in the data matrix. When using this kind of data matrix with some computer programs for survey estimation, such as SUDAAN, each observation with a missing value for any of the variables included in the analysis is excluded. On the other hand, in some programs, such as PC CARP, a complete data matrix is required. This leads to loss of information for the other variables for which data are not missing. Therefore, efforts are often made to get a more complete data set. To attain this goal, different imputation techniques have been devised.

Imputation implies simply that a missing value of the study variable y for a sample element k in the data matrix is substituted by an imputed value $\hat{y}_k$. For example, in some computer packages, a technique called *mean imputation* is available, in which an overall mean, calculated from the response values of the study variable, is inserted in place of the missing values for that variable. However, there are certain disadvantages in this method, as will be demonstrated in Example 4.3. In more advanced methods, auxiliary information available from the target population or from the original sample is utilized to model the missing values realistically. In addition, a single missing value may be replaced by two or more imputed values, as in the method of *multiple imputation*. This method can improve the variance estimation, and it is available in certain computer packages, such as OSIRIS IV and GAUSS/MISS. Also, in PC CARP, an incomplete data matrix can be rendered complete by treating it with the subprogram PRE CARP, which generates a complete data matrix using one of the commonly used imputation techniques called *hot-deck imputation.*

The following simple example deals with an incomplete sample data set obtained using SRS from the *Province'91* population. For this purpose, two artificial unit nonresponses are created: the number of unemployed in two municipalities are given missing values. The results from imputation by various nonresponse modelling techniques are presented, and the behaviour of the point estimates and their average standard errors is discussed for the estimation of the population total of unemployed.

Example 4.3

Imputation of two missing values in a sample selected from the *Province'91* population with SRS. The sample is chosen to be the same as that used in Section 3.5, and it includes $n = 8$ municipalities. Two missing values are created by deleting the values of the study variable UE91 in two municipalities (Kuhmoinen and Toivakka). Certain auxiliary information is given for modelling the missing values.

Three imputation techniques will be applied to the incomplete sample data. The first is the mean imputation method (Model-0). The mean of the respondents $(n^{(r)} = 6)$ is 1049. The two missing values are replaced by this overall mean. The second imputation method, based on dividing the sample into two adjustment-cells, urban and rural municipalities (Model-1), uses the information that both of the nonrespondents are rural municipalities. The mean of the responding rural municipalities $(n_2^{(r)} = 3)$ is 231. This mean value is inserted in the incomplete data matrix in place of the missing values. The third nonresponse modelling method (Model-2) uses the variable HOU85 as auxiliary information.

Table 4.3 A simple random sample from the *Province'91* population including two missing values.

ID	Element LABEL	Study var. UE91	Post STR	Aux. var. HOU85
1	Jyväskylä	4 123	1	26 881
4	Keuruu	760	1	4 896
5	Saarijärvi	721	1	3 730
26	Pihtipudas	331	2	1 946
18	Kuhmoinen	.*	2	1 463
31	Uurainen	219	2	932
30	Toivakka	.*	2	834
15	Konginkangas	142	2	556

* A missing value is denoted as '.'.

The sample data is sorted into descending order according to the values of HOU85. Once the ordered data set has been created, the first available sample value preceding a municipality with a missing value is taken and inserted in the data set in place of the missing value. This method is called hot-deck imputation. According to this method, the imputed value for Toivakka is 219, and for Kuhmoinen 331. The imputed values are displayed in Table 4.4.

Imputation has two different impacts. Firstly, it gives a point estimate for a missing value and, secondly, it has an effect on the standard error of the estimator that we are interested in. The obvious gain from imputation is that the analyst has a complete data matrix for calculation, but if the imputation model gives biased imputed values, the results of analysis may be misleading. All depends on how succesfully the imputation model catches the non-response. This means that in every adjustment cell the propensity to respond is random. The respondent mean is then an unbiased estimate for all the elements belonging to this cell, including the missing values. But this does not remove the problems for variance, because imputed values are also estimates and have their own variance component which is to be added to the variance of the basic estimator.

A brief consideration will be given to the properties of the four imputation models. For better understanding, it is worthwhile to compare the imputed values with the true values given in the last row of Table 4.4. In particular, it can be seen that the overall mean of the respondents used in the mean imputation method seriously overestimates the true value of the study variable. Moreover, because the mean imputation keeps the sum of squares unchanged in the variance formula, but increases the number of observations, the variance of the estimator will be decreased. This leads to underestimation of the variance of the total and produces confidence intervals that are too narrow. The method based on the stratified means assigns the same imputed value to the two municipalities, but the value is much more realistic than the previous one. And in the hot-deck imputation, different imputed values are assigned for the two municipalities, and therefore similar underestimation cannot be realized, unlike using the mean imputation method. In Table 4.5, the

Table 4.4 Single imputation under various imputation methods; a simple random sample from the *Province '91* population.

Model	Imputed value	
	Kuhmoinen	Toivakka
No imputation	. . .	. . .
Model-0 (overall mean)	1049*	1049*
Model-1 (stratified mean)	231*	231*
Model-2 (hot deck)	331*	219*
True value	187	127

* Imputed values

Table 4.5 The imputed data set obtained by various imputation models (the *Province '91* population).

ID	Element	No Imp.	Imputed values by model Model-0	Model-1	Model-2	Full sample
1	Jyväskylä	4123	4123	4123	4123	4123
4	Keuruu	760	760	760	760	760
5	Saarijärvi	721	721	721	721	721
26	Pihtipudas	331	331	331	331	331
18	Kuhmoinen	.	1049*	231*	331*	187
31	Uurainen	219	219	219	219	219
30	Toivakka	.	1049*	231*	219*	127
15	Konginkangas	142	142	142	142	142

* Imputed values
Sampling rate for imputation models is 6/32 = 0.1875
Sampling rate for 'Full sample' is 8/32 = 0.2500

complete data sets are displayed, including the imputed values.

Each of the imputed data sets as is analysed as if it were a complete data set. This includes eight $(n = 8)$ complete observations as intended in the original sampling plan. To gain further insight into the effects of the various imputation methods, the estimate of the total of UE91 and its standard error are calculated from each data set.

The estimates of the population total and the standard errors should be compared with the figures in the last row of Table 4.6. The full sample estimates were calculated from the original complete data set before the two missing values were inserted. The mean imputation or Model-0 leaves the total estimate calculated for the responded cases unchanged $(\hat{t} = \hat{t}_{model-0} = 33\,576)$, but leads to underestimation of the standard error $(12\,650 \leq 13\,282)$. The more advanced nonresponse adjustment models Model-1 and Model-2 result in estimates which are closer to their true values.

The efficiency of an imputation method depends on the modelling of the nonresponse, if the nonresponse is recognized to be systematic, i.e. nonignor-

Table 4.6 Estimates of the total and its standard error by various imputation models (the *Province'91* population).

Model	Estimate	s.e. $(\hat{t})$
No imputation $(n = 6)$	33 576	17 988
Model-0	33 576	13 166
Model-1	27 032	13 721
Model-2	27 384	13 668
Full sample $(n = 8)$	26 440	13 282

Sampling rate for imputation models is 6/32 = 0.1875
Sampling rate for 'Full sample' is 8/32 = 0.2500

able. In the foregoing example, the most realistic imputation was carried out by using the auxiliary information available from the sample or from the target population. The imputation was then completed by using the mean of the appropriate subgroup of the sample, or the value of a real substitute to be inserted in place of a missing value. In choosing an auxiliary variable, it is important that the variable correlates with the variable to be imputed. However, because in these techniques a missing value is replaced by a single imputed value, underestimation of the standard error can result. To avoid this inconvenience, multiple imputation techniques can be applied instead. In these techniques, two or more imputed values are calculated and a missing value is replaced by their mean. This will increase the standard error of an estimator, since a new variance component enters into the variance formulae – the variation between the multiple imputed values.

Summary

Typically nonresponse is present in large-scale sample surveys. This involves missing data in the form of unit nonresponses and item nonresponses which can cause biased estimation and erroneus standard error estimates. Effective activities are important during the data collection to reduce the nonresponse to a nonsignificant level. The remaining nonresponse can be adjusted for by various techniques. For this, the first task is to evaluate if the nonresponse is ignorable or not. If ignorable to a reasonable degree, the adjustments do not contribute noticeably to the estimation results. But if the nonresponse is nonignorable or harmful, standard estimation ignoring nonresponse can lead to seriously biased estimates. Thus, some adjustment method for accounting for missing values is needed. We introduced two practical ways to perform an adjustment by modelling the nonresponse using auxiliary information available in the sampled data set and in the sampling frame. The difference between these methods depends upon the extent to which auxiliary information was utilized. A more complete treatment of the methodology for missing data is to be found in Little and Rubin (1987). Rubin (1987) considers multiple imputation techniques in detail.

Three sources of nonsampling errors were introduced in the beginning of this chapter: nonresponse, measurement errors and outliers. Measurement errors are thoroughly discussed in Biemer *et al.* (1991), and strategies to handle outliers are presented in Barnett and Lewis (1984). The book by Lessler and Kalsbeek (1992), as well as the book by Groves (1989) and that edited by Groves *et al.* (1988), serve as general references for nonsampling errors in surveys.

5

Linearization and Sample Re-use in Variance Estimation

In this and the following three chapters we discuss estimation, testing and modelling methods for complex analytical surveys common for example in social and health sciences. In analytical surveys, variance estimation is needed to obtain standard error estimates of sample means and proportions for the total population and, more importantly, for various subpopulations. In modelling procedures, variance estimates of estimated model coefficients, such as regression coefficients, are needed for proper test statistics. Subpopulation means and proportions are defined as ratio estimators in Section 5.2. Approximation techniques are required for the estimation of the variances of these nonlinear estimators. These techniques supplement those examined for descriptive surveys in Chapters 2 and 3. The linearization method, considered in Section 5.3, is used as the basic approximation method. Alternative methods (balanced half-samples, jackknife and bootstrap) based on sample re-use techniques are examined in Section 5.4, and all the methods are compared numerically in Section 5.5. The approximation methods are demonstrated for the Mini-Finland Health Survey, providing a complex ánalytical survey where stratified cluster sampling is used.

5.1 THE MINI-FINLAND HEALTH SURVEY

The Mini-Finland Health Survey was designed to obtain a comprehensive picture of health and of the need for care in Finnish adults, and to develop methods for monitoring health in the population (Aromaa *et al.* 1989). The sampling design of the survey belongs to the class of two-stage stratified cluster sampling. A variety of data collection methods were used; one aim of the survey was to compare the reliability of these various methods

(Heliövaara *et al.* 1993). A large part of the data were collected in health examinations using a Mobile Clinic Unit, and by personal interviews. Cluster sampling with regional clusters was thus well motivated by cost efficiency.

The target population of the survey was the Finnish population aged 30 years or over. A two-stage stratified cluster-sampling design was used in such a way that one cluster was sampled from each of the 40 geographical strata. The one-cluster-per-stratum design was used to attain a deep stratification of the population of the clusters. The sample of 8000 persons was allocated to achieve an *epsem sample* (equal probability of selection method; see Section 3.2). Recall that an epsem sample refers to a design involving a constant overall element sampling fraction.

Original Sampling Design

The 320 population clusters in the original sampling design consisted of one municipality or, in some cases, two regionally neighbouring municipalities. The clusters were stratified by whether they were urban or rural and the shares of the population in manufacturing industry and agriculture. From the largest towns, 8 self-representing strata were formed. The other 32 strata consisted of several nearly equal-sized clusters and consisted of 40 000–60 000 eligible inhabitants. One cluster was sampled from these noncertainty strata using PPS sampling with a cumulative method where the inclusion probabilities were proportional to the size of the target population in a stratum (see Section 2.5). Second-stage sample sizes were obtained by proportional allocation, resulting in an epsem design. Sample sizes from the sampled clusters varied between 50 and 500 people, the mean being 150. The person-level samples were drawn by systematic sampling in each stratum using a register database as the sampling frame, which covered the relevant population of the sampled clusters.

Modified MFH Survey Sampling Design

The estimation of the between-cluster variance was not possible in the noncertainty strata because only one cluster was drawn from each stratum. The original design was thus modified for variance estimation by using the so-called *collapsed stratum technique*. A total of 16 pseudo-strata were formed from the 32 noncertainty strata so that there were two clusters in each of the new strata. A pair of strata was formed by combining two of the original strata which were approximately equal-sized and had similar values for the stratification variables. In the 8 self-representing strata, two pseudo-clusters were formed by randomly dividing the sample into two approximately equal-sized parts. In the modified design, called the MFH Survey sampling design, there are totals of 24 strata and 48 sample clusters. The MFH design is described in more detail in Lehtonen and Kuusela (1986).

The relatively small number of sample clusters in the MFH Survey sampling design can cause a problem in the estimation of variances and covariances. The number of clusters determines the degrees of freedom available for variance and covariance estimation. These degrees of freedom are defined as the number of sample clusters less the number of strata, i.e. $48-24 = 24$ in the MFH design. This small number can cause instability in variance and co-variance estimates, possibly resulting in difficulties in testing and modelling procedures. The situation is different in the Occupational Health Care Survey and in the Finnish Health Security Survey, where the number of sample clusters is much larger (these surveys will be described in Sections 6.1 and 9.2, respectively). The Second National Assessment of the Comprehensive School (described in Section 9.3) involves a small number of sample clusters, and can thus suffer from similar problems to those in the MFH Survey design. The consequences of instability are discussed in Chapter 6.

Data Collection and Nonresponse

The main phases of the field survey were a health interview, a health examination, which consisted of two phases, and a so-called in-depth examination. The field survey was carried out in 1978–1981. The main methods were interviews, questionnaires, tests of performance, physical and biochemical measurements, observer assessments, and a clinical examination by a doctor. The interview was made by local public health or hospital nurses, and the health examination was carried out by a Mobile Clinic Unit.

Of the 8000 people in the sample, 7703 (96%) completed the health interview, and 7217 (90%) took part in the screening phase of the health examination. Over 6000 persons of those examined during the screening phase had at least one symptom, or finding, or gave a disease history which led to their being asked to attend the clinical phase of health examination; 94% of them attended. Almost 5300 of those examined during the screening phase were asked to attend the doctor's clinical examination; 4840 participated. The data for non-attendants were amended after the field study. Thus, clinical data based on a doctor's examination, or data similar to these data, are available for all 5292 persons invited to the doctor's examinations. The response rates are thus very high for each phase of the survey.

Design Effects

The regional clusters in the MFH Survey sampling design had quite large populations. Because of this, only slight intra-cluster correlations can be expected in most study variables. But there are also variables for which

Table 5.1 Design-effect estimates of sample means or proportions of selected study variables in the MFH Survey data set.

Study variable	deff
Systolic blood pressure	3.2
Chronic morbidity	2.0
Number of physician visits	1.4
Body mass index	1.4
Serum cholesterol	1.2
Number of dental visits	1.0
Number of sick days	0.9

clustering effects are noticeable. Design-effect estimates of sample means or proportions of selected study variables are displayed in Table 5.1, which covers data from the screening phase of the health examination. Design-effect estimates vary between 3.2 and 0.9, the largest estimate being for the mean of a continuous variable, systolic blood pressure. Design-effect estimates in many study variables were close to one, and in some cases less than one, indicating a weak clustering effect.

Demonstration Data Set

In examining variance approximation techniques for subpopulation means and proportions, we use a subgroup of the MFH Survey data consisting of 30–64-year-old males who took part in the screening phase of the health examination and who also belonged to an active labour force or had a past labour

Table 5.2 Age distribution, proportions (%) of chronically ill persons (CHRON) and persons exposed to physical health hazards at work (PHYS), and average of systolic blood pressure (SYSBP) in the MFH survey subgroup of 30–64-year-old males.

Age	Sample n	Sample %	CHRON %	PHYS %	SYSBP Mean
30–34	508	18.8	13.8	12.8	134.0
35–39	384	14.2	21.4	17.4	136.2
40–44	437	16.2	28.4	18.8	138.5
45–49	395	14.6	44.8	18.5	141.9
50–54	379	14.0	52.2	17.4	144.7
55–59	336	12.4	68.5	21.4	151.2
60–64	260	9.6	73.8	21.2	154.3
Total sample	2699	100.0	39.8	17.8	141.8

history. These data consist of 2699 eligible males. The data set includes sampling identifiers STRATUM, CLUSTER and WEIGHT, and two binary response variables, CHRON (presence of chronic illness) and PHYS (suffering or having suffered from physical health hazards at work), and a continuous response variable SYSBP (systolic blood pressure). Information on these data is displayed in Table 5.2. Note that the selected subgroup is of a cross-classes type properly reflecting all essential properties of the MFH Survey sampling design such as the number of strata (24) and the number of sample clusters (48) covered.

Our aim is to estimate the variances of the subpopulation proportion estimator of CHRON and the subpopulation mean estimator of SYSBP by using approximation methods based on linearization and sample re-use. Both response variables indicated relatively strong intra-cluster correlation from the total MFH Survey data. The response variable PHYS is used in a test for two-way tables in Chapter 7. Before turning to these tasks we briefly discuss the issue of weighting in the relevant MFH Survey subgroup.

Poststratification

The MFH Survey data set can be regarded as self-weighting because the design is epsem and adjustment for nonresponse is not necessary. However, for further demonstration of poststratification as considered in Sections 3.3 and 4.1, we develop the poststratification weights, and we compare the unweighted and weighted estimation results. For this, let us suppose for a moment that we are working with a simple random sample (although this is not actually true for the MFH Survey data set).

We construct the poststratification weights using the regional age distributions for both sexes which are available on the population level. We first divide the target population into 30 regional age–sex poststrata with five regions and three age groups. Let us consider the selected MFH Survey subgroup of 30–64-year-old males; the corresponding population and sample frequency distributions and proportions are displayed in Table 5.3. Using these distributions, two different weights are derived for the sample elements in poststratum l: a weight $w_l^* = N_l/n_l$, and a rescaled weight $w_l^{**} = w_l^* \times n/N$, where N_l and n_l denote the population size and sample size in poststratum l, respectively, and N and n are the corresponding sizes of the population and the sample data set.

The weights w_l^* indicate the amount of population elements 'represented' by a single sample element. Over an n-element sample data set, these weights sum up to the relevant population size N. The rescaled weights w_l^{**} sum up to n. In Table 5.3, these weights vary only slightly around their mean value of one, indicating the self-weighting property of the MFH data set. In a strictly self-weighting data set, rare in practice, the weights w_l^* would be constant and the rescaled weights w_l^{**} would be equal to one, for all sample elements.

When using the weights, it is obvious that the weight w_l^* is suitable for proper

Table 5.3 Poststratification weight generation for the MFH Survey subgroup of 30–64-year-old males. Population and sample sizes N_l and n_l, the corresponding proportions P_l and p_l, and the weights w_l^* and w_l^{**} in the 15 poststrata for males.

Poststratum	N_l	n_l	P_l	p_l	w_l^*	w_l^{**}
1	56 658	140	0.058 06	0.051 87	404.70	1.1192
2	32 450	94	0.033 25	0.034 83	345.21	0.9547
3	21 681	66	0.022 22	0.024 45	328.50	0.9085
4	71 324	199	0.073 08	0.073 73	358.41	0.9912
5	41 422	123	0.042 44	0.045 57	336.76	0.9313
6	33 168	93	0.033 99	0.034 46	356.65	0.9863
7	75 172	215	0.077 03	0.079 66	349.64	0.9669
8	45 507	131	0.046 63	0.048 54	347.38	0.9607
9	33 011	97	0.033 82	0.035 94	340.32	0.9412
10	116 822	309	0.119 70	0.114 49	378.06	1.0456
11	62 917	172	0.064 47	0.063 73	365.80	1.0116
12	47 261	157	0.048 43	0.058 17	301.03	0.8325
13	188 252	466	0.192 89	0.172 66	403.97	1.1172
14	88 185	254	0.090 36	0.094 11	347.19	0.9602
15	62 105	183	0.063 64	0.067 80	339.37	0.9386
Total	975 935	2699	1.000 00	1.000 00		

estimation of population totals and the rescaled weight w_l^{**} is convenient in testing and modelling procedures when population totals are not of interest. Popular software for survey analysis, such as SUDAAN and PC CARP, accept either type of weight variable in their analysis programs. However, in the analysis programs of standard statistical software, such as SAS and SPSS, the rescaled weight w_l^{**} is necessary for weighted estimation and testing.

Developing a weight variable for poststratification is more complicated for a non-epsem data set from a complex sampling design, because there may already exist an element weight to compensate for unequal inclusion probabilities. For the simplest case, an adjusted weight to account for nonresponse can be derived by multiplying the sampling weight by the response rate in a poststratum, and then the product can be used as a weight variable in an analysis program. Strictly speaking, however, the variance estimators of poststratified estimates are different from the estimators obtained by using the adjusted weights. Therefore, for example in SUDAAN, PC CARP and WesVarPC, options are available for proper variance estimation with poststratification. However, in practice the differences in variance estimates are usually small.

Let us compare the estimation results from an unweighted and a weighted MFH Survey data set, and using genuine poststratified estimators. For simplicity, we ignore the original stratification and clustering; the MFH sample data set is thus taken as a simple random sample (drawn with replacement) for the unweighted analysis (SRSWR), a stratified simple random sample with non-proportional allocation in the weighted analysis (STRWR), and a poststratified simple random sample in the third case. Weighted estimates are obtained using the weights w_l^* or w_l^{**} in the weight variable, and the poststratification is carried out by supplying the population sizes N_l in each poststratum in the relevant options of the SUDAAN procedure DESCRIPT. The corresponding sample means and standard error estimates of CHRON, PHYS and SYSBP are displayed below:

Study variable	n	SRSWR Mean	s.e.	STRWR Mean	s.e.	Poststratified Mean	s.e.
CHRON	2699	0.398	0.0094	0.386	0.0084	0.386	0.0085
PHYS	2699	0.178	0.0074	0.176	0.0073	0.176	0.0073
SYSBP	2699	141.8	0.3677	141.4	0.3353	141.4	0.3375

The unweighted and poststratified means differ for CHRON and somewhat for SYSBP because of their dependence on the demographic decomposition of the poststrata, especially on age, which is stronger than for PHYS. It should be noted that poststratification can increase efficiency. Poststratification executed as a usual stratified analysis decreases standard error estimates for CHRON and SYSBP. The extra variance due to the poststratification can be seen from the last column (especially for SYSBP) where the standard errors are estimated using the most appropriate variance estimators. However, when compared to the stratified analysis, the differences are still quite small.

5.2 RATIO ESTIMATORS

In the estimation of variances we concentrate on *ratio estimators*, which are the simplest examples of *nonlinear* estimators. Means and proportions estimated in population subgroups, for example, the mean of systolic blood pressure and the proportion of chronically ill persons in the MFH Survey subgroup, are typical nonlinear ratio estimators. Variance estimation is examined under a stratified cluster-sampling design which is epsem like the MFH Survey sampling design. This kind of sampling design is simple for variance estimation and is popular in practice.

Nonlinear Estimators

A linear estimator constitutes a linear function of the sample observations. Totals such as $\hat{t} = N \sum_{k=1}^{n} y_k/n$ are linear estimators when calculated from a simple random sample whose size n is fixed in advance. Under cluster sampling, situations are often encountered in which a fixed-size sample cannot be assumed. This occurs, for example, in one-stage cluster sampling if the cluster sizes B_α vary. Then, in the total estimator $\hat{t}_{rat} = N \sum_{\alpha=1}^{a} y_\alpha / \sum_{\alpha=1}^{a} B_\alpha$ (considered in Section 3.2) where y_α is the sample sum of the response variable in cluster α the denominator should also be taken as a random variate whose value depends on which clusters are drawn. Because of this, $\hat{t}_{rat}$ turns out to be a nonlinear estimator.

The estimator $\hat{t}_{rat}$ is a special case of the ratio estimation considered in Section 3.3, where ratio estimation refers to the estimation of the population total T of a response variable using auxiliary information. There, the estimator $\hat{t}_{rat} = \hat{r} \times T_z$ was derived, where $\hat{r} = \hat{t}/\hat{t}_z$ is a ratio of the total estimators $\hat{t}$ and $\hat{t}_z$ of the response variable of interest and an auxiliary variable z, respectively, and T_z is the known population total of z. For the estimation of the population ratio $R = T/T_z$, the estimator $\hat{r}$ is directly available, and it can be written as $\hat{r} = \sum_{\alpha=1}^{a} y_\alpha / \sum_{\alpha=1}^{a} z_\alpha$. The estimator $\hat{r}$ is called an *estimator of a ratio*, or a *ratio estimator*. In this estimator, the denominator is the sample size, which is not assumed fixed. In practice, subpopulation means and proportions estimated from a subgroup of a sample such that the subgroup sample size is not fixed, as in the MFH Survey subgroup of 30–64-year-old males, provide the most common examples of ratio estimators. We shall especially consider such ratio estimators.

Combined Ratio and Separate Ratio Estimators

Let the population clusters be divided into H strata so that there are M_h clusters in stratum h. A first-stage sample of m_h (≥ 2) clusters is drawn from each stratum h, and a second-stage sample of a total of $n = \sum_{h=1}^{H} n_h$ elements is drawn from the $m = \sum_{h=1}^{H} m_h$ sample clusters. As we often work with subgroups of the sample whose sizes are not fixed in advance, we will use x_h in place of n_h. Note that we do not use z_h to avoid confusion with notation used for an auxiliary variable. We assume that the sample is self-weighting, i.e. the inclusion probability of each of the N population elements is constant over the strata and adjustment for nonresponse is not necessary. Element weights are thus constant for all sample elements. Further, let $y_{h\alpha} = \sum_{\beta=1}^{x_{h\alpha}} y_{h\alpha\beta}$ denote the subgroup sample sum of the response variable in sample cluster α of stratum h, and let $x_{h\alpha}$ denote the corresponding sample size. Two types of ratio estimators are derived by using the sample sums $y_{h\alpha}$ and $x_{h\alpha}$. A *combined ratio* (across-stratum ratio) *estimator* is given by

$$\hat{r} = \frac{\sum\limits_{h=1}^{H} y_h}{\sum\limits_{h=1}^{H} x_h} = \frac{\sum\limits_{h=1}^{H} \sum\limits_{\alpha=1}^{m_h} y_{h\alpha}}{\sum\limits_{h=1}^{H} \sum\limits_{\alpha=1}^{m_h} x_{h\alpha}}, \tag{5.1}$$

which is a ratio estimator of a mean $\bar{Y} = T/N$ or of a proportion $P = N_1/N$ where T is the total of a continuous response variable and N_1 is the count of persons having the value one on a binary response variable in the population subgroup considered. It is essential to note that in the ratio estimator $\hat{r}$ not only the numerator quantities $y_{h\alpha}$ vary between clusters but also the denominator quantities $x_{h\alpha}$ may do so and, therefore, $\hat{r}$ is a nonlinear estimator.

For (5.1) $y_{h\alpha}$ and $x_{h\alpha}$ were first summed over the strata and clusters. A *separate ratio* (stratum-by-stratum ratio) *estimator* is a weighted sum of stratum ratios y_h/x_h. It is given by

$$\hat{r}_s = \sum_{h=1}^{H} W_h \hat{r}_h, \tag{5.2}$$

where $W_h = N_h/N$ are known stratum weights, and

$$\hat{r}_h = \frac{y_h}{x_h} = \frac{\sum\limits_{\alpha=1}^{m_h} y_{h\alpha}}{\sum\limits_{\alpha=1}^{m_h} x_{h\alpha}}, \quad h = 1, ..., H.$$

The separate ratio estimator is often used in descriptive surveys, whereas the combined ratio estimator is more common in complex analytical surveys. We will exclusively use combined ratio estimators in this and subsequent chapters and call them ratio estimators. In the case of a continuous response variable, we put $\hat{r} = \bar{y}$ (a sample mean) and in the case of a binary response, $\hat{r} = \hat{p}$ (a sample proportion). We will often denote the ratio estimator in (5.1) simply as $\hat{r} = y/x$, where $y = \sum_{h=1}^{H} y_h$ and $x = \sum_{h=1}^{H} x_h$. The quantities y and x thus refer to the sample sum of the response variable and the sample size, respectively, in a subgroup of the sample. Note that the above discussion applies equally to an estimator $\hat{r}$ calculated from the whole sample if its size is not fixed by the sampling design.

The ratio estimator $\hat{r}$ is not unbiased but is consistent. The bias of $\hat{r}$ depends on the variability of the cluster sample sizes in the subgroup. The coefficient of variation of the cluster sample sizes $x_{h\alpha}$ can be used as a measure of this variability. If the coefficient of variation is small, the ratio estimator $\hat{r}$ is nearly linear and hence nearly unbiased. The bias is not disturbing if the coefficient of variation is less than, say 0.2.

Various kinds of subgroups can be formed in which the bias properties of ratio estimators can vary. In *cross-classes*, which cut smoothly across the strata and sample clusters, the decrease in the subgroup sample sizes $x_{h\alpha}$ within clusters is proportional to the decrease in the subgroup sample size relative to the total sample size. The coefficient of variation of the subgroup sample sizes hence has the same magnitude as for the total sample. For this kind of a subgroup, basic features of the sampling design are well reflected; for example, the number of strata and sample clusters covered by a cross-class are usually the same as for the entire sample. Alternatively, in *segregated classes* covering only a part of the sample clusters, the coefficient of variation of the subgroup sample sizes can increase substantially. These are, for example, regional subgroups. It should be noted that, in contrast to a cross-classes type domain, a segregated class does not properly reflect the properties of the sampling design, possibly leading to instability problems in variance estimation (see Section 6.2). Between these extremes are *mixed classes*, which perhaps are the most common subgroup types in practice. Demographic subgroups often constitute cross-classes while socioeconomic subgroups tend to be mixed classes. Moreover, a property of design-effect estimates of subpopulation ratio estimators for cross-classes is that they tend to approach unity with decreasing subgroup sample size. This property is not shared by the other types of subgroups.

Variance Estimation of a Ratio Estimator

For the ratio estimator (5.1), not only the cluster-wise variation in the numerator $\sum_{h=1}^{H} y_h$ but also the variation in the denominator $\sum_{h=1}^{H} x_h$ contributes to the total variance. Therefore, variance estimation of a ratio estimator is more complicated than that of a linear estimator. Analytical variance estimators for linear estimators, such as for population totals considered in Chapter 2, were derived according to the special features of each basic sampling technique. For nonlinear estimators, analytical variance estimators can be cumbersome or they may not be available. Other types of variance estimators are thus needed. To be successful, these estimators, and the corresponding computational techniques, should have multi-purpose properties that cover the most common types of complex sampling designs and nonlinear estimators.

Approximative variance estimators can be used for variance estimation of a nonlinear estimator. These variance estimators are not sampling-design-specific, unlike those for linear estimators. Approximative variance estimators are flexible so that they can be applied for different kinds of nonlinear estimators, including the ratio estimator, under a variety of multi-stage designs covering all the different real sampling designs selected for this book. We use the *linearization method* as the basic approximation method. Alternative methods are based on *sample re-use techniques* such as *balanced half-samples, jackknife* and *bootstrap*. Approximative techniques for variance estimation are available in most statistical software products for variance estimation such as SUDAAN (Shah *et al.* 1995), PC CARP (Fuller *et al.* 1989) and OSIRIS (Institute for Social Research 1992).

Certain simplifying assumptions are often made when using approximative variance estimators. In variance estimation under a multi-stage design, each sampling stage contributes to the total variance. For example, under a two-stage design, an analytical variance estimator of a population total is composed of a sum of the between-cluster and within-cluster variance components as shown in Section 3.2. In the simplest use of the approximation methods, a possibly multi-stage design is reduced to a one-stage design, and the clusters are assumed to be drawn with replacement. Variances are then estimated by using the between-cluster variation only. In more advanced uses of the approximation techniques, the variation of all the sampling stages can be properly accounted for.

5.3   LINEARIZATION METHOD

Linearization Method for a Nonlinear Estimator

In the estimation of the variance of a general nonlinear estimator, denoted by $\hat{\theta}$, we adopt the method based on the so-called *Taylor series expansion*. The method is usually called the *linearization* method because we first reduce the original nonlinear quantity to an approximate linear quantity by using the linear terms of the corresponding Taylor series expansion, and then construct the variance formula and an estimator of the variance of this linearized quantity.

Let an s-dimensional parameter vector be denoted by $\mathbf{Y} = (Y_1, ..., Y_s)'$ where Y_j are population totals or means. The corresponding estimator vector is denoted by $\hat{\mathbf{Y}} = (\hat{Y}_1, ..., \hat{Y}_s)'$ where $\hat{Y}_j$ are estimators of Y_j. We consider a nonlinear parameter $\theta = f(\mathbf{Y})$ with a consistent estimator denoted by $\hat{\theta} = f(\hat{\mathbf{Y}})$. A simple example is a subpopulation mean parameter $\theta = \bar{Y} = Y_1/Y_2$ with a ratio estimator $\hat{\theta} = \bar{y} = \hat{Y}_1/\hat{Y}_2 = y/x$, where $y = \sum_{h=1}^{H} y_h$ is the subgroup sample sum of the response variable and $x = \sum_{h=1}^{H} x_h$ is the subgroup sample size, both regarded as random quantities.

Suppose that for the function $f(\mathbf{y})$ continuous second-order derivatives exist in an open sphere containing $\mathbf{Y}$ and $\hat{\mathbf{Y}}$. Using the linear terms of the Taylor series expansion we have an approximative linearized expression,

$$\hat{\theta} - \theta \doteq \sum_{j=1}^{s} \frac{\partial f(\mathbf{Y})}{\partial y_j} (\hat{Y}_j - Y_j), \qquad (5.3)$$

where $\partial f(\mathbf{Y})/\partial y_j$ refers to partial derivation. Using the linearized equation

(5.3), the variance approximation of $\hat{\theta}$ can be expressed by

$$V(\hat{\theta}) = V\left(\sum_{j=1}^{s} \frac{\partial f(\mathbf{Y})}{\partial y_j}(\hat{Y}_j - Y_j)\right)$$

$$= \sum_{j=1}^{s}\sum_{k=1}^{s} \frac{\partial f(\mathbf{Y})}{\partial y_j}\frac{\partial f(\mathbf{Y})}{\partial y_k} V(\hat{Y}_j, \hat{Y}_k), \qquad (5.4)$$

where $V(\hat{Y}_j, \hat{Y}_k)$ denote variances and covariances of the estimators $\hat{Y}_j$ and $\hat{Y}_k$. We have hence reduced the variance of a nonlinear estimator $\hat{\theta}$ to a function of variances and covariances of s linear estimators $\hat{Y}_j$. A variance estimator $\hat{v}(\hat{\theta})$ is obtained from (5.4) by substituting the variance and covariance estimators $\hat{v}(\hat{Y}_j, \hat{Y}_k)$ for the corresponding parameters $V(\hat{Y}_j, \hat{Y}_k)$. The resulting variance estimator is a first-order Taylor series approximation where justification for ignoring the remaining higher-order terms is essentially based on practical experience derived from various complex surveys in which the sample sizes have been sufficiently large.

As an example of the linearization method, let us consider further a ratio estimator. The parameter vector is $\mathbf{Y} = (Y_1, Y_2)'$ with the corresponding estimator vector $\hat{\mathbf{Y}} = (\hat{Y}_1, \hat{Y}_2)'$. The nonlinear parameter to be estimated is $\theta = f(\mathbf{Y}) = Y_1/Y_2$, and the corresponding ratio estimator is $\hat{\theta} = f(\hat{\mathbf{Y}}) = \hat{Y}_1/\hat{Y}_2$. The partial derivatives are

$$\partial f(\mathbf{Y})/\partial y_1 = 1/Y_2 \quad \text{and} \quad \partial f(\mathbf{Y})/\partial y_2 = -Y_1/Y_2^2.$$

Hence we have

$$V(\hat{\theta}) = \sum_{j=1}^{2}\sum_{k=1}^{2} \frac{\partial f(\mathbf{Y})}{\partial y_j}\frac{\partial f(\mathbf{Y})}{\partial y_k} V(\hat{Y}_j, \hat{Y}_k)$$

$$= \frac{1}{Y_2}\frac{1}{Y_2}V(\hat{Y}_1) + \frac{1}{Y_2}\left(-\frac{Y_1}{Y_2^2}\right)V(\hat{Y}_1, \hat{Y}_2)$$

$$+ \left(-\frac{Y_1}{Y_2^2}\right)\frac{1}{Y_2}V(\hat{Y}_2, \hat{Y}_1) + \left(-\frac{Y_1}{Y_2^2}\right)\left(-\frac{Y_1}{Y_2^2}\right)V(\hat{Y}_2)$$

$$= (1/Y_2^2)(V(\hat{Y}_1) + \theta^2 V(\hat{Y}_2) - 2\theta V(\hat{Y}_1, \hat{Y}_2))$$

$$= \theta^2(Y_1^{-2}V(\hat{Y}_1) + Y_2^{-2}V(\hat{Y}_2) - 2(Y_1Y_2)^{-1}V(\hat{Y}_1, \hat{Y}_2)). \qquad (5.5)$$

Basic principles of the linearization method for variance estimation of a nonlinear estimator under complex sampling are due to Keyfitz (1957) and Tepping (1968). Woodruff (1971) suggested simplified computational algorithms for the approximation by transforming an *s*-dimensional situation to a one-dimensional case. A good reference for the method is Wolter (1985). The linearization method can be also used for more complex nonlinear estimators such as correlation and regression coefficients. The linearization method is used in most software products, such as SUDAAN and PC CARP, for variance estimation of ratio estimators and for more complicated nonlinear estimators. We next consider the estimation of the approximative variance of a ratio estimator by using the linearization method.

Linearization Method for a Combined Ratio Estimator

A variance estimator of the ratio estimator $\hat{r} = y/x = \sum_{h=1}^{H} y_h / \sum_{h=1}^{H} x_h$ given by (5.1) should, according to equation (5.5), include the following terms: first, a term accounting for cluster-wise variation of the subgroup sample sums $y_{h\alpha}$, second, a term accounting for cluster-wise variation of the subgroup sample sizes $x_{h\alpha}$, and finally, a term accounting for joint cluster-wise variation of the sample sums $y_{h\alpha}$ and $x_{h\alpha}$, i.e. their covariance. A variance estimator of $\hat{r}$ can thus be obtained from equation (5.5) by substituting the estimators $\hat{v}(y)$, $\hat{v}(x)$ and $\hat{v}(y, x)$ for the corresponding variance and covariance terms $V(y), V(x)$ and $V(y, x)$. Hence we have

$$\hat{v}_{des}(\hat{r}) = \hat{r}^2(y^{-2}\hat{v}(y) + x^{-2}\hat{v}(x) - 2(yx)^{-1}\hat{v}(y, x)), \qquad (5.6)$$

as the *design-based* variance estimator of $\hat{r}$ based on the linearization method, where $\hat{v}(y)$ is the variance estimator of the subgroup sample sum y, $\hat{v}(x)$ is the variance estimator of the subgroup sample size x, and $\hat{v}(y, x)$ is the covariance estimator of y and x.

The variance estimator (5.6) is consistent if the estimators $\hat{v}(y), \hat{v}(x)$ and $\hat{v}(y, x)$ are consistent. The cluster sample sizes $x_{h\alpha}$ should not vary too much for the reliable performance of the approximation based on the Taylor series expansion. The method can be safely used if the coefficient of variation of $x_{h\alpha}$ is less than 0.2. If the cluster sample sizes are equal, the variance and covariance terms $\hat{v}(x)$ and $\hat{v}(y, x)$ are zero and the variance approximation reduces to $\hat{v}_{des}(\hat{r}) = \hat{v}(y)/x^2$. And for a binary response from simple random sampling with replacement, this variance estimator reduces to the binomial variance estimator $\hat{v}_{des}(\hat{p}) = \hat{v}_{bin}(\hat{p}) = \hat{p}(1 - \hat{p})/x$, where $x = n$, the size of the available sample data set.

The variance estimator (5.6) is a large-sample approximation in that a good variance estimate can be expected if not only a large element-level sample is available but a large number of sample clusters is also present. In the case of a small number of sample clusters, the variance estimator can be unstable; this will be examined in Chapter 6.

Strictly speaking, the variance and covariance estimators in (5.6) depend on the actual sampling design. But assuming that at least two sample clusters are drawn from each stratum and by using the with-replacement assumption, i.e. assuming that clusters are drawn independently of each other, we obtain relatively simple variance and covariance estimators which can be generally applied for multi-stage stratified epsem samples:

$$\hat{v}(y) = \sum_{h=1}^{H} m_h \hat{s}_{yh}^2, \quad \hat{v}(x) = \sum_{h=1}^{H} m_h \hat{s}_{xh}^2$$

and

$$\hat{v}(y, x) = \sum_{h=1}^{H} m_h \hat{s}_{yxh},$$

where

$$\hat{s}_{yh}^2 = \sum_{\alpha=1}^{m_h} (y_{h\alpha} - y_h/m_h)^2/(m_h - 1),$$

$$\hat{s}_{xh}^2 = \sum_{\alpha=1}^{m_h} (x_{h\alpha} - x_h/m_h)^2/(m_h - 1),$$

and

$$\hat{s}_{yxh} = \sum_{\alpha=1}^{m_h} (y_{h\alpha} - y_h/m_h)(x_{h\alpha} - x_h/m_h)/(m_h - 1). \tag{5.7}$$

Note that by using the with-replacement approximation, only the between-cluster variation is accounted for. Therefore, the corresponding variance estimators underestimate the true variance. This bias is negligible if the stratum-wise first-stage sampling fractions are small, which is the case when there are a large number of population clusters in each stratum (see Section 3.2).

For the estimation of the between-cluster variance, at least two sample clusters are needed. If the sampling design is such that exactly two clusters are drawn from each stratum the estimators (5.7) can be further simplified:

$$\hat{v}(y) = \sum_{h=1}^{H} (y_{h1} - y_{h2})^2, \quad \hat{v}(x) = \sum_{h=1}^{H} (x_{h1} - x_{h2})^2$$

and

$$\hat{v}(y, x) = \sum_{h=1}^{H} (y_{h1} - y_{h2})(x_{h1} - x_{h2}). \tag{5.8}$$

This kind of design is popular in practice because of the simplicity of the variance and covariance estimators. The modified MFH Survey sampling design is of this type. The linearization method is demonstrated in the MFH

Survey's two-stage design in Example 5.1. There, an application of the method by the SUDAAN software for survey analysis is also given.

Example 5.1

Linearization method in the MFH Survey. We consider the estimation of the variance of a subpopulation proportion estimator $\hat{r} = \hat{p}$ for the binary response variable CHRON (chronic morbidity) and a subpopulation mean estimator $\hat{r} = \bar{y}$ for the continuous response variable SYSBP (systolic blood pressure) by the linearization method. The MFH Survey subgroup covers 30–64-year-old males. The subgroup sample size is $x = 2699$ and the data set is self-weighting. In the modified MFH Survey sampling design, described in Section 5.1, there are $H = 24$ regional strata and $m = 48$ regional sample clusters. Two sample clusters are thus drawn from each stratum. Recall that the subgroup maintains these prop-

Table 5.4 Cluster sample sums $y_{h\alpha}$ of the response variables CHRON and SYSBP and the corresponding cluster sample sizes $x_{h\alpha}$ for the subgroup of 30–64-year-old males in the MFH Survey.

Stratum h	Cluster α	CHRON $y_{h\alpha}$	SYSBP $y_{h\alpha}$	$x_{h\alpha}$	Cluster α	CHRON $y_{h\alpha}$	SYSBP $y_{h\alpha}$	$x_{h\alpha}$
1	1	70	29 056	204	2	74	29 417	210
2	1	12	3 692	26	2	14	4 564	30
3	1	15	7 741	59	2	16	8 585	63
4	1	9	6 277	45	2	14	5 668	43
5	1	10	2 322	17	2	16	3 960	30
6	1	10	3 080	21	2	6	3 252	22
7	1	10	3 966	27	2	4	3 261	24
8	1	12	4 156	28	2	6	2 852	20
9	1	15	6 617	46	2	23	6 616	48
10	1	37	10 552	73	2	25	11 032	77
11	1	11	8 759	60	2	25	9 876	72
12	1	33	9 901	69	2	24	6 828	47
13	1	31	8 624	61	2	27	9 390	66
14	1	22	6 960	48	2	20	7 130	49
15	1	18	6 646	49	2	22	7 094	49
16	1	24	9 841	69	2	37	11 786	83
17	1	19	6 910	48	2	23	6 446	45
18	1	25	10 742	73	2	29	9 026	61
19	1	36	9 350	65	2	34	8 912	62
20	1	9	3 810	26	2	22	7 098	51
21	1	18	6 998	53	2	34	9 970	69
22	1	29	11 146	79	2	41	13 215	94
23	1	22	6 596	48	2	18	6 002	41
24	1	15	3 808	27	2	7	3 148	22
Over both clusters in all strata						1073	382 678	2699

erties of the sampling design because it constitutes a cross-classes type domain. The data set is displayed in Table 5.4.

For the binary response variable CHRON we obtain :

$$y = \sum_{h=1}^{24} \sum_{\alpha=1}^{2} y_{h\alpha} = \sum_{h=1}^{24} (y_{h1} + y_{h2}) = 1073$$

chronically ill males in the sample, and a sample sum of

$$x = \sum_{h=1}^{24} \sum_{\alpha=1}^{2} x_{h\alpha} = \sum_{h=1}^{24} (x_{h1} + x_{h2}) = 2699$$

males in the subgroup. The subpopulation proportion estimate of CHRON is

$$\hat{p} = y/x = 1073/2699 = 0.3976.$$

For the variance estimate $\hat{v}_{des}(\hat{p})$ of $\hat{p}$ we calculate the variance and covariance estimates $\hat{v}(y)$, $\hat{v}(x)$ and $\hat{v}(y, x)$. By using equations (5.8) these are:

$$\hat{v}(y) = \sum_{h=1}^{24} (y_{h1} - y_{h2})^2 = 1545, \quad \hat{v}(x) = \sum_{h=1}^{24} (x_{h1} - x_{h2})^2 = 2527$$

and

$$\hat{v}(y, x) = \sum_{h=1}^{24} (y_{h1} - y_{h2})(x_{h2} - x_{h2}) = 1435.$$

Using these estimates we obtain a variance estimate (5.6):

$$\hat{v}_{des}(\hat{p}) = \hat{p}^2 (y^{-2}\hat{v}(y) + x^{-2}\hat{v}(x) - 2(y \times x)^{-1}\hat{v}(y, x))$$
$$= 0.3976^2 \times (1073^{-2} \times 1545 + 2699^{-2} \times 2527$$
$$- 2 \times (1073 \times 2699)^{-1} \times 1435) = 0.1103 \times 10^{-3}.$$

For the continuous response variable SYSBP we obtain the sample sum

$$y = \sum_{h=1}^{24} \sum_{\alpha=1}^{2} y_{h\alpha} = \sum_{h=1}^{24} (y_{h1} + y_{h2}) = 382\,678.$$

Hence the subpopulation mean estimate of SYSBP is

$$\bar{y} = y/x = 382\,678/2699 = 141.785.$$

For the variance estimate $\hat{v}_{des}(\bar{y})$ of $\bar{y}$ we obtain:

$$\hat{v}(y) = \sum_{h=1}^{24} (y_{h1} - y_{h2})^2 = 50\,469\,516$$

and

$$\hat{v}(y, x) = \sum_{h=1}^{24} (y_{h1} - y_{h2})(x_{h1} - x_{h2}) = 349\,962.$$

Using these estimates we obtain a variance estimate (5.6):

$$\hat{v}_{des}(\bar{y}) = \bar{y}^2 (y^{-2}\hat{v}(y) + x^{-2}\hat{v}(x) - 2(y \times x)^{-1}\hat{v}(y, x))$$

$$= 141.785^2 \times (382\,678^{-2} \times 50\,469\,516 + 2699^{-2} \times 2527$$

$$- 2 \times (382\,678 \times 2699)^{-1} \times 349\,962) = 0.2788.$$

All these variances could be estimated from the cluster-level data set given in Table 5.4. For CHRON we next calculate a binomial variance estimate of $\hat{p}$ corresponding to simple random sampling with replacement, and the corresponding design-effect estimate. The variance estimate is

$$\hat{v}_{bin}(\hat{p}) = \hat{p}(1 - \hat{p})/x = 0.3976 \times (1 - 0.3978)/2699 = 0.0887 \times 10^{-3},$$

where $\hat{v}_{bin}$ is the standard binomial variance estimator. The design-effect estimate is $\hat{d}(\hat{p}) = \hat{v}_{des}(\hat{p})/\hat{v}_{bin}(\hat{p}) = 1.24$. Note that the design-effect estimate is noticeably smaller than that for the total survey data because the subgroup is a cross-class. The design-effect estimate also indicates that intra-cluster correlation in CHRON in the subgroup is only slight. For SYSBP, on the other hand, an access to the individual-level data set is required for the calculation of the variance estimate of $\bar{y}$ with an assumption of simple random sampling with replacement. This turns out to be

$$\hat{v}_{srswr}(\bar{y}) = \sum_{k=1}^{2699} (y_k - \bar{y})^2 /(2699(2699 - 1)) = 0.1352,$$

and hence the design-effect estimate is $\hat{d}(\bar{y}) = 2.06$. The estimate indicates a substantial intra-cluster correlation in the response SYSBP in the subgroup, even the estimate is considerably smaller than that for the total survey data. The coefficient of variation of the subgroup sample size is c.v. $(x) =$ s.e. $(x)/x = 0.019$ which is small enough to justify the use of the Taylor series linearization.

We finally collect the estimation results below.

Study variable	Estimate $\hat{r}$	Standard-error estimate		
		s.e.$_{des}(\hat{r})$	s.e.$_{srs}(\hat{r})$	deff
CHRON	0.3976	0.0105	0.0094	1.24
SYSBP	141.785	0.5280	0.3677	2.06

The estimation of the variance of a ratio-type proportion or mean estimator can be carried out by suitable software for survey analysis. We briefly describe this using the SAS-callable SUDAAN procedure DESCRIPT (see Appendix 1). As a further example of the use of the procedure, we also produce estimation results for three age groups (30–39, 40–49, 50–64 years). We use the SUDAAN sampling design option WR to request variance estimation with the linearization method under the assumption of with-replacement sampling of the clusters (SUDAAN design options are discussed in more detail in Section 8.3 and in Appendix 1). The relevant SUDAAN code for the design-based analysis are:

```
1 PROC DESCRIPT DATA=<dataset> DESIGN=WR DEFF ATLEVEL2=2;
2 NEST       STRATUM CLUSTER;
3 WEIGHT     _ONE_;
4 VAR        CHRON SYSBP;
5 SUBGROUP   AGE;
6 LEVELS     3;
7 TABLES     AGE;
8 RFORMAT    AGE AGEF.;
```

In addition to the design option WR and the keyword DEFF for obtaining design-effect estimates given in the PROC statement, there are two sampling design-specific SUDAAN statements: NEST and WEIGHT. In the NEST statement the stratification and clustering variable names are given, and in the WEIGHT statement the variable name of the weight variable is given. Because in the MFH data set the rescaled weights w_k^{**} are all one, the SUDAAN automatic variable _ONE_ is used to assign a value one for the weight variable for all sample elements. In the VAR statement, the response variable names are supplied. The next four statements define the grouping of the data and the tabulation of the estimates. We use the individual-level data set in this analysis, not the cluster-level one. Running these statements by using SUDAAN we obtain the following output:

SUDAAN Survey Data Analysis Software
Copyright Research Triangle Institute   March 1995
Release 6.40

Number of observations read   : 2699 Weighted count: 2699
Number of observations skipped : 0
(WEIGHT variable nonpositive)
Denominator degrees of freedom : 24

by: Variable, Age.

Variable Age	Mean	SE Mean	DEFF Mean	Sample Size	Count at Level 2
Chronic morbidity					
Total	0.3976	0.0105	1.24	2699	48
30-39	0.1704	0.0126	1.00	892	48
40-49	0.3618	0.0167	1.00	832	48
50-64	0.6359	0.0170	1.22	975	48
Systolic blood pressure					
Total	141.7851	0.5280	2.06	2699	48
30-39	134.9159	0.6929	2.08	892	48
40-49	140.1178	0.8939	2.10	832	48
50-64	149.4923	0.6857	1.03	975	48

The design-effect estimates of the CHRON proportions in the age groups are equal to one for the first two groups, and the last group has a design-effect estimate close to the total male subpopulation estimate. For the variable SYSBP, the design-effect estimates in first two age groups are somewhat larger than that for the total male subpopulation. These large design-effect estimates might indicate instability in the design-based variance estimates $\hat{v}_{des}$. Note that there are $f = m - H = 24$ degrees of freedom available for variance estimation; this quite small number can be the source of possible instability in the variance approximation. However, this would be more apparent in subpopulations which are not cross-classes, such as regional subpopulations. The ratio estimates with their estimated standard errors and design effects could also be obtained by using the SUDAAN procedure RATIO, the PC CARP program, or certain OSIRIS programs.

5.4   SAMPLE RE-USE METHODS

Sample re-use methods can be used as an alternative to the linearization method in variance approximation of a nonlinear estimator $\hat{\theta}$ under complex multi-stage designs. The term *re-use* refers to a procedure where variance estimation is based on repeated utilization of the sampled data set which itself is obtained as a single sample from the population. Therefore, these methods are sometimes called *pseudoreplication* techniques. Pseudoreplication should be distinguished from techniques such as the *random groups methodology*, which rely on true replication where several independent samples are actually drawn from the same population. These methods are excluded here because of their limited practical applicability in complex analytical surveys.

In this section, we consider three particular sample re-use techniques: balanced half-samples, jackknife and bootstrap. They all share the following basic variance estimation procedure (which actually originates from random groups methodology):

1. From the sample data set, we draw K *pseudosamples* by a particular technique with a value of K that is specific to each re-use method.
2. An estimate $\hat{\theta}_k$ mimicking the parent estimator $\hat{\theta}$ is obtained from each of the K pseudosamples.
3. The variance $V(\hat{\theta})$ of the estimator $\hat{\theta}$ is estimated by using the observed variation of the pseudosample estimates $\hat{\theta}_k$, essentially based on squared differences of the form $(\hat{\theta}_k - \hat{\theta})^2$. An average of the K pseudosample estimates $\hat{\theta}_k$ can be used in place of $\hat{\theta}$ to form the squared differences.

The estimator $\hat{\theta}$ is usually a nonlinear estimator, a ratio estimator or an estimator of a regression coefficient. In the linearization method, analytical expressions for partial derivatives of such nonlinear functions were needed in the construction of a variance estimator. This is not so in the sample re-use techniques. In fact, the basic variance estimation procedure described above is independent of the type of the estimator and, therefore, the methods are applicable for any kind of a nonlinear estimator. Pseudoreplication techniques, especially the bootstrap, however, involve much more computation than the linearization method; thus they are flexible but computer-intensive.

The technique of balanced half-samples was introduced by McCarthy (1966, 1969) for variance approximation of a nonlinear estimator under an epsem design, where a large number of strata are formed and exactly two clusters are drawn with replacement from each stratum. For variance estimation in a similar design, McCarthy (1966) also introduced the jackknife method which was originally developed by Quenouille (1956) for bias reduction of an estimator. A key property of the jackknife method is compactly stated as *jack-of-all-trades and master of none*. Both methods have been generalized for more complex designs involving more than two clusters per stratum and without-replacement sampling of clusters. Good introductions to recent developments in balanced half-samples and jackknife techniques are Wolter (1985) and Rao *et al.* (1992).

Bootstrapping was introduced by Efron (1982) for a general nonparametric methodology for various statistical problems: *'Our goal is to understand a collection of ideas concerning the nonparametric estimation of bias, variance and more general measures of error'* (Efron 1982, p. 1). Since then, the technique has been extensively applied, using computer-intensive simulation, for a variety of non-standard variance and confidence-interval approximation problems when working with independent observations. Originating like the jackknife outside the survey sampling framework, the bootstrap technique has been only recently applied for variance estimation of nonlinear estimators in complex surveys. One of the first developments for finite-population without-replacement sampling was McCarthy and Snowden (1985). Extensions of the bootstrap technique are given in Rao and Wu (1988), Rao et al.(1992), covering non-smooth functions such as quantiles, and Sitter (1992). A brief summary of the bootstrap technique for complex surveys is given in Särndal et al. (1992).

We only introduce the basics principles of the sample re-use techniques and concentrate on their practical application within the MFH Survey setting. As an example of a nonlinear estimator we again consider the (combined) ratio estimator $\hat{r} = y/x$, given by (5.1), where $y = \sum_{h=1}^{H} y_h$ is the sum of the cluster-level subgroup sample sums of a response variable and $x = \sum_{h=1}^{H} x_h$ is the corresponding sum of the cluster-level subgroup sample sizes. A two-stage epsem sampling design is assumed such that the clusters are drawn with replacement. The with-replacement assumption involves bias to the approximative variance estimates but the bias is negligible if the first-stage sampling fraction is small. Note that the cluster-level data set used for variance approximation in all the re-use methods is similar to that used in the linearization method.

Balanced half-samples and jackknife techniques for variance approximation of a ratio estimator $\hat{r}$ are examined in a design where exactly two clusters are drawn from each stratum. Note that the MFH Survey sampling design is of this type. The bootstrap technique is applied to a more general design where at least two clusters are drawn from each stratum but the number of sample clusters is constant over the strata. Under these designs, the techniques are here called *balanced repeated replications (BRR), jackknife repeated replications (JRR)* and *bootstrap repeated replications (BOOT)*. Because there are several alternative versions of BRR and JRR suggested in the literature, our aim is also to compare estimation results with each other, and also with the results attained by the linearization method. An overall comparison is given in Section 5.5.

The balanced half-samples and jackknife techniques are available in the OSIRIS IV statistical software, which is one of the pioneering commercial software products for survey analysis. A more recent software package, WesVarPC, also uses BRR and jackknife techniques for variance estimation. However, estimation results for the BRR, JRR and BOOT techniques given in this section are obtained using simple home-made programs coded in the SAS language. One of these programs, namely that for the bootstrap, is given with the accompanying numerical results section in Appendix 3.

Sample re-use methods differ in their asymptotic and other properties, computational requirements, and practicality. Comparative results for the properties of the sample re-use methods for nonlinear estimators from complex sampling are reported by Kish and Frankel (1970, 1974), Bean (1975), Krewski and Rao (1981), Rao and Wu (1985, 1988) and Rao et al. (1992). We discuss the relative merits of the methods in Section 5.5.

The BRR Technique

In its basic form, the technique of *balanced repeated replications* can be applied to variance approximation in epsem designs where exactly two clusters are drawn with replacement from each stratum, and the number of strata is large. We consider using this design, the BRR method for a ratio estimator $\hat{r} = y/x$ which is a subpopulation mean or proportion estimator, where $y = \sum_{h=1}^{H}(y_{h1} + y_{h2})$ and $x = \sum_{h=1}^{H}(x_{h1} + x_{h2})$ and $y_{h\alpha}, x_{h\alpha}$ are the cluster-level sample sums previously given.

The way of forming pseudosamples in the BRR technique starts from the fact that, with H strata and $m_h = 2$ sample clusters per stratum, the total sample can be split into 2^H overlapping half-samples each with H sample clusters. For each half-sample, one of the pairs (y_{11}, x_{11}) and (y_{12}, x_{12}) from the first stratum, one of the pairs (y_{21}, x_{21}) and (y_{22}, x_{22}) from the second stratum, and so forth, is selected. A ratio estimator

$$\hat{r}_k = \frac{y_k}{x_k} = \frac{\sum_{h=1}^{H} \sum_{\alpha=1}^{2} \delta_{h\alpha k} y_{h\alpha k}}{\sum_{h=1}^{H} \sum_{\alpha=1}^{2} \delta_{h\alpha k} x_{h\alpha k}}, \quad k = 1, ..., 2^H \tag{5.9}$$

is derived for each half-sample k, where the weights $\delta_{h\alpha k} = 1$ if the cluster $h\alpha$ is selected in the kth half-sample, and $\delta_{h\alpha k} = 0$ otherwise.

Variance estimator of the mean of $\hat{r}_k$ over all half-samples, namely

$$\bar{\hat{r}} = \sum_{k=1}^{2^H} \hat{r}_k / 2^H, \tag{5.10}$$

and that of the parent estimator $\hat{r}$, can be constructed by using $\hat{r}_k$ obtained from the half-samples. Hence we have:

$$\hat{v}(\bar{\hat{r}}) = \sum_{k=1}^{2^H} (\hat{r}_k - \bar{\hat{r}})^2 / 2^H,$$

and

$$\hat{v}(\hat{r}) = \sum_{k=1}^{2^H} (\hat{r}_k - \hat{r})^2 / 2^H. \tag{5.11}$$

If $\hat{r}$ is a linear estimator, an identity $\hat{r} = \bar{\hat{r}}$ holds, and the two variance estimators in (5.11) are equal. Although for a ratio estimator the identity does not hold, in practice the parent estimate and the mean of the half-sample estimates are usually close and either of the variance estimators (5.11) could be used as a variance estimator for the parent estimator $\hat{r}$. But it is obvious that these variance estimators are not useful in practice because they often presuppose forming a very large number of half-samples, e.g. in the MFH Survey setting about 17 million. To avoid the heavy task of constructing all possible pseudosamples, a subset of them may be selected. But if this subset is chosen at random, a nonzero cross-stratum covariance term will appear in the corresponding variance estimator. In the BRR technique, a subset of K half-samples is selected by a *balanced* method. Balancing involves the selection of the half-samples in such a way that the cross-stratum covariance term is zero. This considerably reduces the number of half-samples needed. In practice, the number K should be selected such that it is at least equal to the number of strata H.

Balanced selection of half-samples is achieved by applying a method developed by Plackett and Burman (1946) for the construction of $K \times K$ orthogonal matrices where K is an integer multiple of 4. An example of such an orthogonal *Hadamard* matrix $\mathbf{B}$ with $K = 12$ such that $\mathbf{B'B} = 12 \times \mathbf{I}$, where $\mathbf{I}$ denotes an identity matrix, is given below. The rows in the matrix refer to the half-samples and the columns to the strata. A $+1$ in a cell (k, h) of the matrix denotes that the first cluster $h1$ in a stratum h is included in the kth half-sample, whilst -1 denotes that the cluster $h2$ is included. Note that complement half-samples can be obtained simply by reversing the signs in the matrix. The number of half-samples, $K = 12$, is thus noticeably smaller than the total amount of possible half-samples, which in this case is $2^{12} = 4096$.

| Half-sample | Stratum h | | | | | | | | | | | |
k	1	2	3	4	5	6	7	8	9	10	11	12
1	+1	−1	+1	−1	−1	−1	+1	+1	+1	−1	+1	−1
2	+1	+1	−1	+1	−1	−1	−1	+1	+1	+1	−1	−1
3	−1	+1	+1	−1	+1	−1	−1	−1	+1	+1	+1	−1
4	+1	−1	+1	+1	−1	+1	−1	−1	−1	+1	+1	−1
5	+1	+1	−1	+1	+1	−1	+1	−1	−1	−1	+1	−1
6	+1	+1	+1	−1	+1	+1	−1	+1	−1	−1	−1	−1
7	−1	+1	+1	+1	−1	+1	+1	−1	+1	−1	−1	−1
8	−1	−1	+1	+1	+1	−1	+1	+1	−1	+1	−1	−1
9	−1	−1	−1	+1	+1	+1	−1	+1	+1	−1	+1	−1
10	+1	+1	−1	−1	+1	+1	+1	−1	+1	+1	−1	−1
11	−1	−1	−1	−1	+1	+1	+1	−1	+1	+1	+1	−1
12	−1	−1	−1	−1	−1	−1	−1	−1	−1	−1	−1	−1

If the actual number of strata is 12, we use the full matrix in the balanced construction of the half-samples. If H is smaller than K, e.g. 10, we can choose any 10 rows of the matrix. In the MFH Survey design we will use $K = 24$, which equals the number of strata. When working with linear estimators, *full orthogonal balance* is reached, which involves equality of a full-sample mean estimate with the estimate obtained as an average of the half-sample estimates, by choosing K as an integer multiple of 4 which is greater than H. Hadamard matrices of orders 2 to 100 are given in Wolter (1985); such matrices can also be easily reproduced by a suitable computer algorithm.

Several BRR variance estimators are suggested in the literature for the variance $V(\hat{r})$ of the parent estimator $\hat{r}$. The variance estimator based on the estimators $\hat{r}_k$ from the K half-samples and the full-sample estimator $\hat{r}$ is

$$\hat{v}_{1.brr}(\hat{r}) = \sum_{k=1}^{K}(\hat{r}_k - \hat{r})^2/K, \qquad (5.12)$$

which is equal to (5.11) based on all 2^H half-samples. As a counterpart to the variance estimator $\hat{v}_{1.brr}(\hat{r})$, an estimator based on estimates $\hat{r}_k^c$ obtained from the K complement half-samples is given by

$$\hat{v}_{2.brr}(\hat{r}) = \sum_{k=1}^{K}(\hat{r}_k^c - \hat{r})^2/K. \qquad (5.13)$$

By using the variance estimators (5.12) and (5.13), a combined variance estimator

$$\hat{v}_{3.brr}(\hat{r}) = (\hat{v}_{1.brr}(\hat{r}) + \hat{v}_{2.brr}(\hat{r}))/2 \qquad (5.14)$$

is derived. Counterparts to the variance estimators (5.12)–(5.14) can be derived based on the averages of $\hat{r}_k$ and $\hat{r}_k^c$. An estimator corresponding to $\hat{v}_{1.brr}$ is hence

$$\hat{v}_{4.brr}(\hat{r}) = \sum_{k=1}^{K}(\hat{r}_k - \hat{\bar{r}})^2/K, \quad \text{where} \quad \hat{\bar{r}} = \sum_{k=1}^{K}\hat{r}_k/K, \qquad (5.15)$$

and that formed by using the complement half-samples is

$$\hat{v}_{5.brr}(\hat{r}) = \sum_{k=1}^{k}(\hat{r}_k^c - \hat{\bar{r}}^c)^2/K, \quad \text{where} \quad \hat{\bar{r}}^c = \sum_{k=1}^{K}\hat{r}_k^c/K. \qquad (5.16)$$

Using $\hat{v}_{4.brr}$ and $\hat{v}_{5.brr}$ we obtain a counterpart to $\hat{v}_{3.brr}$:

$$\hat{v}_{6.brr}(\hat{r}) = (\hat{v}_{4.brr}(\hat{r}) + \hat{v}_{5.brr}(\hat{r}))/2. \qquad (5.17)$$

By using the estimators $\hat{r}_k$ and $\hat{r}_k^c$ from all the half-samples we finally obtain

$$\hat{v}_{7.brr}(\hat{r}) = \sum_{k=1}^{K}(\hat{r}_k - \hat{r}_k^c)^2/4K. \qquad (5.18)$$

For a linear estimator all these variance estimators coincide. However, this is not so for a ratio estimator. For example, there is a relationship between $\hat{v}_{3.brr}$ and $\hat{v}_{7.brr}$:

$$\hat{v}_{3.brr}(\hat{r}) = \hat{v}_{7.brr}(\hat{r}) + \sum_{k=1}^{K}(\hat{\bar{r}} - \hat{r})^2/K,$$

and hence $\hat{v}_{3.brr}(\hat{r}) \geq \hat{v}_{7.brr}(\hat{r})$. By Wolter (1985), $\hat{v}_{7.brr}$ could be regarded as the most natural BRR variance estimator for the parent estimator $\hat{\theta}$. In practice, however, all the estimators should yield nearly equal variance estimates, as appears to be true in the MFH Survey.

Example 5.2

The BRR technique in the MFH Survey. We continue working with variance approximation of ratio-type subpopulation mean and proportion estimators from the MFH Survey data, as considered in the previous section for the linearization method. The binary response variable CHRON (chronic morbidity) and the continuous response variable SYSBP (systolic blood pressure) are used. The subgroup consists of 30–64-year-old males; the subgroup size is 2699. A proportion estimator for CHRON is denoted by $\hat{r} = \hat{p}$ and a mean estimator for SYSBP is denoted by $\hat{r} = \bar{y}$. We calculate all the seven BRR variance estimators for $\hat{p}$ and $\bar{y}$.

Recall that there are $H = 24$ strata and $m = 48$ sample clusters in the modified MFH Survey design, with exactly two clusters drawn from each stratum. Variance estimation by BRR starts with forming the K half-samples and the corresponding complement half-samples. We choose $K = 24$, i.e. the number of strata, and use the whole matrix in forming the half-samples and their complements. Note that for a full orthogonal balance we would choose $K = 28$. We work out a weight matrix from the 24×24 Hadamard matrix to perform the computations, which are based on the cluster-level data set given in Example 5.1.

The parent ratio and mean estimates $\hat{p}$ and $\bar{y}$, and the corresponding means of the half-sample estimates $\hat{p}_k$ and $\bar{y}_k$ with their complement half-sample estimates $\hat{p}_k^c$ and $\bar{y}_k^c$, are first calculated. These are:

$$\hat{p} = 0.3976, \quad \hat{\bar{p}} = \sum_{k=1}^{24}\hat{p}_k/24 = 0.3953 \quad \text{and} \quad \hat{\bar{p}}^c = \sum_{k=1}^{24}\hat{p}_k^c/24 = 0.3997,$$

$$\bar{y} = 141.785, \quad \hat{\bar{y}} = \sum_{k=1}^{24}\bar{y}_k/24 = 141.804 \quad \text{and} \quad \hat{\bar{y}}^c = \sum_{k=1}^{24}\bar{y}_k^c/24 = 141.768.$$

All three CHRON proportion estimates and SYSBP mean estimates are close.

We next calculate the BRR variance estimates (5.12)–(5.18). For CHRON, by using $\hat{p}$ we obtain from the half-samples and their complements:

$$\hat{v}_{1.brr}(\hat{p}) = \sum_{k=1}^{24}(\hat{p}_k - 0.3976)^2/24 = 0.1104 \times 10^{-3},$$

$$\hat{v}_{2.brr}(\hat{p}) = \sum_{k=1}^{24}(\hat{p}_k^c - 0.3976)^2/24 = 0.1103 \times 10^{-3},$$

and

$$\hat{v}_{3.brr}(\hat{p}) = (\hat{v}_{1.brr}(\hat{p}) + \hat{v}_{2.brr}(\hat{p}))/2 = 0.1103 \times 10^{-3}.$$

By using the mean estimates $\bar{\hat{p}}$ and $\bar{\hat{p}}^c$ we obtain the counterparts:

$$\hat{v}_{4.brr}(\hat{p}) = \sum_{k=1}^{24}(\hat{p}_k - 0.3953)^2/24 = 0.1052 \times 10^{-3},$$

$$\hat{v}_{5.brr}(\hat{p}) = \sum_{k=1}^{24}(\hat{p}_k^c - 0.3997)^2/24 = 0.1056 \times 10^{-3},$$

and

$$\hat{v}_{6.brr}(\hat{p}) = (\hat{v}_{4.brr}(\hat{p}) + \hat{v}_{5.brr}(\hat{p}))/2 = 0.1054 \times 10^{-3}.$$

From all the half-samples we finally obtain:

$$\hat{v}_{7.brr}(\hat{p}) = \sum_{k=1}^{24}(\hat{p}_k - \hat{p}_k^c)^2/(4 \times 24) = 0.1103 \times 10^{-3}.$$

For CHRON the first three BRR variance estimates, and the last one, happen to be equal to those obtained by the linearization method. Those based on the mean of the half-sample estimates are somewhat, but not very much smaller.
For SYSBP we obtain the following BRR variance estimates:

$$\hat{v}_{1.brr}(\bar{y}) = \sum_{k=1}^{24}(\bar{y}_k - 141.785)^2/24 = 0.2791,$$

$$\hat{v}_{2.brr}(\bar{y}) = \sum_{k=1}^{24}(\bar{y}_k^c - 141.785)^2/24 = 0.2790,$$

$$\hat{v}_{3.brr}(\bar{y}) = (\hat{v}_{1.brr}(\bar{y}) + \hat{v}_{2.brr}(\bar{y}))/2 = 0.2791,$$

$$\hat{v}_{4.brr}(\bar{y}) = \sum_{k=1}^{24} (\bar{y}_k - 141.804)^2/24 = 0.2787,$$

$$\hat{v}_{5.brr}(\bar{y}) = \sum_{k=1}^{24} (\bar{y}_k^c - 141.768)^2/24 = 0.2788,$$

$$\hat{v}_{6.brr}(\bar{y}) = (\hat{v}_{4.brr}(\bar{y}) + \hat{v}_{5.brr}(\bar{y}))/2 = 0.2787,$$

$$\hat{v}_{7.brr}(\bar{y}) = \sum_{k=1}^{24} (\bar{y}_k - \bar{y}_k^c)^2/(4 \times 24) = 0.2790.$$

For SYSBP, all the BRR variance estimates (and that obtained by the linearization method) are equal to 0.279 when rounded to three digits.

All the BRR variance estimators provided similar results for a ratio estimator, a subpopulation proportion or a mean, for the response variables considered. These results equal those drawn from other comparable empirical studies. Also on theoretical grounds, no definite preference for the BRR variance estimators of a nonlinear estimator can be given. The REPERR program for regression analysis under complex surveys, available in the OSIRIS software, uses the variance estimator $\hat{v}_{3.brr}$ for various parent estimators including the ratio estimator. The WesVarPC program uses the variance estimator $\hat{v}_{1.brr}$ as the basic BRR variance estimator. A BRR variant, called the Fay's method (Judkins 1990), is also available resembling jackknife type estimation.

The JRR Technique

The particular jackknife method based on *jackknife repeated replications* has many features of the BRR technique, since only the method of forming the pseudosamples is different. Application of the JRR technique to a design where more than two sample clusters are drawn from a stratum is more straightforward than for BRR. We, however, consider the JRR technique in the simplest case where the number of sample clusters per stratum is exactly two, and the clusters are assumed to be drawn with replacement, i.e. with a similar design to that required for BRR. JRR variance estimators are derived for a ratio estimator $\hat{r}$ which is a subpopulation proportion or estimator.

We construct the pseudosamples following the method suggested by Frankel (1971). For the first pseudosample, we exclude the first cluster $h1$ from the first stratum and weight the second cluster $h2$ by the value 2, leaving the remaining $H–1$ strata unchanged. By repeating this procedure for all strata, we get a total of H pseudosamples. For a similar set of H complement pseudo samples we change the order of the clusters that are excluded. The JRR variance estimators are derived by using these two sets of pseudosamples.

Like the BRR technique, several alternative JRR variance estimators can be constructed for the parent ratio estimator $\hat{r}$. For these, we first derive the pseudosample estimators for each stratum. Let $\hat{r}_h$ denote a pseudosample estimator based on excluding cluster $h1$ and duplicating cluster $h2$ in stratum h:

$$\hat{r}_h = \frac{2y_{h2} + \sum_{h' \neq h}^{H} \sum_{\alpha=1}^{2} y_{h'\alpha}}{2x_{h2} + \sum_{h' \neq h}^{H} \sum_{\alpha=1}^{2} x_{h'\alpha}}, \quad h = 1, ..., H. \tag{5.19}$$

These estimators are constructed for each pseudosample. From the complement pseudosamples, we obtain corresponding estimators $\hat{r}_h^c$ by excluding cluster $h2$ and duplicating cluster $h1$. Using the pseudosample estimators and the complement pseudosample estimators we can derive the first set of JRR variance estimators for the parent estimator $\hat{r}$. Hence we have

$$\hat{v}_{1.jrr}(\hat{r}) = \sum_{h=1}^{H} (\hat{r}_h - \hat{r})^2, \tag{5.20}$$

and from the complement pseudosamples

$$\hat{v}_{2.jrr}(\hat{r}) = \sum_{h=1}^{H} (\hat{r}_h^c - \hat{r})^2. \tag{5.21}$$

A combined variance estimator is

$$\hat{v}_{3.jrr}(\hat{r}) = (\hat{v}_{1.jrr}(\hat{r}) + \hat{v}_{2.jrr}(\hat{r}))/2. \tag{5.22}$$

Another set of variance estimators can be obtained by using the so-called *pseudovalues* introduced by Quenouille (1956) to reduce the bias of an estimator. In the case considered above, pseudovalues are of the form

$$\hat{r}_h^p = 2\hat{r} - \hat{r}_h, \quad h = 1, ..., H, \tag{5.23}$$

and for the complement pseudosamples they are denoted by $\hat{r}_h^{pc}$. By using the first set of H pseudovalues $\hat{r}_h^p$ we obtain a bias-corrected estimator given by

$$\hat{\bar{r}}^p = \sum_{h=1}^{H} \hat{r}_h^p / H, \tag{5.24}$$

and using the pseudovalues $\hat{r}_h^{pc}$ from the complement pseudosamples we obtain

$$\hat{\bar{r}}^{pc} = \sum_{h=1}^{H} \hat{r}_h^{pc} / H. \tag{5.25}$$

Counterparts to the variance estimators (5.20)–(5.22) can be derived from the pseudovalues and the bias-corrected estimators, giving

$$\hat{v}_{4.jrr}(\hat{r}) = \sum_{h=1}^{H} (\hat{r}_h^p - \hat{\bar{r}}^p)^2, \tag{5.26}$$

and from the complement pseudosamples

$$\hat{v}_{5.jrr}(\hat{r}) = \sum_{h=1}^{H} (\hat{r}_h^{pc} - \hat{\bar{r}}^{pc})^2. \tag{5.27}$$

A combined variance estimator can also be derived:

$$\hat{v}_{6.jrr}(\hat{r}) = (\hat{v}_{4.jrr}(\hat{r}) + \hat{v}_{5.jrr}(\hat{r}))/2. \tag{5.28}$$

Finally, from all the $2H$ pseudosamples we obtain:

$$\hat{v}_{7.jrr}(\hat{r}) = \sum_{h=1}^{H} (\hat{r}_h - \hat{r}_h^c)^2/4. \tag{5.29}$$

A similar way of constructing the JRR variance estimators was used to that given for the BRR technique. For a linear estimator, the bias-corrected JRR estimators reproduce the parent estimator, and all the JRR variance estimators coincide. This is not the case for nonlinear estimators, but in practice all JRR variance estimators should give closely related results. Like BRR, the variance estimator $\hat{v}_{7.jrr}$ could be taken as the most natural estimator of the variance of the parent estimator $\hat{\theta}$.

The JRR technique can be extended to a more general case where more than two clusters are drawn from each stratum, and for without-replacement sampling of clusters. Pseudosamples and their complements are constructed by consecutively excluding a cluster and weighting the remaining clusters appropriately in a stratum (see Wolter 1985, Section 4.6).

Like BRR, we use the JRR technique for variance estimation of a ratio estimator $\hat{r}$ for the MFH Survey design.

Example 5.3

The JRR technique in the MFH Survey. We continue considering the estimation of the variance of a ratio-type subpopulation proportion estimator $\hat{p}$ of CHRON (chronic morbidity) and a subpopulation mean estimator $\bar{y}$ of SYSBP (systolic blood pressure) for 30–64-year-old males. By using the cluster-level data set available, we calculate all the seven JRR variance estimates for $\hat{p}$ and $\bar{y}$.

Because $H=24$ we construct 24 JRR pseudosamples with their complements by the Frankel method. The parent ratio and mean estimates $\hat{p}$ and $\bar{y}$, and the corresponding bias-corrected estimators given by (5.24) and (5.25) based on the pseudovalues $\hat{p}_h^p$, $\hat{p}_h^{pc}$, $\bar{y}_h^p$ and $\bar{y}_h^{pc}$ calculated from the pseudosamples and their complements, are first obtained. These are:

$$\hat{p} = 0.3976, \quad \hat{\bar{p}}^p = \sum_{k=1}^{24} \hat{p}_k^p / 24 = 0.3972 \quad \text{and} \quad \hat{\bar{p}}^{pc} = \sum_{k=1}^{24} \hat{p}_k^{pc} / 24 = 0.3980,$$

$$\bar{y} = 141.785, \quad \hat{\bar{y}}^p = \sum_{k=1}^{24} \bar{y}_k^p / 24 = 141.793 \quad \text{and} \quad \hat{\bar{y}}^{pc} = \sum_{k=1}^{24} \bar{y}_k^{pc} / 24 = 141.777.$$

All three CHRON proportion estimates and SYSBP mean estimates are close. Next we calculate the JRR variance estimates. For a CHRON proportion estimator $\hat{p}$ the first variance estimate (5.20) is

$$\hat{v}_{1.jrr}(\hat{p}) = \sum_{h=1}^{24} (\hat{p}_h - 0.3976)^2 = 0.1099 \times 10^{-3},$$

and from the complement pseudosamples we obtain, using (5.21):

$$\hat{v}_{2.jrr}(\hat{p}) = \sum_{h=1}^{24} (\hat{p}_h^c - 0.3976)^2 = 0.1107 \times 10^{-3}.$$

The combined variance estimate (5.22) is thus

$$\hat{v}_{3.jrr}(\hat{p}) = (\hat{v}_{1.jrr}(\hat{p}) + \hat{v}_{2.jrr}(\hat{p}))/2 = 0.1103 \times 10^{-3}.$$

The second set (5.26)–(5.29) of JRR variance estimates is obtained by using the pseudovalues and the bias-corrected estimators. A counterpart of $\hat{v}_{1.jrr}$ is

$$\hat{v}_{4.jrr}(\hat{p}) = \sum_{h=1}^{24} (\hat{p}_h^p - 0.3972)^2 = 0.1060 \times 10^{-3},$$

and from the complement pseudosamples we have

$$\hat{v}_{5.jrr}(\hat{p}) = \sum_{h=1}^{24} (\hat{p}_h^{pc} - 0.3980)^2 = 0.1067 \times 10^{-3}.$$

The combined variance estimate is

$$\hat{v}_{6.jrr}(\hat{p}) = (\hat{v}_{4.jrr}(\hat{p}) + \hat{v}_{5.jrr}(\hat{p}))/2 = 0.1063 \times 10^{-3}.$$

From all the pseudosamples and their complements we obtain:

$$\hat{v}_{7.jrr}(\hat{p}) = \sum_{h=1}^{24} (\hat{p}_h - \hat{p}_h^c)^2 / 4 = 0.1103 \times 10^{-3}.$$

The JRR variance estimates for the CHRON proportion estimator $\hat{p}$ are quite close, as expected. For the SYSBP mean estimator $\bar{y}$ we obtain the following JRR variance estimates:

$$\hat{v}_{1.jrr}(\bar{y}) = \sum_{h=1}^{24}(\bar{y}_h - 141.785)^2 = 0.2773,$$

$$\hat{v}_{2.jrr}(\bar{y}) = \sum_{h=1}^{24}(\bar{y}_h^c - 141.785)^2 = 0.2803,$$

$$\hat{v}_{3.jrr}(\bar{y}) = (\hat{v}_{1.jrr}(\bar{y}) + \hat{v}_{2.jrr}(\bar{y}))/2 = 0.2788,$$

$$\hat{v}_{4.jrr}(\bar{y}) = \sum_{h=1}^{24}(\bar{y}_h^p - 141.793)^2 = 0.2759,$$

$$\hat{v}_{5.jrr}(\bar{y}) = \sum_{h=1}^{24}(\bar{y}_h^{pc} - 141.777)^2 = 0.2789,$$

$$\hat{v}_{6.jrr}(\bar{y}) = (\hat{v}_{4.jrr}(\bar{y}) + \hat{v}_{5.jrr}(\bar{y}))/2 = 0.2774,$$

$$\hat{v}_{7.jrr}(\bar{y}) = \sum_{h=1}^{24}(\bar{y}_h - \bar{y}_h^c)^2/4 = 0.2788.$$

For SYSBP, the JRR variance estimates of $\bar{y}$ are also very close. All the JRR variance estimators of a proportion estimator and a mean estimator provided closely related numerical results. Therefore, either practical or computational considerations can guide the selection of an appropriate JRR variance estimator. In the OSIRIS program REPERR there are a variety of jackknife methods available for variance approximation of a ratio estimator, including the JRR technique with the variance estimator $\hat{v}_{3.jrr}$. In WesVarPC, the jackknife technique is available with two variants. The first (JK1) is for designs where explicit stratification is not used and the second (JK2) uses the estimator $\hat{v}_{1.jrr}$. As the way of forming pseudosamples in the JRR technique is very simple, a program for JRR variance estimation on a cluster-level data set could be easily written in a flexible statistical language such as SAS.

The BOOT Technique

Similar to the other sample re-use methods, the bootstrap can be used for variance approximation of a nonlinear estimator under a complex sampling design. The method, however, differs from BRR and JRR in many respects, e.g. the generation of pseudosamples is quite different. We consider the bootstrap technique for variance estimation of a ratio estimator under a two-stage stratified epsem design where a constant number of clusters (which may be greater than two) is drawn with replacement from each stratum. We adopt a simple version of the bootstrap, introduced in Rao and Wu (1988) as a *naive bootstrap* for this kind of a design, and call it the *BOOT technique*. An application for the MFH Survey design, supplemented with appropriate SAS code for numerical calculation of the variance estimates, is provided (see Appendix 3).

Let us assume that $m_h = a \; (\geq 2)$ clusters are drawn with replacement from each of the H strata. The number of sample clusters is thus $m = a \times H$. We construct the bootstrap pseudosamples in the following way:

Step 1.   From the a sample clusters in stratum h draw a simple random sample of size a with replacement. This is performed independently in each stratum. The resulting H simple random samples together constitute a *bootstrap sample* of m clusters.

Step 2.   Repeating Step 1 K times, a total of K independent bootstrap samples are obtained.

It is important in Step 1 that the simple random samples in each stratum are drawn with replacement, and the stratum-wise samples are drawn independently. So, a particular sample cluster in a stratum may be included in a bootstrap sample many (even a) times, or not at all.

We consider the BOOT technique for the estimation of the variance of the ratio estimator $\hat{r}$ as previously. A ratio estimator for a bootstrap sample k is denoted by $\hat{r}_k \; (k = 1, ..., K)$. The mean of the bootstrap sample estimates $\hat{r}_k$ provides a *bootstrap estimator*

$$\hat{\bar{r}} = \sum_{k=1}^{K} \hat{r}_k/K. \tag{5.30}$$

A Monte Carlo variance estimator based on $\hat{r}_k$ and the bootstrap estimator (5.30) is first derived for the parent estimator $\hat{r}$:

$$\hat{v}_{mc}(\hat{r}) = \sum_{k=1}^{K} (\hat{r}_k - \hat{\bar{r}})^2/K. \tag{5.31}$$

Unfortunately, this intuitively attractive variance estimator is unacceptable because it is not consistent for the variance of $\hat{r}$ and, moreover, it is not unbiased even for the variance of a linear estimator, as Rao and Wu (1988) have shown. But in the case considered, where a constant number of clusters is drawn from each stratum, an appropriately rescaled Monte Carlo variance estimator provides a consistent variance estimator for the parent estimator $\hat{r}$. Hence the first BOOT variance estimator is

$$\hat{v}_{1.boot}(\hat{r}) = \frac{a}{a-1} \hat{v}_{mc}(\hat{r}) = \frac{a}{a-1} \sum_{k=1}^{K} (\hat{r}_k - \hat{\bar{r}})^2/K. \tag{5.32}$$

By using the parent estimator $\hat{r}$ in place of the bootstrap estimator, another variance estimator is obtained:

$$\hat{v}_{2.boot}(\hat{r}) = \frac{a}{a-1} \sum_{k=1}^{K} (\hat{r}_k - \hat{r})^2 / K. \tag{5.33}$$

It should be noticed that for the naive bootstrap there is no obvious solution to the scaling problem in the case where the number of sample clusters per stratum varies. Rao and Wu (1988) derive a *rescaling bootstrap* for these cases, based on drawing simple random samples of size m_h (≥ 1) clusters with replacement from a stratum. With appropriate selection of m_h, different versions of the bootstrap are provided. Sitter (1992) proposes a generalization of this method, based on resampling without replacement rather than with replacement, and repeating this many times with replacement. Rao *et al.* (1992) redefine the rescaling bootstrap to be also suitable for variance estimation of non-smooth functions such as the median.

In the BOOT technique, to obtain variance estimation results with sufficient precision the number K of bootstrap samples should be large, preferably 500 to 1000. The technique thus requires large processing capabilities and can consume a lot of computer resources. In this, the BOOT technique is more obviously computer-intensive than BRR and JRR. To indicate the sensitivity of numerical results of variance estimation to the number of bootstrap samples, we perform the calculations with various values of K given in the computational example shown in Appendix 3.

Example 5.4

The BOOT technique in the MFH Survey. We apply the BOOT technique for variance approximation of subpopulation proportion and mean estimators $\hat{p}$ (for CHRON) and $\bar{y}$ (for SYSBP), both considered as ratio estimators. The MFH Survey subgroup consists of 2699 males aged 30–64 years. In the MFH Survey design there are $H = 24$ strata each with $a = 2$ sample clusters, so each bootstrap sample constitutes of $m = 2 \times 24 = 48$ clusters. In the generation of the bootstrap samples we use the cluster-level data set. We obtain a bootstrap sample by drawing a simple random sample of two clusters with replacement independently from each stratum. Thus, a cluster in a stratum can appear in a bootstrap sample either 0, 1 or 2 times so that the sample size from a stratum is always two clusters. Note that the number of such samples can become large; e.g. if we have 1000 bootstrap samples, a total of 24 000 independent samples of size two must be drawn. In this example $K = 1000$ bootstrap samples. In Appendix 3, the values $K = 10$ and 100 are also used. The value $K = 10$ is for practical purposes too small and we use it for pedagogical purposes only. Stabilization of the estimates can be seen when the value of K is increased to 1000 bootstrap samples, which therefore provides the most useful variance estimation results.

An estimate $\hat{r}_k$ mimicking the parent estimator $\hat{r}$ is calculated from each of the K bootstrap samples. A bootstrap estimate is then calculated as an average of the $\hat{r}_k$. By using the $\hat{r}_k$, the bootstrap estimate and the parent estimate, we finally obtain BOOT variance estimates $\hat{v}_{1.boot}(\hat{r})$ and $\hat{v}_{2.boot}(\hat{r})$.

With $K = 1000$ bootstrap samples, the parent estimates and the bootstrap estimates (5.30) for CHRON proportion and SYSBP mean are

$$\hat{p} = 0.3976, \text{ and the bootstrap estimate is } \hat{\bar{p}} = 0.3973,$$
$$\bar{y} = 141.785, \text{and the bootstrap estimate is } \hat{\bar{y}} = 141.783.$$

The BOOT variance estimates (5.32) and (5.33) for CHRON proportion $\hat{p}$ are, respectively

$$\hat{v}_{1.boot}(\hat{p}) = 2 \times \sum_{k=1}^{1000} (\hat{p}_k - 0.3973)^2/1000 = 0.1039 \times 10^{-3}$$

and

$$\hat{v}_{2.boot}(\hat{p}) = 2 \times \sum_{k=1}^{1000} (\hat{p}_k - 0.3976)^2/1000 = 0.1040 \times 10^{-3}.$$

The BOOT variance estimates for SYSBP mean $\bar{y}$ are

$$\hat{v}_{1.boot}(\bar{y}) = 2 \times \sum_{k=1}^{1000} (\bar{y}_k - 141.783)^2/1000 = 0.2798$$

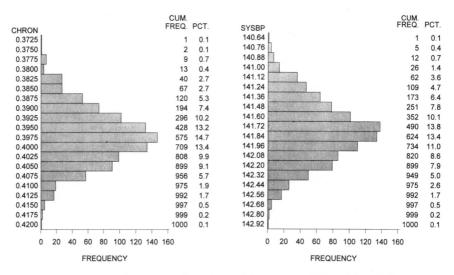

Figure 5.1 Bootstrap histograms for CHRON (a binary variable) and SYSBP (a continuous variable) from the bootstrap estimates $\hat{r}_k$ with $K = 1000$ bootstrap samples.

and

$$\hat{v}_{2.boot}(\bar{y}) = 2 \times \sum_{k=1}^{1000} (\bar{y}_k - 141.785)^2 / 1000 = 0.2798.$$

For a CHRON proportion estimator $\hat{p}$ and a SYSBP mean estimator $\bar{y}$, both BOOT variance estimates are approximately equal. As in the other re-use methods, any definite preference for the type of variance estimator has not been suggested. From a computational point of view, the estimator $\hat{v}_{2.boot}$ is simpler than $\hat{v}_{1.boot}$. A SAS macro for bootstrap variance approximation by (5.32) and (5.33), with more detailed numerical results, is given in Appendix 3. By using this macro the data for the bootstrap histograms in Figure 5.1 were produced.

5.5 COMPARISON OF VARIANCE ESTIMATORS

The linearization method and sample re-use methods were used as basic approximation techniques for variance estimation of a nonlinear ratio estimator. It was assumed that the sample was from a two-stage epsem sampling design with at least two clusters drawn with replacement from each stratum. The linearization method was considered under a design with varying number (≥ 2) of sample clusters per stratum. Basic forms of the balanced half-samples (BRR) and jackknife repeated replications (JRR) techniques involved a design with exactly two sample clusters per stratum, and the number of strata is assumed large. Both methods have been generalized for designs with a varying number (≥ 2) of sample clusters per stratum. The bootstrap technique was considered under a design where a constant number (≥ 2) of clusters were drawn from each stratum. Also the bootstrap has been generalized for the case of a varying number of sample clusters per stratum. Of the approximation methods, the bootstrap tends to require more computer resources. We next compare the numerical results obtained from the MFH Survey for variance approximation by the linearization and sample re-use techniques.

Comparison of the Variance Estimates in the MFH Survey

Using the linearization, BRR, JRR and BOOT techniques we estimated the variance of a subpopulation proportion estimator of a binary response CHRON (chronic morbidity), and the variance of a subpopulation mean estimator of a continuous response SYSBP (systolic blood pressure). Both estimators were ratio-type estimators for the MFH Survey subgroup consisted of 2699 males aged 30–64 years. Detailed results were given in Examples 5.1–5.4. There were a total of 24 strata each with two sample clusters in the MFH Survey sampling design which therefore provides adequate data for demonstrating all the variance approximation methods. A cluster-level data set with 48 observations was used in all techniques.

Variance and design-effect estimates for a CHRON proportion $\hat{p}$ and a SYSBP mean $\bar{y}$ are displayed in Table 5.5. The design-effect estimator is of the form deff= $\hat{v}/\hat{v}_{srswr}$, where $\hat{v}$ is the variance estimator being considered and $\hat{v}_{srswr}$ is the variance estimator corresponding to simple random sampling with replacement.

For CHRON, the variance estimate from linearization, and the first three BRR and JRR estimates and the last one from the BRR and JRR techniques are all nearly equal. When compared with these estimates, the fourth, fifth and sixth BRR and JRR variance estimates, and both of the BOOT variance estimates, are somewhat smaller. Note that the linearization and the last BRR

Table 5.5 Linearization, BRR, JRR, BOOT, and SRSWR variance and design-effect estimates $\hat{v}$ and deff of a CHRON proportion estimate $\hat{p}$ and a SYSBP mean estimate $\bar{y}$ in the MFH Survey subgroup of 30–64-year-old males.

Method	Chronic morbidity $10^{-3} \times \hat{v}(\hat{p})$	deff $(\hat{p})$	Systolic blood pressure $\hat{v}(\bar{y})$	deff $(\bar{y})$
Linearization				
DES	0.1103	1.24	0.2788	2.06
Balanced repeated replications				
1	0.1104	1.24	0.2791	2.06
2	0.1103	1.24	0.2790	2.06
3	0.1103	1.24	0.2791	2.06
4	0.1052	1.18	0.2787	2.06
5	0.1056	1.19	0.2788	2.06
6	0.1054	1.19	0.2787	2.06
7	0.1103	1.24	0.2790	2.06
Jackknife repeated replications				
1	0.1099	1.24	0.2773	2.05
2	0.1107	1.25	0.2803	2.07
3	0.1103	1.24	0.2788	2.06
4	0.1060	1.19	0.2759	2.04
5	0.1067	1.20	0.2789	2.06
6	0.1063	1.20	0.2774	2.05
7	0.1103	1.24	0.2788	2.06
Bootstrap				
1	0.1039	1.17	0.2798	2.07
2	0.1040	1.17	0.2798	2.07
SRSWR	0.0888	1.00	0.1352	1.00

and JRR variance estimates (which could be taken as the most appropriate variance estimates) are equal. For SYSBP, all the BRR variance estimates are nearly equal, and the JRR estimates indicate larger variation. The BOOT variance estimates are somewhat larger than the others. For SYSBP, the linearization and the last BRR and JRR variance estimates are also nearly equal.

The design-effect estimates indicate varying degree of intra-cluster correlation for CHRON and SYSBP. CHRON has noticeably less intra-cluster correlation than SYSBP. For SYSBP, the design-effect estimates indicate only slight variation between techniques.

In conclusion, variance estimates of the ratio estimators obtained by the linearization, BRR, JRR and BOOT techniques do not differ significantly from each other, for both response variables. Therefore, software availability might guide the selection of a technique in practice. Of the commercial software products which are widely used, SUDAAN and PC CARP exclusively use the linearization method. OSIRIS programs PSALMS, PSRATIO, PSTABLE and PSTOTAL also use linearization, whereas in the program REPERR both BRR and JRR techniques are implemented. For variance approximation of a ratio estimator using the bootstrap, the SAS macro given in Appendix 3 can be used in certain cases.

Other Properties of the Variance Approximation Methods

The variance approximation techniques based on linearization, BRR and JRR have been extensively evaluated in the literature since 1970 by empirical investigations and simulation studies, on more theoretical arguments. However, the bootstrap has been a target of comparison with the other methods only recently. We briefly refer to some of the most important results.

Kish and Frankel (1974) empirically studied the relative performances of linearization, BRR and JRR under an epsem one-stage stratified design with two clusters drawn with replacement from each stratum. They showed first that for a linear estimator, the variance estimators coincided and were the same as a standard textbook variance estimator. Properties of the variance estimators were different for nonlinear estimators such as ratio estimators, regression coefficients and correlation coefficients. The linearization method provided the most stable variance estimates whilst BRR gave the least stable, but none of the estimators gave an overall best performance when many criteria were considered. Kish and Frankel concluded that the linearization technique might be the best choice for ratio estimators, and sample re-use techniques for other nonlinear estimators.

Krewski and Rao (1981) showed that linearization, BRR and JRR have similar first-order asymptotic properties. Rao and Wu (1985) considered higher-order properties and showed that linearization and JRR provide equal second-order properties under a design where two clusters are drawn with replacement from each stratum. Rao and Wu (1988) considered the bootstrap and showed that the first-order properties of their rescaling bootstrap variance estimator coincide with those of linearization, BRR and JRR. Second-order properties, however, differ. The rescaling bootstrap also indicated greater instability than either the linearization or the JRR. Rao *et al.* (1992) studied the performances of jackknife, BRR and bootstrap for variance estimation of the median and noticed no considerable differences between the methods.

5.6   CHAPTER SUMMARY AND FURTHER READING

Summary

Proper estimation of the variance of a ratio estimator is important in the analysis of complex surveys. First, variance estimates are needed to derive standard errors and confidence intervals for nonlinear estimators such as a ratio estimator. The estimation of the variance of ratio mean and ratio proportion estimators was carried out under an epsem two-stage stratified cluster-sampling design, where the sample data set was assumed self-weighting so that adjustment for nonresponse was not necessary. The demonstration data set from the modified sampling design of the Mini-Finland Health Survey (MFH Survey) fulfilled these conditions.

A ratio-type estimator $\hat{r} = y/x$ was examined for the estimation of the sub-population mean and proportion in the important case of a subgroup of the sample whose size x was not fixed by the sampling design. Therefore, the denominator quantity x in $\hat{r}$ is a random variable, involving its own variance and covariance with the numerator quantity y. In addition to the variance of y, these variance and covariance terms contributed to the variance estimator of a ratio estimator calculated with the linearization method. This method was considered in depth because of its wide applicability in practice and popularity in software products for survey analysis.

We also introduced alternative methods for variance estimation of a ratio estimator based on sample re-use methods. The techniques of balanced half-samples (BRR) and jackknife (JRR) are traditional sample re-use methods, but the bootstrap (BOOT) has been applied for complex surveys only recently. Being computer-intensive, they differ from the linearization technique but are, as such, readily applicable for different kinds of nonlinear estimators. With-replacement sampling of clusters was assumed for all the approximation methods. With this assumption, the variability of a ratio estimate was evaluated using the between-cluster variation only, leading to relatively simple variance estimators. The design effect was used extensively as a measure of the contribution of the clustering on a variance estimate, relative to the variance estimate based on simple random sampling with replacement.

The MFH Survey sampling design was selected for variance estimation because of its simplicity: there were exactly two sample clusters in each stratum in the modified sampling design. A subgroup of the MFH Survey data set covering 30–64-year-old males was used with all the variance approximation methods. This specific subgroup was chosen instead of the entire MFH Survey sample because the total sample size was fixed by the sampling design, but for the subpopulation considered the sample size was a random variate, thus providing a good target for demonstrating variance estimation with approximative methods. The selected subgroup constitutes a cross-classes type domain mimicking properly all essential properties of the MFH Survey sampling design such as inclusion of elements from all of the 24 strata and 48 sample clusters. This would not be the case if, for example, a regional subgroup were chosen where only a part of the strata and sample clusters would be covered.

The variance approximation methods provided similar results in variance estimation of a proportion estimator of a binary response variable CHRON (chronic morbidity), which was a slightly intra-cluster correlated variable, and for a mean estimator of a continuous response variable SYSBP (systolic blood pressure) having stronger intra-cluster correlation. Because no theoretical arguments are available for choosing between the approximative variance estimators, technical factors such as software availability often guides the selection of an appropriate method in practice.

Computer programs are available for variance approximation of ratio estimators in complex surveys. The SUDAAN and PC CARP software use the linearization method. In the OSIRIS software, the linearization, BRR and JRR techniques are available. The WesVarPC software uses exclusively pseudoreplication methods based on BRR and jackknife techniques. With SUDAAN and PC CARP, appropriate sampling-design options can be used to account for the contribution of various stages of a multi-stage design to variance estimates, supplementing the simple variance estimators examined here. In addition to ratio estimators, variance approximation of more complex nonlinear estimators is also possible with these software packages.

The assumption of an epsem design is relaxed in the next chapter where the estimation of the covariance matrix of a vector of ratio estimators is considered. There, appropriate element weights are derived and used in the estimation under a non-epsem design.

Further Reading

In-depth consideration of the estimation of variance of a ratio, and other nonlinear estimators, can be found in Wolter (1985). Supplementary sources on the topic, in addition to those already mentioned, are Kalton (1977, 1983), Verma *et al.* (1980), Rust (1985) and Rao (1988). Thorough discussion on the concept of design effect is given in Kish (1995).

6

Covariance-matrix Estimation of Ratio Estimators

Variance estimation is extended in this chapter to the estimation of the covariance matrix of several ratio estimators which are each calculated for a particular population subgroup. Such a covariance-matrix estimate includes not only variance estimates of the ratio estimators, but also the covariances of separate ratio estimators. Covariance-matrix estimates of such ratio estimators as subpopulation proportions and means are needed, for example, to conduct logit modelling and other types of modelling procedures. Thus, this chapter partly serves as background material for Chapter 8. The other extension given in this chapter is to consider non-epsem complex designs. This is done by incorporating appropriate element weights in the estimators.

All the approximation methods for variance estimation of a ratio estimator under a complex sampling design, introduced in Chapter 5, would also be available for the covariance-matrix estimation. We choose the linearization method because of its practical importance. Covariance-matrix estimation using linearization is considered in Section 6.2. In Section 6.3, certain special covariance-matrix estimators are examined, aimed adjusting for extravariation based on the so-called *effective sample sizes*. For this, the concept of design effect of a ratio estimator is extended to a *design-effects matrix* of a vector of several ratio estimators. The design-effects matrix is also is used when assessing the contribution from clustering on a covariance-matrix estimate. The numerical examples given in this chapter come from the Occupational Health Care Survey where stratified cluster sampling is used.

6.1 THE OCCUPATIONAL HEALTH CARE SURVEY

In this section we describe the sampling design, data collection, and properties of the available survey data of the Occupational Health Care Survey (OHC

Survey). The sampling design of the OHC Survey is an example of stratified cluster sampling where both one-stage and two-stage sampling are used. Thus, the OHC Survey sampling design is slightly more complex than the MFH Survey. Moreover, in the OHC Survey sampling design a large number of sample clusters are available, and the design produces noticeable clustering effects for several response variables. Therefore, this sampling design is very suitable for examining covariance-matrix estimation in a complex survey. The OHC Survey will be used for further examples given in Chapters 7 and 8.

In Finland, as in many industrialized countries, the provision of occupational health (OH) services is regulated by legislation. An Act on Occupational Health Services came into force in 1979 to guide the development of OH services. All employers, with a few minor exceptions, would be required to provide OH services for their employees so that the activities would focus on the main work-related health hazards. Through the National Sickness Insurance Scheme, employers are reimbursed by the Social Insurance Institution for 55% (in 1992) of the costs of OH services. For employees, the OH services are free of charge. Nowadays, over 80% of the total employee population is covered by OH services. Being responsible for the financing of OH services the Social Insurance Institution has had a great interest in following the development of OH services and the functioning of the OHC Act. Several sample surveys have been carried out for this purpose, with a major one, the OHC Survey, conducted in 1985. An overview and the main results of the OHC Survey are given in Kalimo *et al.* (1991).

Sampling Design

The OHC Survey can be characterized as a multi-purpose analytical sample survey similar to the MFH Survey. The OHC Survey was aimed at assessing implementation of the activities prescribed by the OHC Act, at discovering how well the essential goals of the legislation had been attained, and at defining how OH services could be further developed. The survey focused on establishments in all industries except farming and forestry, on the employers and employees, and on the units which provided the OH services for the sites surveyed. There were about 2 million employees and over 100 000 industrial establishments in the target populations.

In the study design, the industrial establishment was the primary unit of sampling and data collection. Because in Finland there are nation-wide registers available for a sampling frame covering fairly well the target establishments, cluster sampling was a natural choice to be used with establishments as the clusters, i.e. primary sampling units. In contrast to the MFH sampling design, the principal motivation for cluster sampling in the OHC Survey was subject matter rather than cost efficiency.

Within the establishment sampling frame, sizes of PSUs varied widely, from one-person workplaces to enterprises with a thousand or more workers. This property of varying cluster sizes should be taken into account when considering the person-level sample size for data collection. Therefore the population of clusters was stratified by cluster size and by using two-stage sampling in strata that covered large sites. In addition to size, type of industry of establishment was used to form six explicit strata. One-stage sampling was used in strata covering establishments with a maximum of 100 employees; otherwise, two-stage sampling was used with approximately 50 employees sampled from each large site. This would produce an estimated total sample of about 17 000 employees in a sample of 1542 establishments. Stratum-wise allocation of the clusters, based on prior knowledge of their expected mean sizes, was carried out so that the employee sample would be nearly epsem, giving approximately equal inclusion probabilities for the employees. The sampling design is described in more detail in Lehtonen (1988).

Data Collection and Nonresponse

Structured questionnaires were used to collect data from employers, employees and OH units. During the data collection it turned out that a number of sample establishments, mainly small ones, had closed down, and the final number of establishments for the appropriate questionnaire was 1362. The response rate was 88%. Furthermore, 82% (13 355) of the employees from the 1195 responding establishments completed the personnel questionnaire. Finally, 93% of the OH units of the responding establishments completed the appropriate questionnaire; this produced information on 760 out of a total of 816 establishments covered by OHC. The numbers of establishments and employees in the resulting survey data for each stratum are displayed in Table 6.1.

Analyses based on logit models indicated statistically significant variation in the response rates of the establishment questionnaire, depending on certain structural features of the establishments such as size, type of industry and organizational type. Predicted response rates for the appropriate questionnaire (based on a logit model with size, type of industry, organizational type, and interaction of the two last mentioned as the model terms) are displayed in Figure 6.1. Small size, belonging to the construction industry, and having only a single site all increased the probability of nonresponse.

Nonresponse was quite small in large establishments and was independent of the type of industry or organizational type. It was also noted that establishments covered by OH services, and for which the regulations of the OHC Act were obligatory, responded most frequently to the appropriate questionnaire. Also, establishments for which the regulations of the Act were obligatory had an approximately equal response rate whether or not they were covered by OH services. Nonresponse was highest in those smallest single-site establish-

Table 6.1 The number of establishments and employees by stratum in the OHC Survey data.

Stratum	Size	Number of Establishments	Employees	Average cluster sample size
1	1–10	696	1730	2.5
2	11–100	176	4143	23.5
3	101–500	52	2396	46.1
4	501+	21	976	46.5
5	(all sizes)	109	1396	12.8
6	(all sizes)	141	2714	19.2
Total sample		1195	13355	11.2

Type of industry :
  Strata 1–4: All except those in strata 5 and 6
  Stratum 5: Construction industry
  Stratum 6: Public services

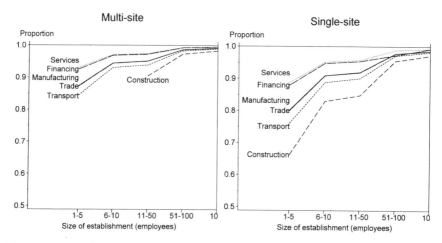

Figure 6.1 Predicted response rates in the establishment questionnaire (based on a logit model) by size and type of industry of establishment, in establishments of multi-site enterprises and in single-site establishments.

ments which operated in the construction industry and were not covered by OH services.

Weighting for nonresponse was required in the cluster-level analyses, for example, for the estimation of coverage of the OHC. The weight was constructed so that stratum-wise variation in inclusion probabilities of the PSUs was also compensated for. At the employee level the sampling design was nearly epsem, and the total number of employees at the small nonresponse establishments was relatively small. Therefore, adjustment for nonresponse in the element-level analyses was not so critical as at the cluster level. This was so, for example, in inferences concerning employee-level target populations on establishments covered by OH services.

Design Effects

A subgroup of establishments with a minimum of 10 employees will make up the OHC Survey data set used for demonstration purposes in examples. The data set includes a total of 250 clusters in five strata, and a total of 7841 employees. The data set can be regarded as approximately self-weighting. Cluster sample sizes in this subgroup vary from 10 to about 60 workers. Note that the subgroup is of a segregated classes type. These data, for selected response variables, are displayed in Table 6.2.

The number of sample clusters, i.e. establishments, is large (250) and this is favourable for covariance-matrix estimation. The sample establishments tend to be homogeneous with respect to certain subject-level response variables, resulting in positive intra-cluster correlations. For example, in a manufactur-

Table 6.2   The available OHC Survey data by sex and age of respondent, and proportions (%) of chronically ill persons (CHRON) and persons exposed to physical health hazards of work (PHYS), and the mean of the standardized first principal component of nine psychic (psychological or mental) symptoms (PSYCH).

Sex	Age	Sample n	%	CHRON %	PHYS %	PSYCH Mean
Males		4485	57.2	29.3	46.0	−0.104
Females		3356	42.8	29.2	19.4	0.139
Males	15–24	504	6.4	15.5	52.8	−0.300
	25–34	1355	17.3	19.8	50.8	−0.160
	35–44	1453	18.5	27.1	42.9	−0.073
	45–54	847	10.8	44.2	41.9	−0.033
	55–64	326	4.2	61.3	39.3	0.102
Females	15–24	418	5.3	16.0	19.1	0.095
	25–34	993	12.7	18.9	18.9	0.132
	35–44	1002	12.8	26.5	17.9	0.104
	45–54	681	8.7	43.5	18.5	0.168
	55–64	262	3.3	61.8	29.4	0.301
Both sexes	15–24	922	11.8	15.7	37.5	−0.121
	25–34	2348	29.9	19.4	37.4	−0.036
	35–44	2455	31.3	26.9	32.7	−0.000
	45–54	1528	19.5	43.8	31.5	0.056
	55–64	588	7.5	61.6	34.9	0.191
Total sample		7841	100.0	29.2	34.6	0.000

Table 6.3 Averages of design-effect estimates of proportion estimates of selected groups of binary response variables in the OHC Survey data set (number of variables in parentheses).

Study variable	Mean deff
Physical working conditions (12)	6.5
Psycho-social working conditions (11)	3.3
Psycho-somatic symptoms (8)	2.0
Psychic symptoms (9)	1.8

ing firm, working conditions tend to be similar for most of the workers, these conditions being different from those of an office establishment which in turn are also internally homogeneous. This produces design-effect estimates of means and proportions noticeably greater than one, especially for subject-level response variables measuring workplace-related matters such as physical or psycho-social working conditions. In some other variables, intra-cluster correlations were smaller, e.g. in variables describing overall psychic (psychological or mental) strain and psycho-somatic symptoms. Design effects for selected response variables are displayed in Table 6.3.

The average design-effect estimates are noticeably large especially in response variables strongly associated with working conditions. The averages are closer to one in the variables which cannot be considered as being work-related. For further analyses, three response variables are selected: the variables PHYS (physical health hazards of work) and CHRON (chronic morbidity) which are binary, and the variable PSYCH (psychic strain) which is continuous. PHYS has strong intra-cluster correlation with a large overall design-effect estimate of 7.2. The overall design-effect estimates of CHRON and PSYCH are 1.8 and 2.0, respectively. Moreover, PHYS is apparently work-related; this is not as clear for CHRON and PSYCH.

6.2 LINEARIZATION METHOD FOR COVARIANCE-MATRIX ESTIMATION

Weighted Ratio Estimator

We previously considered the case of a single ratio estimator. A *vector* of ratio estimators consists of u ratio estimators, where $u \geq 2$ is the number of population subgroups called *domains*. The domains are formed by cross-classifying one or more categorical predictors such as sex, age group, socioeconomic factors, or regional variables. Our aim is to estimate consistently the domain ratio parameters and the corresponding covariance matrix of the ratio estimators under a given complex sampling design. For this, we construct a *weighted ratio estimator* to be used for the domain ratios. For a binary response variable, we work with weighted domain proportions, and for a continuous response with weighted domain means.

Let the population of N elements be divided into u non-overlapping subpopulations or domains. The unknown population ratio vector is a column vector denoted by $\mathbf{R} = (R_1, \dots, R_u)'$. It consists of u domain ratio parameters $R_j = T_j/N_j$, where T_j denotes the population domain total of a response variable and N_j denotes the domain size, $\sum_{j=1}^{u} N_j = N$. In the binary case, the ratio parameter vector is denoted by $\mathbf{p} = (p_1, \dots, p_u)'$, consisting of proportion parameters $p_j = N_{j1}/N_j$, where N_{j1} is the population total of a binary response in domain j. And in the continuous case, the parameter vector is denoted by $\bar{\mathbf{Y}} = (\bar{Y}_1, \dots, \bar{Y}_u)'$, where $\bar{Y}_j$ are domain mean parameters $\bar{Y}_j = T_j/N_j$. A sample of n elements is drawn using stratified cluster sampling such that m_h clusters are drawn from each of the $h = 1, \dots, H$ strata with a total $m = \sum_{h=1}^{H} m_h$ of sample clusters, where $H \geq 1, m \geq 2H$ and $m > u$. In two-stage cluster sampling, a sample of $n_{h\alpha}$ elements is drawn from sample cluster α in stratum h, $\sum_{h=1}^{H} \sum_{\alpha=1}^{m_h} n_{h\alpha} = n$. If sampling is performed in one stage, all the elements of the selected sample clusters are taken in the element level sample.

In complex surveys, epsem designs with an equal inclusion probability for each population element are often used because they are convenient for statistical analysis. We considered such designs in Chapter 5, and the MFH and OCH Survey sampling designs are taken as being epsem. In practice, however, element inclusion probabilities can vary between the strata, and, even in epsem designs, weighting may be necessary to adjust for nonresponse to attain consistent estimation. Also to cover these cases, we derive a weighted ratio estimator which is more generally applicable than that previously considered for epsem samples.

For a self-weighting data set, an epsem sampling design is required and unit nonresponse is considered ignorable (see Chapter 4 and Section 5.2). If the data set is not self-weighting, an appropriate weight variable should be generated for statistical analyses. A weight variable assigns a positive value for each element of the data set such that unequal element inclusion probabilities and nonresponse are adjusted. Basically, as shown in Chapter 4, the weight w_k for a sample element k is the reciprocal of the probability of inclusion and response, i.e. $w_k = 1/(\pi_k \theta_k)$. In epsem designs, π_k is a constant. In *non-epsem* designs, unequal inclusion probabilities may arise, for example, due to non-proportional allocation. For nonresponse adjustment, the sample data set can be divided into a number of adjustment cells, and the response rate θ_c is assumed constant within cell c but is allowed to vary between the cells. The cells are formed by using auxiliary variables which are also available for nonresponse cases. When using poststratification, adjustment cells are formed by using auxiliary information on the population level (see Sections 3.3 and 5.1 and Chapter 4). Note that the weight is a constant in a self-weighting data set because π_k and θ_k are constants.

As shown in Section 5.1, there are two main approaches for a weight variable. In a descriptive survey where the population total on a study variable is estimated, a weight variable is constructed such that the sum of all n element weights w_k^* provides a consistent estimate $\hat{N}$ of the population size N. This type of weighting was extensively used in Chapters 2 and 3. In analytical surveys, where such totals are rarely estimated, it is customary to rescale the weights so that their sum equals the size n of the available sample data set. Although either kind of a weight variable can be used in software available for survey analysis such as SUDAAN or PC CARP, rescaled weights w_k^{**}, that sum up to n, are often more convenient for statistical analyses requiring a weight variable.

When using weights w_k^*, a vector $\hat{\mathbf{r}} = (\hat{r}_1, \ldots, \hat{r}_u)'$ of combined ratio estimators is constructed consisting of domain ratio estimators $\hat{r}_j = \hat{t}_j/\hat{N}_j$, where $\hat{t}_j$ is a weighted total estimator of the population total T_j of the response variable in domain j and $\hat{N}_j$ is the weighted size of domain j, and $\sum_{j=1}^{u} \hat{N}_j = \hat{N}$, the sum of all n sample weights. As a result, the weighted estimators t_j and $\hat{N}_j$ are consistent for the corresponding population analogues T_j and N_j, so the domain ratio estimator $\hat{r}_j$ is consistent for the domain ratio R_j in a given complex sampling design.

The weighted totals $\hat{t}_j$ and $\hat{N}_j$ in the previous domain ratio estimators $\hat{r}_j$ are scaled to sum to the population level. For analytical purposes, we rescale the weights so that they sum to n, the size of the sample data set. Thus to derive an estimator $\hat{r}_j$ we use the scaled weighted analogues y_j and x_j of $\hat{t}_j$ and $\hat{N}_j$ such that $y_j = (n/\hat{N})\hat{t}_j$ and $x_j = (n/\hat{N})\hat{N}_j$ with $\sum_{j=1}^{u} x_j = n$. The domain ratio estimator $\hat{r}_j$ can thus be written in the form

$$\hat{r}_j = \frac{y_j}{x_j} = \frac{\sum_{h=1}^{H} \sum_{\alpha=1}^{m_h} y_{jh\alpha}}{\sum_{h=1}^{H} \sum_{\alpha=1}^{m_h} x_{jh\alpha}} = \frac{\sum_{h=1}^{H} \sum_{\alpha=1}^{m_h} \sum_{\beta=1}^{x_{h\alpha}} w_{jh\alpha\beta}^{**} y_{jh\alpha\beta}}{\sum_{h=1}^{H} \sum_{\alpha=1}^{m_h} \sum_{\beta=1}^{x_{h\alpha}} w_{jh\alpha\beta}^{**}}, \quad j = 1, \ldots, u, \qquad (6.1)$$

where $y_{jh\alpha}$ is the weighted sample sum of the response variable for the elements falling in domain j in sample cluster α of stratum h, and $x_{jh\alpha}$ is the corresponding weighted domain sample size. The rescaled weights $w_{jh\alpha\beta}^{**}$ in (6.1) therefore sum up to n.

For a binary response, the ratio estimator $\hat{\mathbf{r}}$ with elements of the form (6.1) is a proportion estimator vector denoted by $\hat{\mathbf{p}} = (\hat{p}_1, \ldots, \hat{p}_u)'$ which consists of domain ratio estimators $\hat{p}_j = y_j/x_j = \hat{n}_{j1}/\hat{n}_j$, where $\hat{n}_{j1}$ is the weighted sample sum of the binary response for sample elements belonging to the domain j and $\hat{n}_j$ is the weighted domain size such that $\sum_{j=1}^{u} \hat{n}_j = n$. Under an epsem design and, moreover, if the data set is self-weighting, a simple unweighted estimator $\hat{\mathbf{p}}^U = (\hat{p}_1^U, \ldots, \hat{p}_u^U)'$ of $\mathbf{p}$ is obtained, where $\hat{p}_j^U = n_{j1}/n_j$ is a consistent estimator of the domain parameter p_j, n_{j1} is the sample sum of the binary response in domain j and n_j is the corresponding domain sample size such that $\sum_{j=1}^{u} n_j = n$. In this case $\hat{\mathbf{p}}$ and $\hat{\mathbf{p}}^U$ coincide. Note that if the data set is not self-weighting the estimator $\hat{\mathbf{p}}^U$ is not consistent for $\mathbf{p}$.

For a continuous response variable, we denote the weighted ratio estimator vector $\bar{\mathbf{y}} = (\bar{y}_1, \ldots, \bar{y}_u)'$, where the domain sample means $\bar{y}_j = y_j/x_j$ are consistent for the corresponding population domain means $\bar{Y}_j = T_j/N_j$. The corresponding unweighted counterpart is $\bar{\mathbf{y}}^U = (\bar{y}_1^U, \ldots, \bar{y}_u^U)'$.

It may be noted that the data actually needed for the ratio estimators $\hat{r}_j$ consist of m cluster-level scaled weighted sample sums $y_{jh\alpha}$ and $x_{jh\alpha}$. Indeed, the analysis of such data can be performed by using the cluster-level data set of size m and access to the element-level data set of size n is not necessarily required. In practice, however, when using software for survey analysis, the weighted sample sums y_j and x_j are estimated from an element-level data set by using the rescaled element weights $w_{jh\alpha\beta}^{**}$.

Covariance-matrix Estimation

The unknown population covariance matrix $\mathbf{V}/n$ of the ratio estimator vector $\hat{\mathbf{p}}$ has u rows and u columns, thus it is a $u \times u$ matrix. $\mathbf{V}/n$ is symmetric such that the lower and upper triangles of the matrix are identical. Variances of the domain ratio estimators are placed on the main diagonal of $\mathbf{V}/n$ and covariances of the corresponding domain ratio estimators on the off-diagonal part of the matrix. There is a total of $u \times (u+1)/2$ distinct parameters in $\mathbf{V}/n$ that need to be estimated.

The variance and covariance estimators $\hat{v}_{des}(\hat{r}_j)$ and $\hat{v}_{des}(\hat{r}_j, \hat{r}_k)$, being respectively the diagonal and off-diagonal elements of a consistent covariance-matrix estimator $\hat{\mathbf{V}}_{des}$ of the asymptotic covariance matrix $\mathbf{V}/n$ of the ratio estimator vector $\hat{\mathbf{r}} = (\hat{r}_1, \ldots, \hat{r}_u)'$, are derived using the linearization method considered in Section 5.3. The variance and covariance estimators of the sample sums y_j and x_j in a variance estimator $\hat{v}_{des}(\hat{r}_j)$ of $\hat{r}_j = y_j/x_j$, and the covariance estimators of the sample sums y_j, y_k, x_j and x_k in the covariance estimators $\hat{v}_{des}(\hat{r}_j, \hat{r}_k)$ of $\hat{r}_j$ and $\hat{r}_k$ in separate domains, are straightforward generalizations of the corresponding variance and covariance estimators given in Section 5.3 for the variance estimator of a single ratio estimator $\hat{r}$. We therefore do not show these formulae.

Like the scalar case, the variance and covariance estimators of $\hat{r}_j$ and $\hat{r}_k$ are based on the with-replacement assumption and the variation accounted for is the between-cluster variation. This causes bias in the estimates, but the bias can be assumed negligible if the first-stage sampling fraction is small.

The variance and covariance estimators of y_j, x_j, y_k and x_k are finally collected into the corresponding $u \times u$ covariance-matrix estimators $\hat{\mathbf{V}}_{yy}$, $\hat{\mathbf{V}}_{xx}$

and $\hat{\mathbf{V}}_{yx}$. By using these estimators, the design-based covariance-matrix estimator of $\hat{\mathbf{r}}$ based on the linearization method is given by

$$\hat{\mathbf{V}}_{des} = \mathrm{diag}(\hat{\mathbf{r}})(\mathbf{Y}^{-1}\hat{\mathbf{V}}_{yy}\mathbf{Y}^{-1} + \mathbf{X}^{-1}\hat{\mathbf{V}}_{xx}\mathbf{X}^{-1}$$

$$- \mathbf{Y}^{-1}\hat{\mathbf{V}}_{yx}\mathbf{X}^{-1} - \mathbf{X}^{-1}\hat{\mathbf{V}}_{xy}\mathbf{Y}^{-1})\mathrm{diag}(\hat{\mathbf{r}}), \qquad (6.2)$$

where

$\mathrm{diag}(\hat{\mathbf{r}}) = \mathrm{diag}(\hat{r}_1, \ldots, \hat{r}_u) = \mathrm{diag}(y_1/x_1, \ldots, y_u/x_u)$
$\mathbf{Y} = \mathrm{diag}(\mathbf{y}) = \mathrm{diag}(y_1, \ldots, y_u)$
$\mathbf{X} = \mathrm{diag}(\mathbf{x}) = \mathrm{diag}(x_1, \ldots, x_u)$
$\hat{\mathbf{V}}_{yy}$ is the covariance-matrix estimator of the sample sums y_j and y_k
$\hat{\mathbf{V}}_{xx}$ is the covariance-matrix estimator of the sample sums x_j and x_k
$\hat{\mathbf{V}}_{yx}$ is the covariance-matrix estimator of the sums y_j and x_k, and
$\hat{\mathbf{V}}_{xy} = \hat{\mathbf{V}}'_{yx}$

and the operator 'diag' generates a diagonal matrix with the elements of the corresponding vector as the diagonal elements and with off-diagonal elements zero. Note that in a linear case, all elements of the covariance-matrix estimators $\hat{\mathbf{V}}_{xx}$, $\hat{\mathbf{V}}_{yx}$ and $\hat{\mathbf{V}}_{xy}$ are zero.

In the estimation of the elements of $\hat{\mathbf{V}}_{des}$, at least two clusters are assumed to be drawn with replacement from each of the H strata. In the special case of $m_h = 2$ clusters routinely used in survey sampling, the estimators can be simplified in a similar manner to that done in Section 5.3.

As a simple example, let the number of domains be $u = 2$. The elements of the covariance-matrix estimator

$$\hat{\mathbf{V}}_{des} = \begin{bmatrix} \hat{v}_{des}(\hat{r}_1) & \hat{v}_{des}(\hat{r}_1, \hat{r}_2) \\ \hat{v}_{des}(\hat{r}_2, \hat{r}_1) & \hat{v}_{des}(\hat{r}_2) \end{bmatrix}$$

are the following: Variance estimators:

$$\hat{v}_{des}(\hat{r}_j) = \hat{r}_j^2 (y_j^{-2}\hat{v}(y_j) + x_j^{-2}\hat{v}(x_j) - 2(y_jx_j)^{-1}\hat{v}(y_j, x_j)), \quad j = 1, 2.$$

Covariance estimator:

$$\hat{v}_{des}(\hat{r}_1, \hat{r}_2) = \hat{r}_1\hat{r}_2((y_1y_2)^{-1}\hat{v}(y_1, y_2) + (x_1x_2)^{-1}\hat{v}(x_1, x_2)$$

$$- (y_1x_2)^{-1}\hat{v}(y_1, x_2) - (y_2x_1)^{-1}\hat{v}(y_2, x_1)).$$

The estimator $\hat{v}_{des}(\hat{r}_2, \hat{r}_1)$ is equal to $\hat{v}_{des}(\hat{r}_1, \hat{r}_2)$ because of symmetry of $\hat{\mathbf{V}}_{des}$. If the estimators $\hat{r}_j$ are taken as linear estimators then the denominators x_j are assumed fixed. In this case, the variance and covariance estimates $\hat{v}(x_j)$ and $\hat{v}(y_j, x_j)$ are zero, and $\hat{v}_{des}(\hat{r}_j) = \hat{v}(y_j)/x_j^2$. And for a binary response in the binomial case, this estimator reduces to $\hat{v}_{bin}(\hat{p}_j) = \hat{p}_j(1 - \hat{p}_j)/n_j$.

It is important to note that $\hat{\mathbf{V}}_{des}$ is distribution-free so that it requires no specific distributional assumptions about the sampled observations. This allows an estimate $\hat{\mathbf{V}}_{des}$ to be nondiagonal unlike all the other covariance-matrix estimates of a ratio estimator vector considered in this chapter. The nondiagonality of $\hat{\mathbf{V}}_{des}$ is because the ratio estimators $\hat{r}_j$ and $\hat{r}_k$ from distinct domains can have nonzero correlations. In contrast, all the other covariance-matrix estimators have zero correlation by definition.

One source of nonzero correlation of the estimators $\hat{r}_j$ and $\hat{r}_k$ from separate domains comes basically from the clustering of the sample. Varying degrees of correlation can be expected depending on the type of the domains. If the domains cut smoothly across the sample clusters, distinct members in a given sample cluster may fall in separate domains j and k such as cross-classes like demographic or related factors. Large correlations can then be expected if the clustering effect is noticeable. In contrast, if the domains are totally segregated in such a way that all members of a given sample cluster fall in the same domain, zero correlations of distinct estimates $\hat{r}_j$ and $\hat{r}_k$ are obtained. This happens if the predictors used in forming the domains are cluster-specific unlike cross-classes where factors are essentially individual-specific. If, for example, households are clusters, typical cluster-specific factors are net income of the household and family size, whereas age and sex of a family member are individual-specific. Mixed-type domains, often met in practice, are intermediate, so that nonzero correlations are present in some dimensions of the table with zero correlations in the others.

Detecting Instability

The covariance-matrix estimator (6.2) is consistent for the asymptotic co-variance matrix $\mathbf{V}/n$ under the given complex sampling design so that, with a fixed cluster sample size, it is assumed to converge to $\mathbf{V}/n$ by increasing the number m of sample clusters. But with small m, an estimate $\hat{\mathbf{V}}_{des}$ can become unstable, i.e. near-singular. This can also happen if the number of domains u is large, which may require the estimation of several hundred distinct variance and covariance terms. The instability of a covariance-matrix estimate causes numerical problems when the inverse of the matrix is formed, which can severely disturb the reliability of testing and modelling procedures.

A near-singularity or *instability problem* is present if the degrees of freedom f for the estimation of the asymptotic covariance matrix $\mathbf{V}/n$ are small. For standard complex sampling designs f can be taken as the number of sample clusters less the number of strata, i.e. $f = m - H$. A stable $\hat{\mathbf{V}}_{des}$ can be expected if f is large relative to the number u of domains or, more specifically, relative to the residual degrees of freedom of the model to be fitted. In practice, instability problems are not expected if a large number of sample clusters are available, and if u is also much smaller than m.

The statistic *condition number* can be used as a measure of instability of $\hat{\mathbf{V}}_{des}$. It is defined as the ratio $\text{cond}(\hat{\mathbf{V}}_{des}) = \hat{\lambda}_{max}/\hat{\lambda}_{min}$, where $\hat{\lambda}_{max}$ and $\hat{\lambda}_{min}$ are the largest and smallest eigenvalues of $\hat{\mathbf{V}}_{des}$, respectively. If this statistic is large, e.g. in the hundreds or thousands, an instability problem is present. If the statistic is small, e.g. noticeably less than 50, no serious instability problems can be expected. Unfortunately, this statistic is not a routine output in software products from survey analysis. In the following table, condition numbers of $\hat{\mathbf{V}}_{des}$ with various values of u are displayed for the proportion estimator vector of the binary response variable CHRON (chronic morbidity) from the MFH and OHC Survey designs. The domains for each survey are formed by sex of respondent and equal-sized age groups.

No. of domains	MFH	OHC
4	6.5	2.8
8	10.6	3.5
12	39.8	3.6
20	421.5	5.6
24	423684	6.6
40	–	9.9

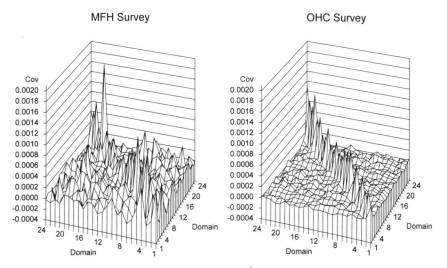

Figure 6.2   The covariance-matrix estimates $\hat{\mathbf{V}}_{des}$ of $u=24$ domain proportion estimates of CHRON in the MFH and OHC Survey designs.

Note that in the MFH Survey $f = 24$, and in the OHC Survey $f = 245$. Therefore, in the MFH Survey, the largest possible value of u is 24, and with this value the corresponding $\hat{\mathbf{V}}_{des}$ becomes very unstable. With values of u less than 12 the estimate remains quite stable. In the OHC Survey, condition numbers slightly increase with increasing u, but $\hat{\mathbf{V}}_{des}$ indicates stability with all values of u. These properties of the covariance-matrix estimates $\hat{\mathbf{V}}_{des}$ can also be depicted graphically. In Figure 6.2, the estimates $\hat{\mathbf{V}}_{des}$ for CHRON proportions with $u = 24$ domains from the MFH and OHC Survey designs are displayed. For the MFH Survey, the instability in $\hat{\mathbf{V}}_{des}$ is indicated by high 'peaks' in the off-diagonal part of the matrix. The stability of $\hat{\mathbf{V}}_{des}$ in the OHC Survey design is also clearly seen.

For binary responses, certain smoothing methods for $\hat{\mathbf{V}}_{des}$ have been proposed to reduce instability. These include Singh's (1985) method of dimensionality reduction of an unstable covariance-matrix estimate, and the techniques of correlation and covariance smoothing of $\hat{\mathbf{V}}_{des}$ suggested by Lehtonen (1990). Also, Morel (1989) proposed a smoothing method applicable in logistic regression.

Example 6.1

Covariance-matrix estimation with the linearization method. Using the OHC Survey data we carry out a detailed calculation of the covariance-matrix estimate $\hat{\mathbf{V}}_{des}$ of a proportion estimate $\hat{\mathbf{p}}$ of the binary response PHYS (physical health hazards of work), and of a mean estimate $\bar{y}$ of the continuous response PSYCH (the first standardized principal component of nine psychic symptoms), in the simple case of $u = 2$ domains formed by the variable sex. $\hat{\mathbf{V}}_{des}$ is thus a 2×2 matrix, and the domains are of cross-class type. A part of the data set needed for the covariance-matrix estimation is displayed in Table 6.4. Note that these data are cluster-level, consisting of $m = 250$ clusters in five strata. Thus, the degrees of freedom $f = 245$. The employee-level sample size is $n = 7841$.

The ratio estimator is $\hat{\mathbf{r}} = (\hat{r}_1, \hat{r}_2)' = (y_1/x_1, y_2/x_2)'$, where $\hat{r}_1$ and $\hat{r}_2$ are given by (6.1). For the binary response PHYS we denote the ratio estimator as $\hat{\mathbf{p}} = (\hat{p}_1, \hat{p}_2)'$, and for the continuous response PSYCH, $\overline{\mathbf{Y}} = (\bar{y}_1, \bar{y}_2)'$. The following figures for PHYS are calculated from Table 6.4:

Sums of the cluster-level sample sums $y_{jh\alpha}(= y_{ji})$ and $x_{jh\alpha}(= x_{ji})$:

$$\hat{n}_{11} = y_1 = 2061 \quad \text{and} \quad \hat{n}_1 = x_1 = 4485 \text{ (males)}$$

$$\hat{n}_{21} = y_2 = 650 \quad \text{and} \quad \hat{n}_2 = x_2 = 3356 \text{ (females)}$$

Proportion estimates for PHYS, i.e. the elements of $\hat{\mathbf{p}} = (\hat{p}_1, \hat{p}_2)'$:

$$\hat{p}_1 = y_1/x_1 = 2061/4485 = 0.4595 \text{ (males)}$$

Table 6.4 Cluster-level sample sums y_{1i} and y_{2i} of the response variables PHYS and PSYCH with the corresponding cluster sample sizes x_{1i} and x_{2i} in sample clusters $i = 1, \ldots, 250$ in two domains formed by sex (the OHC Survey).

i	Stratum h	Cluster α	PHYS y_{1i}	y_{2i}	PSYCH y_{1i}	y_{2i}	x_{1i}	x_{2i}
1	2	1	11	3	−0.1434	−0.0322	36	22
2	2	2	18	4	−0.1925	0.1867	57	21
3	2	3	4	5	0.0045	0.3674	9	15
4	2	4	2	2	0.7135	−0.3679	12	15
5	2	5	1	0	−0.1681	0.1235	27	8
6	2	6	1	0	−0.2673	0.1504	19	21
7	2	7	9	4	0.0099	0.2099	23	27
8	2	8	4	2	0.3681	0.0155	16	31
9	2	9	0	0	−0.5033	0.0755	6	6
10	2	10	3	0	−0.3176	−0.2516	8	8
11	2	11	2	7	0.9746	0.1903	6	67
12	2	12	7	3	−0.3361	0.5572	22	31
13	2	13	4	1	−0.2329	−0.2181	9	7
14	2	14	0	0	−0.2032	0.5893	13	16
15	2	15	1	23	0.4137	0.2565	4	56
.								
.								
.								
245	6	44	14	2	0.1984	−0.4271	23	7
246	6	45	2	1	−0.1049	0.3905	7	7
247	6	46	4	7	−0.2961	0.5018	7	13
248	6	47	0	1	−0.8073	0.9278	3	9
249	6	48	2	0	0.0006	−0.3484	16	13
250	6	49	13	1	−0.1273	−0.1466	26	4
Total sample			2061	650	−26.7501	33.7983	4485	3356

and

$$\hat{p}_2 = y_2/x_2 = 650/3356 = 0.1937 \text{ (females)}$$

We next construct the diagonal 2×2 matrics $\text{diag}(\hat{\mathbf{p}})$, $\mathbf{Y}$ and $\mathbf{X}$ for the calculation of the estimate $\hat{\mathbf{V}}_{des}$ for the PHYS proportion estimator $\hat{\mathbf{p}}$:

$$\text{diag}(\hat{\mathbf{p}}) = \begin{bmatrix} 0.4595 & 0 \\ 0 & 0.1937 \end{bmatrix}, \quad \mathbf{Y} = \begin{bmatrix} 2061 & 0 \\ 0 & 650 \end{bmatrix}$$

and

$$\mathbf{X} = \begin{bmatrix} 4485 & 0 \\ 0 & 3356 \end{bmatrix}.$$

The covariance-matrix estimates $\hat{\mathbf{V}}_{yy}$, $\hat{\mathbf{V}}_{xx}$ and $\hat{\mathbf{V}}_{yx}$, also obtained from the cluster-level data displayed in Table 6.4, are the following:

$$\hat{\mathbf{V}}_{yy} = \begin{bmatrix} 15722.50 & -130.45 \\ -130.45 & 3261.71 \end{bmatrix}$$

$$\hat{\mathbf{V}}_{xx} = \begin{bmatrix} 34560.23 & -7315.43 \\ -7315.43 & 34099.04 \end{bmatrix}$$

and

$$\hat{\mathbf{V}}_{yx} = \begin{bmatrix} 18973.88 & -5907.69 \\ -1098.11 & 6051.14 \end{bmatrix} = \hat{\mathbf{V}}'_{xy}.$$

By using these matrices we finally calculate for PHYS proportions the co-variance-matrix estimate $\hat{\mathbf{V}}_{des}$ given by (6.2). Hence we have

$$\hat{\mathbf{V}}_{des} = \begin{bmatrix} \hat{v}_{des}(\hat{p}_1) & \hat{v}_{des}(\hat{p}_1, \hat{p}_2) \\ \hat{v}_{des}(\hat{p}_2, \hat{p}_1) & \hat{v}_{des}(\hat{p}_2) \end{bmatrix} = 10^{-4} \begin{bmatrix} 2.775 & 0.576 \\ 0.576 & 1.951 \end{bmatrix}.$$

For example, using the estimates calculated, the variance estimate $\hat{v}_{des}(\hat{p}_1)$ is obtained as

$$\hat{v}_{des}(\hat{p}_1) = 0.4595^2 \times (2061^{-2} \times 15\,722.50 + 4485^{-2} \times 34\,560.23$$
$$- 2 \times (2061 \times 4485)^{-1} \times 18\,973.88) = 0.000\,2775.$$

Correlation of $\hat{p}_1$ and $\hat{p}_2$ is 0.25, which is quite large and indicates that the domains actually constitute cross-classes. The condition number of $\hat{\mathbf{V}}_{des}$ is $\text{cond}(\hat{\mathbf{V}}_{des}) = 1.9$, indicating stability of the estimate due to a large f and small u.

For PSYCH, the following figures are calculated from Table 6.4:

Sums of the cluster-level sample sums $y_{jh\alpha}$ and $x_{jh\alpha}$:

$$y_1 = -26.7501 \quad \text{and} \quad x_1 = 4485 \text{ (males)}$$

$$y_2 = 33.7983 \quad \text{and} \quad x_2 = 3356 \text{ (females)}$$

Mean estimates for PSYCH, i.e. the elements of $\bar{\mathbf{y}} = (\bar{y}_1, \bar{y}_2)'$:

$$\bar{y}_1 = y_1/x_1 = -0.1008 \text{ (males)}$$

and

$$\bar{y}_2 = y_2/x_2 = 0.1347 \text{ (females)}$$

The diagonal 2×2 matrices $\text{diag}(\bar{\mathbf{y}})$, $\mathbf{Y}$ and $\mathbf{X}$ are constructed in the same way as for PHYS. The covariance-matrix estimate $\hat{\mathbf{V}}_{xx}$ is equal to that for PHYS, and the covariance-matrix estimates $\hat{\mathbf{V}}_{yy}$ and $\hat{\mathbf{V}}_{yx}$ are:

$$\hat{\mathbf{V}}_{yy} = \begin{bmatrix} 6765.34 & 1036.34 \\ 1036.34 & 6585.20 \end{bmatrix},$$

$$\hat{\mathbf{V}}_{yx} = \begin{bmatrix} -3139.98 & 2129.01 \\ -2051.46 & 2259.73 \end{bmatrix} = \hat{\mathbf{V}}'_{xy}.$$

By using these matrices we calculate for PSYCH means the covariance-matrix estimate $\hat{\mathbf{V}}_{des}$:

$$\hat{\mathbf{V}}_{des} = \begin{bmatrix} \hat{v}_{des}(\bar{y}_1) & \hat{v}_{des}(\bar{y}_1, \bar{y}_2) \\ \hat{v}_{des}(\bar{y}_2, \bar{y}_1) & \hat{v}_{des}(\bar{y}_2) \end{bmatrix} = 10^{-4} \begin{bmatrix} 3.223 & 0.427 \\ 0.427 & 5.856 \end{bmatrix}.$$

Results from the design-based covariance-matrix estimation for PHYS proportions and PSYCH means including the standard-error estimates $\text{s.e.}_{des}(\hat{r}_j)$ are displayed below.

j	Sex	PHYS $\hat{p}_j$	$\text{s.e.}_{des}(\hat{p}_j)$	PSYCH $\bar{y}_j$	$\text{s.e.}_{des}(\bar{y}_j)$	$\hat{n}_j$
1	Males	0.460	0.0167	−0.1008	0.0180	4485
2	Females	0.194	0.0140	0.1347	0.0242	3356
	Total sample	0.346	0.0144	0.0000	0.0158	7841

Variance and covariance estimates $\hat{\mathbf{V}}_{yy}$, $\hat{\mathbf{V}}_{xx}$ and $\hat{\mathbf{V}}_{yx}$ can be calculated using the cluster-level data set displayed in Table 6.4 by, for example, the SAS procedure CORR as used here. The matrix operations in the formula of $\hat{\mathbf{V}}_{des}$ can be executed by any suitable software for matrix algebra such as SAS/IML or PC GAUSS. In practice, however, it is convenient to estimate $\hat{\mathbf{V}}_{des}$ using an element-level data set by software for survey analysis such as SUDAAN, PC CARP or OSIRIS. Generally, in the case of u domains formed by several categorical predictors, a linear ANOVA model can be used by fitting, with an appropriate sampling design option, for the response variable a full-interaction model excluding the intercept. The model coefficients are then equal to the domain proportion or mean estimates, and the covariance-matrix estimate of the model coefficients provides the covariance-matrix estimate $\hat{\mathbf{V}}_{des}$ of the proportions or means.

Let us consider the estimation of $\hat{\mathbf{V}}_{des}$ for PHYS proportions and PSYCH means in two domains using the SUDAAN procedure REGRESS. Sampling design options needed for the design-based estimation are those used in Example 5.1. Note that the WR design option involves the linearization method for covariance-matrix estimation. The weight variable is equal to one in this case. The estimation is based on an element-level data set. Note that the design-effect estimates of PHYS proportion estimates $\hat{p}_j$ are also displayed, and will be used in the next section. The relevant REGRESS statements for the binary response PHYS are:

```
1  PROC REGRESS DATA=<dataset> DESIGN=WR;
2  NEST        STRATUM CLUSTER;
3  WEIGHT      W;
4  SUBGROUP    SEX;
5  LEVELS      2;
6  MODEL       PHYS=SEX/NOINT;
7  RFORMAT     SEX SEXF.;
```

This gives the following piece of output for the response PHYS:

```
Number of observations
  read                          : 7841  Weighted count: 7841
Observations used in the
  analysis                      : 7841  Weighted count: 7841
Observations with
  missing values                :    0  Weighted count:    0
Denominator degrees of freedom  :  245
```

Multiple R-Square for the dependent variable PHYS: 0.076490

by: Independent Variables and Effects.

Independent Variables and Effects	Beta Coeff.	DEFF Beta	SE Beta
Sex			
Males	0.4595	5.0109	0.0167
Females	0.1937	4.1911	0.0140

Covar Beta
by: Independent Variables and Effects, Independent Variables and Effects.

Independent Variables and Effects	Independent Variables and Effects	
	Sex Males	Sex Females
Sex		
Males	2.7752E-04	
Females	5.7620E-05	1.9506E-04

6.3   ADJUSTING FOR EXTRA-BINOMIAL VARIATION

Extravariation is present in a domain ratio estimator when its true variance exceeds the corresponding variance derived with an assumption of simple random sampling. We consider a domain proportion estimator for which extravariation is called *extra-binomial* variation. The source of extra-binomial variation comes from positive intra-cluster correlation of the binary response variable. We previously measured the extra-binomial variation by computing the design-effect estimate of a proportion estimate. A *design-effects matrix* provides a tool for measuring extra-binomial variation in a vector of domain proportion estimators.

For a design-effects matrix estimator, we derive the binomial covariance-matrix estimator of a proportion estimator vector. A design-effects matrix is obtained by using the binomial and the corresponding design-based covariance-matrix estimators. Design-effect estimators taken from the diagonal of the design-effects matrix are used to derive the covariance-matrix estimators that account for extra-binomial variation in a vector of domain proportions. These covariance-matrix estimators are based on the so-called *effective sample sizes*.

Binomial Covariance-matrix Estimator

For a binary response, we assume a binomial sampling model for a proportion vector $\hat{\mathbf{p}}$ so that the weighted number of successes in each domain j is assumed to be generated by a binomial distribution and the generation processes are assumed independent between the u domains. The covariance-matrix estimator $\hat{\mathbf{V}}_{bin}(\hat{\mathbf{p}})$ of a proportion estimator $\hat{\mathbf{p}}$ is a diagonal matrix with diagonal elements derived from the binomial distribution, given by

$$\hat{v}_{bin}(\hat{p}_j) = \hat{p}_j(1 - \hat{p}_j)/\hat{n}_j, \quad j = 1, \ldots, u. \tag{6.3}$$

For the unweighted proportion vector $\hat{\mathbf{p}}^U$, the corresponding estimate, denoted by $\hat{\mathbf{V}}_{bin}(\hat{\mathbf{p}}^U)$, is obtained by using element weights equal to one.

Design-effects Matrix Estimator

Using the design-based covariance-matrix estimator $\hat{\mathbf{V}}_{des}(\hat{\mathbf{p}})$ and the binomial counterpart $\hat{\mathbf{V}}_{bin}(\hat{\mathbf{p}})$, the corresponding design-effects matrix estimator is derived for the domain proportion estimator vector $\hat{p}$, given by

$$\hat{\mathbf{D}} = \hat{\mathbf{V}}_{bin}^{-1}(\hat{\mathbf{p}})\hat{\mathbf{V}}_{des}(\hat{\mathbf{p}}), \tag{6.4}$$

where $\hat{\mathbf{V}}_{bin}^{-1}$ is the inverse of $\hat{\mathbf{V}}_{bin}$. The design-effect estimators $\hat{d}_j$ of $\hat{p}_j$ are the diagonal elements of the design-effects matrix estimator, hence the name *design-effects matrix*. The eigenvalues $\hat{\delta}_j$ of the design-effects matrix are often called the *generalized design-effects*. The sum of the design-effect estimates equals the sum of the eigenvalues, whose sum can be obtained from the sum of the diagonal elements of $\hat{\mathbf{D}}$, i.e. its trace. And the design-effect estimates and the corresponding eigenvalues are equal only in the special case where the estimate $\hat{\mathbf{V}}_{des}$ is also diagonal. All this holds in the case where the first covariance-matrix estimate in (6.4) is a diagonal matrix, such as $\hat{\mathbf{V}}_{bin}$. But in more complicated situations with proportions, where this is not true, the design-effects are not the diagonal elements of $\hat{\mathbf{D}}$ nor is the sum of design-effects equal to the sum of the eigenvalues. These more complicated design-effects matrices are called *generalized design-effects matrices* and will be discussed in Chapters 7 and 8.

The design-effect estimators of the proportion estimators $\hat{p}_j$ are of the form

$$\hat{d}_j = \hat{v}_{des}(\hat{p}_j)/\hat{v}_{bin}(\hat{p}_j), \quad j = 1, \ldots, u, \tag{6.5}$$

where the variance estimators $\hat{v}_{des}$ are diagonal elements of $\hat{\mathbf{V}}_{des}$. The design effect-estimates $\hat{d}_j$ measure the extra-binomial variation in the proportion estimates $\hat{p}_j$ due to the effect of clustering. Extra-binomial variation is present if design-effect estimates are greater than one.

A similar design-effects matrix estimator can be derived for the unweighted proportion estimator vector $\hat{\mathbf{p}}^U$ also. In the resulting design-effects matrix estimator, all the contributions of complex sampling on covariance-matrix estimation are reflected, such as unequal inclusion probabilities, clustering and adjustment for nonresponse, whereas in the design-effects matrix estimator (6.4), only the contribution of the clustering is accounted for. If adopting as a rule the use of a consistent proportion estimator $\hat{\mathbf{p}}$, then working with weighted observations, and thus with (6.4), would be reasonable. Then, the crucial role of adjusting for the clustering effect in the analysis of complex surveys would also be emphasized.

Covariance-matrix Estimation Using Effective Sample Sizes

The concept of *effective sample size* was introduced by Leslie Kish (1965), being defined as the size $\bar{n}$ of a simple random sample required to provide an equally precise estimate for a population mean, or equal size of a test for a null hypothesis, as that attained with the actual sample size n. For a mean estimator, the effective sample size is simply $\bar{n} = n/\hat{d}$, where $\hat{d}$ is the design-effect estimate of the sample mean. The effective sample size $\bar{n}$ is smaller than the actual sample size n if $\hat{d}$ is greater than one. Therefore, the effective sample sizes can be used to adjust for extra variation.

Because u distinct domain design-effect estimates $\hat{d}_j$ are available, we have two possibilities deriving the effective sample sizes for the u domains. For *constant deff adjustment*, each weighted domain sample size $\hat{n}_j$ is divided by the mean $\hat{d} = \sum_{j=1}^{u} \hat{d}_j/u$ of the design-effect estimates of $\hat{p}_j$ to obtain the effective sample sizes denoted by $\bar{n}_j$. The corresponding covariance-matrix estimator $\hat{\mathbf{V}}_{1.eff}(\hat{\mathbf{p}})$ of the domain proportion estimator $\hat{\mathbf{p}}$ can be obtained from (6.3) by inserting $\bar{n}_j$ in place of $\hat{n}_j$.

The estimator $\hat{\mathbf{V}}_{1.eff}$ for $\hat{\mathbf{p}}$ is consistent only if the assumption of a constant design-effect in all domains holds, and this might be unrealistic in practice. However, with segregated classes, it has been noted that the estimator $\hat{\mathbf{V}}_{1.eff}$ can behave adequately if the domain design-effect estimates are nearly equal.

For *domain-specific deff adjustment* the corresponding covariance-matrix estimator $\hat{\mathbf{V}}_{2.eff}(\hat{\mathbf{p}})$ is obtained from the equation $\hat{\mathbf{V}}_{2.eff} = \text{diag}(\hat{\mathbf{V}}_{des})$, i.e. the variance estimates of $\hat{p}_j$ for this adjustment are the diagonal elements $\hat{v}_{des}(\hat{p}_j)$ of the design-based covariance-matrix estimate $\hat{\mathbf{V}}_{des}$. Thus, the estimator $\hat{\mathbf{V}}_{2.eff}$ is consistent only if $\mathbf{V}/n$ is diagonal, which holds when working with domains formed by segregated classes. The estimator $\hat{\mathbf{V}}_{1.eff}$ can also be used as a smoothed version of $\hat{\mathbf{V}}_{2.eff}$ in unstable situations where the variance estimates in the diagonal of the estimate $\hat{\mathbf{V}}_{des}$ can be expected to be imprecise. Unexpectedly wild variation in the domain design-effect estimates $\hat{d}_j$ might indicate this kind of instability.

It is important to note that all the covariance-matrix estimators $\hat{\mathbf{V}}_{bin}$, $\hat{\mathbf{V}}_{1.eff}$ and $\hat{\mathbf{V}}_{2.eff}$ of $\hat{\mathbf{p}}$ are diagonal unlike the design-based counterpart $\hat{\mathbf{V}}_{des}$. Moreover, they rely essentially on an assumption of a binomial distribution; the key difference depends on which domain sample sizes are used in the variance estimators. If the design-based covariance-matrix estimate $\hat{\mathbf{V}}_{des}$ appears to be nearly diagonal, the covariance-matrix estimators $\hat{\mathbf{V}}_{1.eff}$ and $\hat{\mathbf{V}}_{2.eff}$ can serve as approximations in adjusting for extra-binomial variation in $\hat{\mathbf{p}}$.

Example 6.2

Adjusting for extra-binomial variation in a proportion estimator vector. We study further the covariance-matrix estimation of a domain proportion vector $\hat{\mathbf{p}}$ of the binary response PHYS started in Example 6.1, where the design-based estimate $\hat{\mathbf{V}}_{des}$ was calculated by the linearization method. The binomial covariance-matrix estimate $\hat{\mathbf{V}}_{bin}$ is first calculated and, using a design-effects matrix estimate $\hat{\mathbf{D}}$, we also obtain the covariance-matrix estimates $\hat{\mathbf{V}}_{1.eff}$ and $\hat{\mathbf{V}}_{2.eff}$ for the adjustment of extra-binomial variation. The self-weighting OHC Survey data set is used, as displayed in Table 6.4 with two domains formed by sex of respondent. The estimates $\hat{\mathbf{V}}$ and $\hat{\mathbf{D}}$ are thus 2×2 matrices. There are 4485 males and 3356 females in the available data set.

For PHYS, by computing the elements of the binomial covariance-matrix estimate

$$\hat{\mathbf{V}}_{bin}(\hat{\mathbf{p}}) = \begin{bmatrix} \hat{v}_{bin}(\hat{p}_1) & 0 \\ 0 & \hat{v}_{bin}(\hat{p}_2) \end{bmatrix} = \begin{bmatrix} \hat{p}_1(1-\hat{p}_1)/\hat{n}_1 & 0 \\ 0 & \hat{p}_2(1-\hat{p}_2)/\hat{n}_2 \end{bmatrix}$$

of the proportion vector $\hat{\mathbf{p}}$ we obtain

$$\hat{p}_1(1-\hat{p}_1)/\hat{n}_1 = 0.4595(1-0.4595)/4485 = 0.000\,0554 \text{ (males)}$$

and

$$\hat{p}_2(1-\hat{p}_2)/\hat{n}_2 = 0.1937(1-0.1937)/3356 = 0.000\,0465 \text{ (females)}.$$

Inserting these variance estimates in $\hat{\mathbf{V}}_{bin}$ we have

$$\hat{\mathbf{V}}_{bin}(\hat{\mathbf{p}}) = 10^{-4}\begin{bmatrix} 0.554 & 0 \\ 0 & 0.465 \end{bmatrix}.$$

It is important to note that the covariance-matrix estimate $\hat{\mathbf{V}}_{bin}$ is diagonal because the proportion estimates $\hat{p}_1$ and $\hat{p}_2$ are assumed to be uncorrelated. The effect of clustering is not accounted for, even in the variance estimates, in the estimate $\hat{\mathbf{V}}_{bin}$. Therefore, with positive intra-cluster correlation, the binomial variance estimates $\hat{v}_{bin}(\hat{p}_j)$ tend to be underestimates of the corresponding variances. This appears when calculating the design-effects matrix estimate $\hat{\mathbf{D}} = \hat{\mathbf{V}}_{bin}^{-1}\hat{\mathbf{V}}_{des}$ of the estimate $\hat{\mathbf{p}}$:

$$\hat{\mathbf{D}}(\hat{\mathbf{p}}) = \begin{bmatrix} 18\,058.295 & 0 \\ 0 & 21\,489.421 \end{bmatrix} \times 10^{-4}\begin{bmatrix} 2.775 & 0.576 \\ 0.576 & 1.951 \end{bmatrix}$$

$$= \begin{bmatrix} 5.01 & 1.04 \\ 1.24 & 4.19 \end{bmatrix}.$$

The design-effect estimates $\hat{d}_j$ on the diagonal of $\hat{\mathbf{D}}$ are thus

$$\hat{d}(\hat{p}_1) = \hat{v}_{des}(\hat{p}_1)/\hat{v}_{bin}(\hat{p}_1) = 0.000\,2775/0.000\,0554 = 5.01 \text{ (males)}$$

and

$$\hat{d}(\hat{p}_2) = \hat{v}_{des}(\hat{p}_2)/\hat{v}_{bin}(\hat{p}_1) = 0.000\,1951/0.000\,0465 = 4.19 \text{ (females)}.$$

These estimates are quite large, indicating a strong clustering effect for the response PHYS. This results in severe underestimation of standard errors of the estimates $\hat{p}_j$ when the binomial covariance-matrix estimate $\hat{\mathbf{V}}_{bin}$ is used. In addition to the design-effect estimates, the eigenvalues of the design-effect matrix, i.e. the generalized design-effects, can be calculated. These are $\hat{\delta}_1 = 5.81$

and $\hat{\delta}_2 = 3.39$. It may be noted that the sum of the design-effect estimates is 9.20 which is equal to the sum of the eigenvalues. The mean of the design-effect estimates is 4.60 which indicates a strong average clustering effect over the sex groups. However, the mean is noticeably smaller than the overall design-effect estimate $\hat{d} = 7.2$ for the proportion estimate $\hat{p}$ calculated from the whole sample. This is due to the property of design-effect estimates that, when compared against the overall design-effect estimate, they tend to get smaller in cross-class-type domains. The sex-specific design-effect estimates for PHYS can be obtained from the SUDAAN procedure REGRESS output displayed in Example 6.1.

To adjust for extra-binomial variation in an estimate $\hat{p}$, the covariance-matrix estimates $\hat{\mathbf{V}}_{1.eff}$ based on the constant deff adjustment and $\hat{\mathbf{V}}_{2.eff}$ based on the domain-specific deff adjustment are calculated. We first obtain the corresponding effective sample sizes:

Constant deff adjustment:

$$\bar{n}_1 = \hat{n}_1/\hat{d}. = 4485/4.602 = 974.6 \text{ (males)}$$

and

$$\bar{n}_2 = \hat{n}_2/\hat{d}. = 3356/4.602 = 729.2 \text{ (females)}.$$

Domain-specific deff adjustment:

$$\tilde{n}_1 = \hat{n}_1/\hat{d}_1 = 4485/5.011 = 895.0 \text{ (males)}$$

and

$$\tilde{n}_2 = \hat{n}_2/\hat{d}_2 = 3356/4.193 = 800.4 \text{ (females)}.$$

The sums of the effective sample sizes are $\bar{n} = 1703.8$ and $\tilde{n} = 1695.4$, both of which are much smaller than the actual sample size $n = 7841$ because of the strong clustering effect for PHYS.

We next calculate the variance estimates, i.e. the diagonal elements of the corresponding covariance-matrix estimates $\hat{\mathbf{V}}_{1.eff}(\hat{\mathbf{p}})$ and $\hat{\mathbf{V}}_{2.eff}(\hat{\mathbf{p}})$ for $\hat{\mathbf{p}}$. The diagonal elements of $\hat{\mathbf{V}}_{2.eff}$ are equal to those of the design-based estimate $\hat{\mathbf{V}}_{des}$ calculated in Example 6.1. For the constant deff adjustment they are:

$$\hat{v}_{1.eff}(\hat{p}_1) = \hat{p}_1(1 - \hat{p}_1)/\bar{n}_1 = 0.4595 \times 0.5405/974.6$$
$$= 0.000\,2548 \text{ (males)}$$

and

$$\hat{v}_{1.eff}(\hat{p}_2) = \hat{p}_2(1 - \hat{p}_2)/\bar{n}_2 = 0.1937 \times 0.8063/729.2$$
$$= 0.000\,2142 \text{ (females)}.$$

These variance estimates are noticeably larger than the corresponding binomial variance estimates obtained using the unadjusted domain sample sizes, indicating apparent extra-binomial variation.

It is relevant to question the adequacy of the binomial covariance-matrix estimates for PHYS domain proportions based on effective sample sizes. The design-based covariance-matrix estimate $\hat{\mathbf{V}}_{des}$ appeared clearly nondiagonal. This is due to the fact that the domains are cross-classes and intra-cluster correlation for PHYS is strong. Therefore, there is not only extra-binomial variation present but also nonzero covariance between separate domains. The estimators $\hat{\mathbf{V}}_{1.eff}$ and $\hat{\mathbf{V}}_{2.eff}$ therefore cannot be taken to be consistent for $\mathbf{V}/n$. As a consequence, when using effective sample sizes in modelling procedures on the PHYS proportions, more adequate estimation and test results are attained relative to those from the method that ignores the extra-binomial variation, but, unfortunately, they are overly conservative relative to the asymptotically valid design-based method. On the other hand, the method based on effective sample sizes may be the best one can do if the element-level (or cluster-level) data set is not available. This can be the case, for example, in secondary data analyses from published tables aimed at accounting for extra-variation in testing procedures, which is possible if the domain design-effect estimates are provided.

Estimation results for PHYS proportions (including results from Example 6.1) are collected below.

Standard error estimates:

j	Sex	$\hat{p}_j$	s.e.$_{des}$	s.e.$_{bin}$	s.e.$_{1.eff}$	s.e.$_{2.eff}$	$\hat{d}_j$
1	Males	0.460	0.0167	0.0074	0.0160	0.0167	5.01
2	Females	0.194	0.0140	0.0068	0.0146	0.0140	4.19
	Total Sample	0.346	0.0144	0.0054	0.0144	0.0144	7.17

Domain sample sizes:

j	Sex	$\hat{n}_j$	$\bar{n}_j$	$\tilde{n}_j$
1	Males	4485	975	895
2	Females	3356	729	800
	Total sample	7841	1704	1695

6.4 CHAPTER SUMMARY AND FURTHER READING

Summary

Several domain ratios, collected in a vector of ratios, were estimated using relative element weights in a combined ratio estimator derived for each domain.

This produced consistent estimation of the ratios under a non-epsem complex sampling design. Use of the linearization method gave consistent estimation of the covariance matrix of the weighted domain ratio estimator vector. It was demonstrated that positive intra-cluster correlation of a response variable not only increases the variance estimates, but can also introduce nonzero correlations between ratio estimates from separate domains; the asymptotically valid covariance-matrix estimator was derived to account for the extravariation and nonzero correlations. The estimator was essentially nondiagonal with nonzero off-diagonal covariance terms that occurred especially when working with cross-classes-type domains. This kind of a covariance-matrix estimator is needed for asymptotically valid modelling procedures with logit and linear models.

A covariance-matrix estimate calculated by the linearization method might be unstable in such small-sample situations where the number of sample clusters is small. Instability can cause problems in standard-error estimation and in testing and modelling procedures. Techniques are available for detecting instability, based, for example, on a statistic condition number and on graphical inspection of a covariance-matrix estimate.

When working with segregated-type domains, where the correlations of proportion estimates from separate domains are zero, the asymptotic covariance matrix of the proportion estimator vector is diagonal, making it possible to apply the method of effective sample sizes in the covariance-matrix estimation to adjust for extra-binomial variation. Only the design-effect estimates of the domain proportion estimates are needed, making this method technically simple and applicable also in the secondary analyses of published tables. More generally, the method can be used to adjust for extravariation in practice if the design-based covariance-matrix estimate appears to be nearly diagonal.

Further Reading

The estimation of the asymptotic covariance matrix of a domain ratio estimator vector is considered in Hidiroglou and Paton (1987) and Skinner *et al.* (1989, Section 2.13). Smoothed estimates for unstable situations are derived in Singh (1985), Kumar and Singh (1987) and Lehtonen (1990). The method of effective sample sizes is introduced in Scott (1986) and applied in Rao and Scott (1992). Brier (1980), Williams (1982) and Wilson (1989) consider the accounting for extra-binomial variation using the beta-binomial sampling model. The role of weighting for unequal inclusion probabilities and for adjustment for nonresponse in the analysis of complex surveys has deserved its considerable attention in the literature. Important contributions are Little (1991, 1993), Kish (1992) and Pfeffermann (1993).

7

Analysis of One-way and Two-way Tables

One-way and two-way frequency tables commonly occur in the analysis of complex surveys. Such tables are formed by tabulating the available survey data by a categorical variable, or by cross-classifying two categorical variables with the aim being to test hypotheses of goodness of fit, homogeneity or independence. For example, goodness of fit of the age distribution of the MFH Survey subgroup of 30–64-year-old males can be studied relative to the respective population age distribution. Or the OHC Survey data set may be tabulated by sex of respondent and a binary response variable CHRON (chronic morbidity) in a 2×2 table, with a null hypothesis of homogeneity of CHRON proportions in males and females stated. Further, we may consider an independence hypothesis of response variables CHRON and a categorical variable formed by classifying PSYCH (first principal component of psychic—psychological or mental—symptoms) into a number of classes. Under simple random sampling, valid inferences for these hypotheses can be based on a standard Pearson chi-squared test statistic. But with more complex designs, the testing procedures are more complicated because of clustering effects.

For homogeneity and independence hypotheses on an $r \times c$ frequency table from simple random sampling, the Pearson test statistic is asymptotically chi-squared with $(r-1)(c-1)$ degrees of freedom. But this standard asymptotic property is not valid for a frequency table from a complex survey based on cluster sampling. Positive intra-cluster correlation of the variables used in forming the table causes the test to be overly liberal relative to nominal significance levels. Therefore, the observed values of the test statistic can be too large, which can lead to erroneous inferences.

For valid inferences in complex surveys, certain corrections to the Pearson test statistic have been suggested such as Rao–Scott adjustments or, alternatively, test statistics such as the Wald test statistic can be used which automatically account for the clustering. Both approaches are demonstrated with an introductory example for a simple goodness-of-fit test in Section 7.1. The goodness-of-fit test is further considered in Section 7.2. The basics of testing for two-way tables are presented in Section 7.3. In Section 7.4, test statistics for a homogeneity hypothesis in a two-way table are examined, and in Section 7.5, a test of independence of two categorical variables is considered. The OHC and MFH Surveys involving clustered designs, described respectively in Sections 5.1 and 6.1, are used in the examples.

7.1   INTRODUCTORY EXAMPLE

Binomial Test and Effective Sample Size

Let us consider a hypothetical example of a simple goodness-of-fit test, basically originating from Sudman (1976), but applied here for the OHC Survey setting. A sample of $m = 50$ clusters is drawn from a large population of clusters which are industrial establishments. Let us assume that in each sample cluster $i = 1, ..., 50$, there are $n_i = 20$ employees. The element sample size is thus $n = 1000$. Given appropriate data under this sampling design, one might want to study whether the coverage of occupational health care (OHC), i.e. the unknown population proportion p of workers having access to occupational health (OH) services, is 80% based on prior knowledge from the previous year. The null hypothesis $H_0 : p = p_0 = 0.8$ can thus be stated. Let the significance level for this test be chosen as $\alpha = 5\%$.

A survey estimate $\hat{p} = n_1/n = 0.84$ is obtained, where $n_1 = 840$ is the number of sample workers having access to OH services. The binomial test is chosen, to be referred to the standard normal $N(0,1)$ distribution, with a large-sample test statistic

$$Z = |\hat{p} - p_0|/\sqrt{p_0(1 - p_0)/n}, \qquad (7.1)$$

where the denominator is the standard error of the estimate $\hat{p}$ under the null hypothesis. We calculate the value of Z with an assumption of simple random sampling with replacement, and also using a design-based approach that takes the clustering into account. In this simple case, the standard error of $\hat{p}$, needed for the calculation of an observed value of Z, is for both approaches based on a binomial assumption but with different sample sizes.

In a test based on the assumption of simple random sampling, we ignore the clustering and use the actual sample size $n = 1000$ in the standard error formula. The observed value of the test statistic (7.1) is hence

$$Z_{bin} = |\hat{p} - p_0|/\sqrt{p_0(1 - p_0)/1000} = 3.162 > Z_{0.025} = 1.96,$$

where $\sqrt{0.8(1 - 0.8)/1000} = 0.0126$ is the corresponding standard error of

$\hat{p}$. The result obviously suggests rejecting the null hypothesis when compared against the appropriate critical value from a standard $N(0,1)$ distribution.

It appeared that if an establishment is covered by OHC, then each worker at that site has equal access to OH services, which is an important piece of information that was ignored in the previous test. In fact, taking more than one person from a sample establishment does not increase our knowledge of the coverage of OHC at that site. Therefore, the effective sample size is $\bar{n} = 50$ in contrast to the assumed 1000 in the previous test. Recall that the concept of effective sample size refers to the size of a simple random sample, which gives an equally precise estimate for an unknown parameter p as that given by a sample of $n = 1000$ persons from the actual cluster sample design.

By using the effective sample size we have for a design-based test:

$$Z_{des} = |\hat{p} - p_0| / \sqrt{p_0(1 - p_0)/50} = 0.707,$$

where $\sqrt{0.8(1 - 0.8)/50} = 0.0566$, which is much larger than the corresponding standard error from the previous test. Therefore, the observed value of Z_{des} is smaller than that of Z_{bin}, and our test now suggests that the null hypothesis should not be rejected. We shall next study this example in a slightly more general setting, and introduce alternative test statistics in which the effect of clustering can be successfully removed.

Pearson Test Statistic and Rao–Scott Adjustment

The binomial test statistic Z_{bin} appeared to be liberal when compared to the design-based counterpart Z_{des}. This is because, with Z_{bin}, the clustering is not taken into account. Let us examine the asymptotic behaviour of the test statistic Z_{bin} more closely by constructing the corresponding *Pearson test statistic* X_P^2. For this, the following frequency table is used, where n_j are the observed cell frequencies and p_{0j} are the hypothesized cell proportions:

j	n_j	p_{0j}
1	840	0.8
2	160	0.2
All	1000	1.0

In a finite-population framework, let the unknown cell proportions be $p_j = N_j/N$, based on a population of N elements, where N_j is the number of population elements in cell j. The p_j can also be taken as the unknown cell probabilities under a superpopulation framework. The Pearson test statistic for the simple goodness-of-fit hypothesis $H_0 : p_j = p_{0j}, j = 1, 2$, is given by

$$X_P^2 = \sum_{j=1}^{2} (n_j - np_{0j})^2 / (np_{0j}) = n \sum_{j=1}^{2} (\hat{p}_j - p_{0j})^2 / p_{0j}, \qquad (7.2)$$

where the proportions $\hat{p}_j = n_j/n$ are estimates of the parameters p_j with n_j being the sample value of N_j. In the case of two cells, $\hat{p}_2 = 1 - \hat{p}_1$ and $p_{02} = 1 - p_{01}$, and an analogy exists between the statistics Z_{bin} and X_P^2:

$$X_P^2 = n \sum_{j=1}^{2} (\hat{p}_j - p_{0j})^2 / p_{0j} = (\hat{p} - p_0)^2 / (p_0(1 - p_0)/n) = Z_{bin}^2,$$

where $\hat{p} = \hat{p}_1$ and $p_0 = p_{01}$. With two cells, there is one degree of freedom for the goodness-of-fit test statistic X_P^2 because of one constraint (the proportions must sum up to 1) and no parameters need to be estimated.

Rao and Scott (1981) have given general results about the asymptotic distribution of the Pearson test statistic X_P^2. With two cells, the test statistic X_P^2 is asymptotically distributed as a random variate dW, where W is distributed as a chi-squared random variate χ_1^2 with one degree of freedom, and d denotes the design effect of the proportion estimate $\hat{p}$. The design effect can be obtained from $d = V_{des}(\hat{p})/V_{bin}(\hat{p})$, where $V_{des}(\hat{p}) = p_0(1 - p_0)/\bar{n}$ is the design variance of the estimate $\hat{p}$, $\bar{n}$ denotes the effective sample size, and $V_{bin}(\hat{p}) = p_0(1 - p_0)/n$ is the standard binomial variance counterpart. Hence, in this case, the design effect reduces to $d = n/\bar{n}$, which also confirms that the effective sample size is $\bar{n} = n/d$.

If the sample of employees had actually been drawn with simple random sampling directly from the employee population, we would have $d = 1$ because V_{des} and V_{bin} would then be equal. In this case, for two cells, the Pearson test statistic X_P^2 would be asymptotically chi-squared with one degree of freedom. But if the sample is actually drawn under cluster sampling, positive intra-cluster correlation gives a design effect d greater than one. Due to this, the statistic X_P^2 is no longer asymptotically chi-squared with the appropriate degrees of freedom.

Being now aware of the consequences of positive intra-cluster correlation on the asymptotic distribution of the Pearson test statistic X_P^2, the next step is to derive a valid testing procedure. Because, in general, accounting for intra-cluster correlation cannot be incorporated in the formula for X_P^2, an external correction to X_P^2 must be made. For this purpose, first note that the asymptotic expectation of X_P^2 is $E(X_P^2) = d$, which under positive intra-cluster correlation is greater than the nominal expected value of one. Since $E(X_P^2/d) = E(\chi_1^2) = 1$, we can construct a simple Rao–Scott correction to X_P^2 by dividing the observed value of the test statistic by the design effect. The resulting test statistic adjusted for the clustering effect is given by

$$X_P^2(d) = X_P^2/d \qquad (7.3)$$

and is asymptotically chi-squared with one degree of freedom in the case of two cells.

An analogous adjustment can be made to the corresponding *likelihood ratio* (LR) *test statistic* X_{LR}^2 of goodness of fit, which in the case of two cells is

$$X_{LR}^2 = 2n \sum_{j=1}^{2} \hat{p}_j \log (\hat{p}_j/p_{0j}) = 2n \log(\hat{p}(1 - \hat{p})/(p_0(1 - p_0))). \qquad (7.4)$$

Under simple random sampling, the statistic X_{LR}^2 is also asymptotically chi-squared with one degree of freedom when the null hypothesis is true. For clustered designs, the corresponding adjusted test statistic is

$$X_{LR}^2(d) = X_{LR}^2/d, \qquad (7.5)$$

which is asymptotically chi-squared with one degree of freedom.

We next compute the values of the Pearson and LR test statistics, with their Rao–Scott adjustments, for the OHC Survey setting. For the adjustments, the observed design effect is required, and this is

$$d = V_{des}(\hat{p})/V_{bin}(\hat{p}) = 0.0032/0.000\,16 = 20,$$

which can also be calculated as $d = n/\bar{n} = 1000/50 = 20$.

For the Pearson test statistic we obtain:

$$X_P^2 = (0.84 - 0.80)^2/(0.80 \times 0.20/1000) = 10.00$$

with a p-value of 0.0016. The value of the Rao–Scott corrected Pearson test statistic is hence

$$X_P^2(d) = X_P^2/d = Z_{bin}^2/d = 3.162^2/20 = 10.00/20 = 0.50,$$

which has a p-value of 0.4795. It can be noticed also that $Z_{des}^2 = 0.707^2 = 0.50$, i.e. $Z_{des}^2 = X_P^2(d)$ as expected. For the LR test statistic and the corresponding Rao–Scott correction we obtain

$$X_{LR}^2 = 2 \times 1000 \times \log(0.84 \times 0.16/(0.80 \times 0.20)) = 10.56,$$

with a p-value of 0.0012, and

$$X_{LR}^2(d) = X_{LR}^2/d = 10.560/20 = 0.528$$

with a p-value of 0.4675.

The observed design effect $d = 20$ is unusually large since the positive intra-cluster correlation is complete. The intra-cluster correlation coefficient is thus $\omega = 1$, calculated from the equation $d = 1 + (\bar{m} - 1)\omega$, where $\bar{m} = 20$ is the average cluster size. In practice, intra-cluster correlations are usually positive but less than one, and design-effect estimates $\hat{d}$ are correspondingly greater than one. A typical $\hat{d}$ is less than 3, corresponding to an estimated positive intra-cluster correlation coefficient $\hat{\omega} < 0.1$ with $\bar{m} = 20$.

Neyman and Wald Test Statistics

As an alternative to the Pearson test statistic, a *Neyman test statistic* X_N^2 of a simple goodness-of-fit hypothesis can be calculated. In the case of two cells it reduces to

$$X_N^2 = n \sum_{j=1}^{2} (\hat{p}_j - p_{0j})^2 / \hat{p}_j = (\hat{p} - p_0)^2 / (\hat{p}(1 - \hat{p})/n) \qquad (7.6)$$

which differs from the Pearson statistic since the estimated proportions $\hat{p}_j$ are inserted in the denominator in place of the hypothetical ones p_{0j}. With simple random sampling, the Neyman test statistic is asymptotically chi-squared with one degree of freedom in the case of two cells. But under cluster sampling the Neyman test statistic should be adjusted in a similar manner to that used for the Pearson test statistic. The Rao–Scott adjusted Neyman test statistic is hence

$$X_N^2(\hat{d}) = X_N^2/\hat{d} = \hat{d}^{-1}(\hat{p} - p_0)^2 / (\hat{p}(1 - \hat{p})/n). \qquad (7.7)$$

The estimated design effect is calculated by the formula $\hat{d} = \hat{v}_{des}(\hat{p})/\hat{v}_{bin}(\hat{p})$, where $\hat{v}_{des}$ is the design-based variance estimate of $\hat{p}$ corresponding to the actual sampling design and $\hat{v}_{bin}$ is the binomial counterpart.

We next calculate the values of the Neyman test statistic and its Rao–Scott correction. For this, the estimated design-effect is used. The design-based variance estimate of $\hat{p}$ is first obtained:

$$\hat{v}_{des}(\hat{p}) = \sum_{i=1}^{m} (\hat{p}_i - \hat{p})^2 / (m(m-1)) = \sum_{i=1}^{50} (\hat{p}_i - 0.84)^2 / (50 \times 49) = 0.002\,743,$$

where m is the number of sample clusters, $\hat{p}_i$ is the coverage of OHC in sample cluster i, and $\hat{p}$ is the corresponding estimate in the whole sample. It should be noted that $\hat{p}_i$ is either 0 or 1. A design-effect estimate can be calculated using a binomial variance estimate which is

$$\hat{v}_{bin}(\hat{p}) = \hat{p}(1 - \hat{p})/n = 0.000\,134,$$

giving an estimated design effect $\hat{d} = 0.002\,743/0.000\,134 = 20.4$. Alternatively, the design effect can be estimated as $\hat{d} = \hat{v}_{des}(\hat{p})/V_{bin}(\hat{p}) = 17.1$.

The observed value of the Neyman test statistic is

$$X_N^2 = (0.84 - 0.80)^2/(0.84 \times 0.16/1000) = 11.90$$

with a *p*-value of 0.0006. For the Rao–Scott corrected Neyman test statistic we obtain

$$X_N^2(\hat{d}) = X_N^2/\hat{d} = 11.9/20.4 = 0.583$$

with a *p*-value of 0.4451. Note that the observed values of the Neyman test statistic and the corresponding Rao–Scott adjustment are somewhat larger than the values of the Pearson statistic and its Rao–Scott adjustment.

The Neyman test statistic X_N^2 is a special case of the *Wald* (1943) *test statistic* of goodness of fit. The Wald statistic differs from the Pearson, LR and Neyman test statistics by automatically accounting for intra-cluster correlation. This can be seen in the formula of the design-based Wald statistic, which in the case of two cells reduces to

$$X_{des}^2 = (\hat{p} - p_0)^2/\hat{v}_{des}, \qquad (7.8)$$

where $\hat{v}_{des}$ is the design-based variance estimate of $\hat{p}$. The statistic X_{des}^2 is asymptotically chi-squared with one degree of freedom in the cluster-sampling design considered, without any auxiliary corrections. For a simple random sample, the variance estimator $\hat{v}_{bin}$ is used in (7.8) in place of $\hat{v}_{des}$ and so the Neyman test statistic X_N^2 and the resulting Wald statistic, denoted by X_{bin}^2, coincide. Obviously, for a clustered design, X_{bin}^2 also requires an adjustment similar to that of the Neyman statistic.

When calculating the value of the design-based Wald statistic we obtain

$$X_{des}^2 = (0.84 - 0.80)^2/0.002\,743 = 0.583,$$

which is equal to the value of the Rao–Scott corrected Neyman statistic, as expected. This demonstrates the flexibility of the Wald statistic. Using an appropriate variance estimate reflecting the complexities of the sampling design, we have an asymptotically valid test statistic without any auxiliary corrections. This can be seen as an obvious advantage over the Rao–Scott corrected statistics but, as we shall see later, in more general cases when working with more than two cells, there are certain drawbacks to the design-based Wald statistic caused by possible instability in the variance estimates in some small-sample situations.

Finally, we display the test results from the test statistics (7.2)–(7.8) below:

Test statistic	df	Observed value	p-value
Pearson			
X_P^2	1	10.00	0.0016
$X_P^2(d)$ (adjusted)	1	0.500	0.4795
Likelihood ratio			
X_{LR}^2	1	10.56	0.0012
$X_{LR}^2(d)$ (adjusted)	1	0.528	0.4675
Neyman			
$X_N^2(= X_{bin}^2)$	1	11.90	0.0006
$X_N^2(\hat{d})$ (adjusted)	1	0.583	0.4451
Wald			
X_{des}^2	1	0.583	0.4451

The two main approaches to accounting for the clustering effect in the test statistics demonstrated in this example, namely the Rao–Scott adjusting methodology used for the Pearson, likelihood ratio and Neyman test statistics, and the design-based Wald statistic, are readily applicable for more general one-way tables, and for two-way tables where the number of rows and columns is greater than two. We next consider a more general case for a simple goodness-of-fit test, and give details of alternative test statistics. Then, the tests for a homogeneity hypothesis and a hypothesis of independence are considered for a two-way table. In the testing procedures we will concentrate on the design-based Wald statistic and on various Rao–Scott adjustments to the Pearson and Neyman test statistics.

7.2   SIMPLE GOODNESS-OF-FIT TEST

A valid testing procedure for a goodness-of-fit hypothesis in the case of more than two cells is more complicated than the simple case of two cells. This is true both for the design-based Wald statistic and for the Rao–Scott adjustments to the Pearson and Neyman test statistics. We next discuss these testing procedures in some detail.

The design-based Wald statistic provides a natural testing procedure for a simple goodness-of-fit hypothesis since it is generally asymptotically correct in complex surveys. The Wald statistic can be expected to work adequately in practice if a large number of sample clusters are present, which is the case, for example, in the OHC Survey. But the test statistic can suffer from problems of instability if the number of sample clusters is too small. Then, observed values of the statistic can be obtained which are too large. Fortunately, effects of instability on the test statistic can be reduced by an F-correction. Another generally asymptotically valid testing procedure is based on a second-order Rao–Scott adjustment to the Pearson and Neyman test statistics. It is important to be able to obtain a full design-based covariance-matrix estimate for both of these testing procedures, and this presupposes access to the element-level data set.

There are situations met in practice where there is no access to the element-level data set. For example, in secondary analyses on published tables, an estimate of the full design-based covariance matrix is rarely provided. Therefore, a Wald statistic, or a second-order Rao–Scott adjustment, cannot be used. But certain approximate first-order adjustments are possible, if appropriate design-effect estimates are reported. Although adjustments based on these design-effect estimates are asymptotically valid only under special conditions, they can in many situations be used as a better alternative to the uncorrected Pearson or Neyman test statistics.

A goodness-of-fit hypothesis for $u \geq 2$ cells can be written as $H_0 : p_j = p_{0j}$, $j = 1, ..., u$, where $p_j = N_j/N$ are the unknown cell proportions and p_{0j} are the hypothesized cell proportions. The null hypothesis can be conveniently written, using the corresponding vectors, as $H_0 : \mathbf{p} = \mathbf{p}_0$, where $\mathbf{p} = (p_1, ..., p_{u-1})'$ is the vector of the unknown cell proportions, and $\mathbf{p}_0 = (p_{01}, ..., p_{0,u-1})'$ is the vector of the hypothesized proportions. The consistently estimated vector of cell proportions, based on a sample of n elements, is denoted by $\hat{\mathbf{p}} = (\hat{p}_1, ..., \hat{p}_{u-1})'$, where $\hat{p}_j = \hat{n}_j/n$. The $\hat{n}_j$ are scaled weighted cell frequencies accounting for unequal element inclusion probabilities and adjustment for nonresponse, such that $\sum_{j=1}^{u} \hat{n}_j = n$ (see Section 6.2). The $\hat{p}_j$ are ratio estimators if n is not fixed in advance, typically when working with a population subgroup as is assumed here. Note that only $u - 1$ elements are included in each of the vectors $\mathbf{p}$, $\mathbf{p}_0$ and $\hat{\mathbf{p}}$ because the proportions are constrained to sum up to one, thus, for example, $\hat{p}_u = 1 - \sum_{j=1}^{u-1} \hat{p}_j$.

Design-based Wald Statistic

A design-based Wald statistic X_{des}^2 of the simple goodness-of-fit hypothesis was previously introduced for the case of two cells with clustered sampling designs as an alternative to the adjusted Pearson statistic. In the case of more than two cells, the design-based Wald statistic of goodness of fit is slightly more complicated:

$$X_{des}^2 = (\hat{\mathbf{p}} - \mathbf{p}_0)\hat{\mathbf{V}}_{des}^{-1}(\hat{\mathbf{p}} - \mathbf{p}_0), \qquad (7.9)$$

where $\hat{\mathbf{V}}_{des}$ denotes a consistent covariance-matrix estimator of the true covariance matrix V/n of the proportion estimator vector $\hat{\mathbf{p}}$. An estimate $\hat{V}_{des}$ can be obtained by the linearization method using, for example, the SUDAAN software. The statistic X^2_{des} is asymptotically chi-squared with $u - 1$ degrees of freedom if the null hypothesis is true, thus providing a valid testing procedure for complex surveys. In practice, X^2_{des} can be expected to work reasonably if the number of sample clusters is large and the number of cells is relatively small, because then we can expect a stable estimate $\hat{V}_{des}$. Note that the statistic (7.8) is a special case of the statistic (7.9).

Unstable Situations

If there is a small number m of sample clusters available, an instability problem in the estimate $\hat{V}_{des}$ may be encountered because there may only be a few degrees of freedom $f = m - H$ for the estimate (see Section 6.2). Consequences of instability of an estimate $\hat{V}_{des}$ to the Wald statistic X^2_{des} can be severe, making the statistic overly liberal. One of the most widely used techniques to overcome instability is to make a degrees-of-freedom correction to the Wald statistic, giving rise to a new statistic which is assumed F-distributed. There are two alternative F-corrected Wald statistics. The first one is given by

$$F_{1.des} = \frac{f - u + 2}{f(u - 1)} X^2_{des},\qquad (7.10)$$

which is treated as an F-distributed random variate with $u - 1$ and $f - u + 2$ degrees of freedom, and the second is

$$F_{2.des} = X^2_{des}/(u - 1),\qquad (7.11)$$

which is in turn referred to the F-distribution with $u - 1$ and f degrees of freedom. Note that if $u = 2$, both corrections reproduce the original statistic. The effect of an F-correction to X^2_{des} can be easily seen in the case of just two cells. If f is small, then a p-value for X^2_{des} from the F-distribution with one and f degrees of freedom is larger than that from the chi-squared distribution with one degree of freedom, but when f increases then the difference vanishes. Thus, the corrections are ineffective if f is large. But for a small f, they can effectively correct the liberality in the uncorrected Wald statistic; this is true also where $u > 2$.

Thomas and Rao (1987) provide comparative results of the performances of various test statistics of a simple goodness of fit under instability, based on simulation. Although they noticed that the F-corrected Wald statistic $F_{1.des}$ did not indicate overall best performance relative to its competitors, it behaved relatively well in standard situations where instability was not very severe.

The F-corrected Wald statistics are widely applied in practice, and are also implemented in software products for survey analysis.

Pearson Test Statistic and Rao–Scott Adjustments

As noted in the introductory example, test statistics based on an assumption of simple random sampling require adjustments for the clustering effects to meet the desired asymptotic properties. Let us first consider the Pearson test statistic X_P^2. The statistic can be compactly written in a matrix form

$$X_P^2 = n \sum_{j=1}^{u} (\hat{p}_j - p_{0j})^2 / p_{0j} = n(\hat{\mathbf{p}} - \mathbf{p}_0)' \mathbf{P}_0^{-1} (\hat{\mathbf{p}} - \mathbf{p}_0), \qquad (7.12)$$

where $\mathbf{P}_0 = \text{diag}(\mathbf{p}_0) - \mathbf{p}_0\mathbf{p}_0'$ and $\mathbf{P}_0/n$ is the $(u-1) \times (u-1)$ multinomial covariance matrix of $\hat{\mathbf{p}}$ under the null hypothesis, and the operator $\text{diag}(\mathbf{p}_0)$ generates a diagonal matrix with diagonal elements p_{0j}. The covariance matrix $\mathbf{P}_0/n$ is a generalization of the case of $u = 2$ cells to the case of more than two cells. Note that the matrix formula of X_P^2 mimics that of the Wald statistic (7.9), the only difference being that $\mathbf{P}_0/n$ is used instead of $\hat{\mathbf{V}}_{des}$. In the case of two cells, X_P^2 reduces to the simple formula $X_P^2 = (\hat{p}_1 - p_{01})^2/(p_{01}(1 - p_{01})/n)$ previously considered, where the denominator corresponds to a binomial variance derived under the null hypothesis.

To examine the asymptotic distribution of the Pearson statistic X_P^2 we generalize the previous results from the case of two cells to the case of $u > 2$ cells. In this case, X_P^2 is asymptotically distributed as a weighted sum $\delta_1 W_1 + \delta_2 W_2 + \cdots + \delta_{u-1} W_{u-1}$ of $u - 1$ independent chi-squared random variables W_j each with one degree of freedom. The weights δ_j are eigenvalues of a *generalized design-effects matrix* $\mathbf{D} = \mathbf{P}_0^{-1}\mathbf{V}$, where $\mathbf{V}/n$ is the true covariance matrix of the proportion estimator vector $\hat{\mathbf{p}}$ based on the actual sampling design. These eigenvalues are also called *generalized design-effects*. Note that, in general, they do not coincide with the design effects d_j.

If the actual sampling design is simple random sampling, then the generalized design-effects δ_j are all equal to one because the true and assumed covariance matrices $\mathbf{V}/n$ and $\mathbf{P}_0/n$ coincide and, therefore, the generalized design-effects matrix is an identity matrix. The weighted sum $\sum_{j=1}^{u-1} \delta_j W_j$ then reduces to $\sum_{j=1}^{u-1} W_j$, i.e. a sum of $u - 1$ independent chi-squared random variates χ_1^2 whose distribution obviously is χ^2 with $u - 1$ degrees of freedom. Thus, under simple random sampling, the Pearson statistic X_P^2 is asymptotically chi-squared with $u - 1$ degrees of freedom.

If the actual sampling design is more complex by involving clustering, then the true $\mathbf{V}/n$ and the assumed $\mathbf{P}_0/n$ do not necessarily coincide, and in this case, the generalized design-effects δ_j are not equal to one. The δ_j tend to be greater than one on average due to the clustering effect and thus, the asymptotic distribution of the random variate $\sum_{j=1}^{u-1} \delta_j W_j$ is not assumed to be a chi-squared distribution with $u - 1$ degrees of freedom. Therefore, the Pearson test statistic X_P^2 requires corrections similar to those used in the case of two cells. However, there are now more possibilities for an adjusted Pearson statistic, namely the so-called *first-order* and *second-order* Rao–Scott adjustments developed by Rao and Scott (1981). The aim in the first-order adjustment is to correct the asymptotic expectation of the Pearson statistic, and the second-order adjustment also involves an asymptotically correct variance. Technically, both adjustments are based on eigenvalues of an estimated generalized design-effects matrix $\hat{\mathbf{D}}$.

We first consider a simple *mean deff adjustment* to X_P^2, due to Fellegi (1980) and Holt *et al.* (1980), and the first-order Rao–Scott adjustment. These adjustments are aimed at situations where the full design-based estimate $\hat{\mathbf{V}}_{des}$ is not available. If this estimate is provided, a more exact second-order adjustment is preferable.

The mean deff adjustment is based on the estimated design effects $\hat{d}_j$ of the proportions $\hat{p}_j$. An adjusted statistic to (7.12) is calculated by dividing the observed value of the Pearson statistic by the average design effect:

$$X_P^2(\hat{d}_\cdot) = X_P^2/\hat{d}_\cdot, \qquad (7.13)$$

where $\hat{d}_\cdot = \sum_{j=1}^{u} \hat{d}_j/u$ is an estimator of the mean $\bar{d}$ of the unknown design effects d_j. We estimate the design effects by $\hat{d}_j = \hat{v}_{des}(\hat{p}_j)/(\hat{p}_j(1 - \hat{p}_j)/n)$, where $\hat{v}_{des}(\hat{p}_j)$ are design-based variance estimators of the proportion estimators $\hat{p}_j$. This adjustment thus requires that the design-effect estimates of the u cell proportion estimates are available. Positive intra-cluster correlation gives a mean $\hat{d}_\cdot$ greater than one, and so the mean deff adjustment tends to remove the liberality in X_P^2. The mean deff adjustment can also be executed by calculating the effective sample size $\bar{n} = n/\hat{d}_\cdot$ and then inserting $\bar{n}$ into equation (7.12) of X_P^2 in place of n.

The mean deff adjustment is approximate so that it does not involve exact correction to the asymptotic expectation of X_P^2, because the mean of the design effects is generally not equal to the mean of the generalized design-effects. Under the null hypothesis, the asymptotic expectation of X_P^2 is $E(X_P^2) = \sum_{j=1}^{u-1} \delta_j$, so $E(X_P^2/\bar{\delta}) = E(\chi_{u-1}^2) = u - 1$, where the mean of the eigenvalues is $\bar{\delta} = \sum_{j=1}^{u-1} \delta_j/(u - 1)$. This argument leads to a *first-order Rao–Scott adjustment* to X_P^2 given by

$$X_P^2(\hat{\delta}_\cdot) = X_P^2/\hat{\delta}_\cdot, \qquad (7.14)$$

where $\hat{\delta}_\cdot$ is an estimate of the mean $\bar{\delta}$ of the unknown eigenvalues. This mean can be estimated using the design-effect estimates by the equation

$$(u - 1)\hat{\delta}_\cdot = \sum_{j=1}^{u} \frac{\hat{p}_j}{p_{0j}} (1 - \hat{p}_j)\hat{d}_j$$

without estimating the eigenvalues themselves. Alternatively, $\hat{\delta}_.$ can be obtained from the generalized design-effects matrix estimate $\hat{\mathbf{D}} = n\mathbf{P}_0^{-1}\hat{\mathbf{V}}_{des}$ by the equation $\hat{\delta}_. = \text{tr}(\hat{\mathbf{D}})/(u-1)$, i.e. by dividing the trace of $\hat{\mathbf{D}}$ by the degrees of freedom. The adjusted statistic $X_P^2(\hat{\delta}_.)$ is asymptotically chi-squared with $u-1$ degrees of freedom only if the eigenvalues δ_j are all equal, but the statistic is noted to work reasonably in practice if the variation in the estimated eigenvalues $\hat{\delta}_j$ is small. Because only design-effect estimates of $\hat{p}_j$ are needed, the statistic is also suitable for secondary analyses from published tables if the design-effect estimates are supplied. The first-order Rao–Scott adjustment $X_P^2(\hat{\delta}_.)$ is more exact than the corresponding mean deff adjustment $X_P^2(\hat{d}_.)$, which can be taken as a conservative alternative to $X_P^2(\hat{\delta}_.)$.

The first-order Rao–Scott adjustment (7.14) aimed at successfully correcting the Pearson test statistic X_P^2 so that the asymptotic expectation would be equal to the degrees of freedom. If the variation in the estimated eigenvalues $\hat{\delta}_j$ is noted to be large, then a correction to the variance of X_P^2 is also required. This is achieved by a *second-order Rao–Scott adjustment* based on the Satterthwaite (1946) method. The second-order adjusted Pearson statistic is given by

$$X_P^2(\hat{\delta}_., \hat{a}^2) = X_P^2(\hat{\delta}_.)/(1 + \hat{a}^2), \tag{7.15}$$

where an estimator of the squared coefficient of variation a^2 of the unknown eigenvalues δ_j is

$$\hat{a}^2 = \sum_{j=1}^{u-1} \hat{\delta}_j^2/((u-1)\hat{\delta}_.^2) - 1.$$

An estimator of the sum of the squared eigenvalues is given by

$$\sum_{j=1}^{u-1} \hat{\delta}_j^2 = \text{tr}(\hat{\mathbf{D}}^2) = n^2 \sum_{j=1}^{u} \sum_{k=1}^{u} \hat{v}_{des}^2(\hat{p}_j, \hat{p}_k)/p_{0j}p_{0k},$$

where $\hat{v}_{des}(\hat{p}_j, \hat{p}_k)$ are variance and covariance estimators of $\hat{p}_j$ and $\hat{p}_k$. The degrees of freedom must also be adjusted for this statistic; $X_P^2(\hat{\delta}_., \hat{a}^2)$ is asymptotically chi-squared with Satterthwaite adjusted degrees of freedom $df_S = (u-1)/(1 + \hat{a}^2)$. Note that the full covariance-matrix estimate $\hat{\mathbf{V}}_{des}$ is required in the second-order adjustment, whereas in the first-order adjustment only the variance estimates $\hat{v}_{des}$ were needed.

In unstable situations, an F-correction to the first-order Rao–Scott adjustment (7.14) may be beneficial. It is given by

$$FX_P^2(\hat{\delta}_.) = X_P^2/((u-1)\hat{\delta}_.) \tag{7.16}$$

The statistic is referred to the F-distribution with $u-1$ and f degrees of freedom. Thomas and Rao (1987) noted this statistic as better than the uncorrected first-order adjustment in unstable situations.

Neyman (Multinomial Wald) Statistic

The Neyman test statistic X_N^2 was previously used as an alternative to the Pearson statistic. The Neyman statistic corresponds to a Wald statistic derived using an assumption of a multinomial distribution on $\hat{\mathbf{p}}$. The Neyman statistic is

$$X_N^2 = n\sum_{j=1}^{u}(\hat{p}_j - p_{0j})^2/\hat{p}_j = n(\hat{\mathbf{p}} - \mathbf{p}_0)'\hat{\mathbf{P}}^{-1}(\hat{\mathbf{p}} - \mathbf{p}_0), \qquad (7.17)$$

where $\hat{\mathbf{P}} = \text{diag}(\hat{\mathbf{p}}) - \hat{\mathbf{p}}\hat{\mathbf{p}}'$ and $\hat{\mathbf{P}}/n$ is the estimated (empirical) multinomial covariance matrix. Note that this equation mimics equations (7.9) and (7.12) of the design-based Wald statistic and the Pearson statistic; the only difference is that $\hat{\mathbf{P}}/n$ is used instead of $\hat{\mathbf{V}}_{des}$ or $\mathbf{P}_0/n$. Under simple random sampling, X_N^2 is asymptotically chi-squared with $u-1$ degrees of freedom, but for more complex designs the statistic requires adjustments similar to those used for the Pearson statistic. We thus have a mean deff adjustment for X_N^2 given by $X_N^2(\hat{d}.) = X_N^2/\hat{d}.$, a first-order Rao–Scott adjustment $X_N^2(\hat{\delta}.) = X_N^2/\hat{\delta}.$, a second-order Rao–Scott adjustment $X_N^2(\hat{\delta}., \hat{a}^2) = X_N^2(\hat{\delta}.)/(1 + \hat{a}^2)$, and an F-corrected first-order Rao–Scott adjustment $FX_N^2(\hat{\delta}.) = X_N^2(\hat{\delta}.)/(u-1)$.

Test Statistic and Distributional Properties

Our discussion so far indicates that the asymptotic properties of a test statistic depend on the sampling design assumptions specific to the statistic, and on the actual sampling design. More specifically, let $\mathbf{D} = \mathbf{P}^{-1}\mathbf{V}$ be a design-effects matrix where $\mathbf{P}/n$ is the covariance matrix corresponding to the assumed sampling design, and $\mathbf{V}/n$ is the true covariance matrix based on the actual design. Asymptotic distribution of a test statistic depends on the eigenvalues of such a design-effects-matrix. If all the eigenvalues are equal to one, a test statistic of goodness of fit is asymptotically chi-squared with $u-1$ degrees of freedom.

For the Pearson test statistic, the assumed covariance matrix $\mathbf{P}/n$ was a multinomial $\mathbf{P}_0/n$. If the actual design was also simple random sampling, then the true $\mathbf{V}/n$ and assumed $\mathbf{P}/n$ would coincide and all the eigenvalues would be equal to one. But if the actual design is more complex, the covariance matrices do not coincide, and the eigenvalues differ from the nominal value of one. Thus, an adjustment to X_P^2 is required.

For the design-based Wald statistic the situation is different, because the assumed and actual sampling designs coincide. Thus, the covariance matrices $\mathbf{P}/n$ and $\mathbf{V}/n$ in $\mathbf{D}$ are equal by definition. So, if the actual design is simple random sampling, we put $\mathbf{P}/n = \mathbf{V}/n = \mathbf{P}_0/n$, and if the actual design is more complex, involving clustering and stratification, we put $\mathbf{P}/n = \mathbf{V}/n$. In both cases, the eigenvalues of the corresponding design-effects matrix are equal to one and no adjustment to X_{des}^2 is required.

Residual Analysis

If a goodness-of-fit test does not support the null hypothesis, a residual analysis can be performed to study the deviations from H_0. For a simple random sample, the standardized residuals are of the form

$$\hat{e}_j = (\hat{p}_j - p_{0j})/\text{s.e.}_{srs}(\hat{p}_j), \quad j = 1, ..., u, \qquad (7.18)$$

where $\text{s.e.}_{srs}(\hat{p}_j)$ is the square root of the corresponding diagonal element of the multinomial covariance-matrix estimate $\hat{\mathbf{P}}/n$. A large absolute value of $\hat{e}_j$ indicates deviation from H_0. But in complex surveys, these standardized residuals can be too large because the multinomial standard errors tend to underestimate the true standard errors. We therefore derive the design-based standardized residuals by using the corresponding design-based standard errors $\text{s.e.}_{des}(\hat{p}_j)$. Hence we have

$$\hat{e}_j = (\hat{p}_j - p_{0j})/\text{s.e.}_{des}(\hat{p}_j), \quad j = 1, ..., u. \qquad (7.19)$$

Clearly, if design-effect estimates are noticeably larger than one, smaller standardized residuals are obtained by (7.19) relative to the multinomial counterparts. The design-based standardized residuals can be taken as approximate standard normal variates under the null hypothesis, so they can be referred to critical values from the $N(0,1)$ distribution.

Example 7.1

Goodness-of-fit test of the age distribution for the MFH Survey. We consider a goodness-of-fit test for the age distribution of the MFH Survey subgroup of males aged 30–64-years, relative to the respective population age distribution. We have chosen the MFH design to demonstrate also the effects of a small number of sample clusters ($m = 48$) on test results. Sample and population age distributions with the estimated cell design effects of the proportion estimates are displayed in Table 7.1. The standardized design-based residuals are also included in the table.

Because the cell proportions are constrained to sum up to one, there are

Table 7.1 Estimated and hypothesized age distributions, design-effect estimates of the age proportions, and standardized residuals in the MFH Survey subgroup of 30–64-year-old-males.

Age	n_j	Estimated $\hat{p}$	Hypothesized p_{0j}	Deff $\hat{d}_j$	Residuals $\hat{e}_j$
30–44	1329	0.492	0.521	1.51	−2.45
45–54	774	0.287	0.277	1.70	0.88
55–64	596	0.221	0.202	0.43	3.64
Total sample	2699	1.000	1.000		

$u - 1 = 2$ degrees of freedom for the tests. The null hypothesis is stated as $H_0 : p_j = p_{0j}$ with $j = 1,2,3$. The values of the unadjusted Pearson and Neyman test statistics, and the values of the mean deff adjustment and the first-order Rao–Scott adjustment to the Pearson statistic, can be calculated from Table 7.1 using the sample and population proportions $\hat{p}_j$ and p_{0j} and the design-effect estimates $\hat{d}_j$. But the second-order Rao–Scott adjustment and the Wald statistic require a full estimate $\hat{\mathbf{V}}_{des}$ of the proportion estimates. This estimate was obtained using the linearization method. For complete information, we supply the full 3×3 covariance matrix estimate

$$\hat{\mathbf{V}}_{des} = 10^{-5} \times \begin{bmatrix} 13.9481 & -12.0731 & -1.8750 \\ -12.0731 & 12.9158 & -0.8427 \\ -1.8750 & -0.8427 & 2.7177 \end{bmatrix}.$$

The full estimate $\hat{\mathbf{V}}_{des}$ can be calculated, for example with the SUDAAN procedure CATAN, by first defining three indicator variables Y1, Y2 and Y3 (with values 1 or 2) for cell identification and then requesting a full interaction linear model without intercept for each variable pair (by using CATAN model statement, e.g. Y1*Y2=Y1*Y2), and then combining the resulting three 2×2 covariance-matrix estimates of the nonzero beta coefficients. For comparison, we display also the multinomial counterparts $\mathbf{P}_0/n = (\mathrm{diag}(\mathbf{p}_0) - \mathbf{p}_0\mathbf{p}_0')/2699$ and $\hat{\mathbf{P}}/n = (\mathrm{diag}(\hat{\mathbf{p}}) - \hat{\mathbf{p}}\hat{\mathbf{p}}')/2699$. These are

$$\mathbf{P}_0/n = 10^{-5} \times \begin{bmatrix} 9.2464 & -5.3471 & -3.8993 \\ -5.3471 & 7.4202 & -2.0731 \\ -3.8993 & -2.0731 & 5.9724 \end{bmatrix},$$

and

$$\hat{\mathbf{P}}/n = 10^{-5} \times \begin{bmatrix} 9.2603 & -5.2317 & -4.0286 \\ -5.2317 & 7.5817 & -2.3500 \\ -4.0286 & -2.3500 & 6.3786 \end{bmatrix}.$$

The covariance-matrix estimates $\mathbf{P}_0/n$ and $\hat{\mathbf{P}}/n$ can be used in the calculation of the design effects matrix estimate $\hat{\mathbf{D}}$ and the Pearson and Neyman test statistics (7.12) and (7.17). Note that in the calculation of X^2_{des} in (7.9), and X^2_P and X^2_N, we need not use the full matrices but take the 2×2 submatrices from the estimates $\hat{\mathbf{V}}_{des}$, $\mathbf{P}_0/n$ and $\hat{\mathbf{P}}/n$ corresponding to the two elements of the vectors $\hat{\mathbf{p}}$ and $\mathbf{p}_0$. Of course, the Pearson and Neyman statistics can be calculated as well by using the standard formulae which were also given in equations (7.12) and (7.17).

For the adjusted Pearson and Neyman test statistics we obtain:

$$\hat{d}_. = \sum_{j=1}^{3} \hat{d}_j/3 = 1.21$$

$$\hat{\delta}_. = \sum_{j=1}^{3} \hat{p}_j p_{0j}^{-1}(1 - \hat{p}_j)\hat{d}_j/2 = 1.17$$

$$1 + \hat{a}^2 = 2699^2 \sum_{j=1}^{3}\sum_{k=1}^{3}(\hat{v}^2_{des}(\hat{p}_j, \hat{p}_k)/p_{0j}p_{0k})/(2 \times 1.17^2) = 1.37$$

$$df_S = (u - 1)/(1 + \hat{a}^2) = 1.46.$$

Using these estimates we obtain:

Neyman (multinomial Wald) statistic:

$$X^2_N = 9.96 \text{ with 2 df (degrees of freedom) and a } p\text{-value } 0.007$$

Pearson statistic:

$$X^2_P = 10.15 \text{ with 2 df and a } p\text{-value } 0.006$$

Mean deff adjustment to the Pearson statistic:

$$X^2_P(\hat{d}_.) = 10.15/1.21 = 8.38 \text{ with 2 df and a } p\text{-value } 0.015$$

First-order Rao–Scott adjustment to the Pearson statistic:

$$X^2_P(\hat{\delta}_.) = 10.15/1.17 = 8.66 \text{ with 2 df and a } p\text{-value } 0.013$$

F-corrected first-order Rao–Scott adjustment:

$$FX^2_P(\hat{\delta}_.) = 8.66/2 = 4.33 \text{ with 2 and 24 df and a } p\text{-value } 0.025$$

Second-order Rao–Scott adjustment to the Pearson statistic:

$$X^2_P(\hat{\delta}_., \hat{a}^2) = 8.66/1.37 = 6.30 \text{ with } 2/1.37 = 1.46 \text{ df and a } p\text{-value } 0.023$$

Design-based Wald statistic:

$$X^2_{des} = 15.28 \text{ with 2 df and a } p\text{-value } 0.001$$

F-corrected Wald statistics:

$$F_{1.des} = (24 - 3 + 2)/(24 \times 2) \times 15.28 = 7.32 \text{ with 2 and 23 df and a}$$
p-value 0.003

$$F_{2.des} = 15.28/2 = 7.64 \text{ with 2 and 24 df and a } p\text{-value } 0.003.$$

Of the test statistics introduced, the second-order Rao–Scott adjustment and the Wald statistic with an F-correction could be expected to provide the most adequate test results. The mean deff adjustment and the first-order Rao–Scott adjustment are aimed to be used only if the design-effect estimates in Table 7.1 were available but not the covariance-matrix estimate $\hat{\mathbf{V}}_{des}$.

The test results indicate that the uncorrected Pearson and Neyman statistics give liberal results relative to the adjusted Pearson tests, as expected. Of the adjusted tests, the second-order Rao–Scott adjustment and the F-corrected first-order Rao–Scott adjustment are most conservative. The design-based Wald test, however, is unexpectedly liberal, and the F-corrections involve no apparent improvement in this case. The liberality may be due to the relatively few degrees of freedom ($f = 24$) for the estimate $\hat{\mathbf{V}}_{des}$ which might be unstable. Actually, the eigenvalues of the relevant 2×2 submatrix of $\hat{\mathbf{V}}_{des}$ are 0.000 2552 and 0.000 0135, and thus the condition number is 18.9, though this does not indicate serious instability.

Which one of the seven test statistics aimed at accounting for the clustering effects should be chosen in the MFH Survey where the degrees of freedom for $\hat{\mathbf{V}}_{des}$ are small? Assuming first that an estimate $\hat{\mathbf{V}}_{des}$ is provided, the second-order Rao–Scott adjustment would be chosen because of the apparent nondiagonality of $\hat{\mathbf{V}}_{des}$, and because the second-order correction is not expected to be seriously sensitive to instability problems. Although also asymptotically valid, the design-based Wald statistic, and its F-corrections, would be excluded in this case because of obvious liberality. It should be noticed that in other testing situations where the number of sample clusters is larger, the design-based Wald statistic will be a reasonable alternative. If an estimate $\hat{\mathbf{V}}_{des}$ is not available but the appropriate design-effect estimates are provided, the F-corrected first-order Rao–Scott adjustment would be chosen and this seems also to successfully reduce the effect of instability.

The test results do not support the conclusion that the sample and population age distributions were equal. A residual analysis for the design-based standardized residuals $\hat{e}_j$ indicates that the largest deviance is in the third age group, and the standardized residual exceeds the 1% critical value 2.33 from the $N(0,1)$ distribution. The residuals are smaller than the multinomial counterparts, except in the last age group which has a design-effect estimate noticeably smaller than one.

Rejection of H_0 suggests that it might be reasonable to weight the MFH Survey data set to better match the sample age distribution with the population age distribution. In Section 5.1 we demonstrated this by developing the appropriate poststratification weights. It was noted that this weighting caused some, but not large, differences in the weighted estimates, relative to the unweighted ones, in response variables which were apparently age-dependent.

7.3 PRELIMINARIES FOR TESTS FOR TWO-WAY TABLES

In a two-way table, a *test of homogeneity* is appropriate to study whether the class proportions of a categorical response variable are equal over a set of classes of a categorical predictor variable. A *test of independence* is stated when studying whether there is nonzero association between two categorical response variables. The two tests thus conceptually differ in the formulation of the hypotheses and in the interpretation of test results. Under a simple random sample, a multinomial-based test such as the Pearson test can be used with an identical formula of a test statistic, for both hypotheses. For more complex designs, involving clustering, we also separate the tests technically, and derive different adjustments for the corresponding test statistics. We first introduce the preliminaries of the tests with a simple example from the MFH Survey.

Test of Independence

Let us first consider the test of independence in the simplest case of a two-way table. From the MFH Survey demonstration data set of size $n = 2699$ persons we have the following frequency table with two categorical variables, PHYS (physical health hazards of work, 0: none, 1: some) and SYSBP (systolic blood pressure, ≤ 159 or > 159):

| | SYSBP | | |
PHYS	≤ 159	> 159	Total
0	1857	362	2219
1	390	90	480
Total	2247	452	2699

For an *independence hypothesis*, our question is whether the two variables are associated or not. This leads to the null hypothesis

$$H_0 : \quad p_{jk} = p_{j+}p_{+k}, \quad j, k = 1, 2,$$

where p_{jk} are unknown population cell proportions and p_{j+} and p_{+k} are the corresponding row and column marginal proportions in an N element population with cell frequencies N_{jk}. We thus have:

$$p_{jk} = N_{jk}/N \quad \text{and} \quad p_{11} + p_{12} + p_{21} + p_{22} = 1,$$

$$p_{j+} = p_{j1} + p_{j2} \quad \text{and} \quad p_{+k} = p_{1k} + p_{2k}.$$

Because of the constraints on the cell and marginal proportions, the null hypothesis reduces to $H_0 : p_{11} = p_{1+}p_{+1}$, with one degree of freedom for the test.

For the independence hypothesis, the table of observed cell and marginal proportions $\hat{p}_{jk} = \hat{n}_{jk}/n$, and $\hat{p}_{j+} = \hat{p}_{j1} + \hat{p}_{j2}$ and $\hat{p}_{+k} = \hat{p}_{1k} + \hat{p}_{2k}$, can now be derived using the observed cell frequencies $\hat{n}_{jk}$:

	SYSBP		
PHYS	≤ 159	> 159	Total
0	0.6880	0.1341	0.8222
1	0.1445	0.0333	0.1778
Total	0.8325	0.1675	1

Note that the cell proportions sum up to one over the table. A Pearson test statistic for the hypothesis of independence is

$$X_P^2(I) = n \sum_{j=1}^{2} \sum_{k=1}^{2} \frac{(\hat{p}_{jk} - \hat{p}_{j+}\hat{p}_{+k})^2}{\hat{p}_{j+}\hat{p}_{+k}} = \frac{n(\hat{p}_{11} - \hat{p}_{1+}\hat{p}_{+1})^2}{\hat{p}_{1+}(1 - \hat{p}_{1+})\hat{p}_{+1}(1 - \hat{p}_{+1})},$$

which is a scaled measure of the squared differences of the observed proportions from their expected values under the null hypothesis of independence. For a standard inference on the null hypothesis, the Pearson statistic is referred to the chi-squared distribution with one degree of freedom. Calculated from the table above, the observed value of $X_P^2(I)$ is 1.68 with a p-value 0.195, clearly suggesting acceptance of the null hypothesis of independence.

Test of Homogeneity

For the independence hypothesis, both of the classification variables were actually taken as response variables. It is also possible to look at the frequency table from another point of view. If we consider SYSBP as a response variable and PHYS as a predictor variable, for a *homogeneity hypothesis* our question then asks if the distributions of SYSBP in the two classes of PHYS are equal. This leads to a null hypothesis

$$H_0 : p_{1k} = p_{2k}$$

for both values of $k = 1,2$. When compared to the independence hypothesis, we now have different population proportions for which it holds

$$p_{11} + p_{12} = 1 \quad \text{and} \quad p_{21} + p_{22} = 1.$$

Because of these constraints, the null hypothesis reduces to $H_0 : p_{11} = p_{21}$, and again there is one degree of freedom for the test.

For the homogeneity hypothesis, the table of observed cell proportions $\hat{p}_{1k} = \hat{n}_{1k}/\hat{n}_1$ and $\hat{p}_{2k} = \hat{n}_{2k}/\hat{n}_2$, where $\hat{n}_1 = \hat{n}_{11} + \hat{n}_{12}$ and $\hat{n}_2 = \hat{n}_{21} + \hat{n}_{22}$ are row marginal frequencies and observed marginal proportions are $\hat{p}_{j+} = 1$ and $\hat{p}_{+k} = (\hat{n}_{1k} + \hat{n}_{2k})/n$, is the following:

PHYS	SYSBP ≤ 159	> 159	Total
0	0.8369	0.1631	1
1	0.8125	0.1875	1
All	0.8325	0.1675	1

Note that both of the row margins $\hat{p}_{1+}$ and $\hat{p}_{2+}$ are equal to one. A Pearson test statistic for the hypothesis of homogeneity is now given as

$$X_P^2(H) = \sum_{j=1}^{2} \sum_{k=1}^{2} \frac{\hat{n}_j(\hat{p}_{jk} - \hat{p}_{+k})^2}{\hat{p}_{+k}} = \frac{(\hat{p}_{11} - \hat{p}_{21})^2}{\hat{p}_{+1}(1 - \hat{p}_{+1})/\hat{n}_1 + \hat{p}_{+2}(1 - \hat{p}_{+2})/\hat{n}_2},$$

which is again a measure of the squared differences of the observed proportions from their expected values, under the null hypothesis of homogeneity. For inference on the null hypothesis, this Pearson statistic is also referred to the chi-squared distribution with one degree of freedom. Although the formulae of $X_P^2(H)$ and $X_P^2(I)$ were written differently, the observed value, 1.68, for $X_P^2(H)$ is the same as in the test of independence, and the conclusion–accept the null hypothesis – also remains true.

Cell Design Effects

The Pearson tests of independence and homogeneity were executed assuming a simple random sample. But would the conclusions remain if we account for the clustering effect? This can be examined by calculating the design-effect estimates of the estimated cell and marginal proportions of the observed tables for the independence and homogeneity hypotheses. The following piece of output from the SUDAAN procedure CROSSTAB would then be helpful. Cell and marginal design effects for the independence hypothesis are in the first DEFF column, and those for the homogeneity hypothesis are in the second DEFF column.

PHYS SYSBP	Sample Size	Total Percent	Row Percent	DEFF Total Percent	DEFF Row Percent
Total					
Total	2699	100.00	100.00	.	.
-159	2247	83.25	83.25	0.9800	0.9800
160-	452	16.75	16.75	0.9800	0.9800
0					
Total	2219	82.22	100.00	1.6659	.
-159	1857	68.80	83.69	1.5025	0.8778
160-	362	13.41	16.31	0.8146	0.8778
1					
Total	480	17.78	100.00	1.6659	.
-159	390	14.45	81.25	1.4330	1.1480
160-	90	3.33	18.75	1.3383	1.1480

It is obvious that if the design-effect estimates are greater than one on average, then more conservative adjusted test statistics would be obtained, relative to the unadjusted ones, and therefore, the conclusion of accepting the null hypotheses would remain. The mean of the cell design-effect estimates for the independence hypothesis is $\hat{d}_.(I) = 1.27$, giving the mean deff adjusted Pearson statistic $X_P^2(I, \hat{d}_.) = 1.32$ with a p-value 0.251. And the mean of the cell design-effect estimates for the homogeneity hypothesis is $\hat{d}_.(H) = 1.01$, giving the mean deff adjusted Pearson statistic $X_P^2(H, \hat{d}_.) = 1.66$ with a p-value 0.198. These design-based tests involve no new inferential conclusions but, more importantly, they demonstrate that, because of different adjustments, the adjusted Pearson test statistics accounting for the clustering effect do not give numerically equal results although the unadjusted ones do. Difference between the adjustments to $X_P^2(I)$ and $X_P^2(H)$ also holds for the Rao–Scott corrections, and the design-based Wald test statistics of independence and homogeneity hypotheses would not coincide, either.

The test results also indicate that in the case of the MFH Survey, intra-cluster correlation has a greater effect on the test of independence than on the test of homogeneity. This might be so because we are working with cross-classes-type subgroups, and in part due to the few degrees of freedom available for the variance estimates. It should be noticed that the situation can also reverse: it has been noted in many surveys that the inflation due to clustering is often less for tests of independence than for tests of homogeneity (Rao and Thomas 1988). This holds especially in cases where the classes of the predictor variable are of segregated-type regions.

For the analysis of more general $r \times c$ tables from complex surveys, a design-based Wald statistic with an F-correction, and a second-order Rao–Scott adjustment to the standard Pearson and Neyman test statistics, can be constructed for tests of homogeneity and independence as in the case of the simple goodness-of-fit test. In secondary analyses from published tables, the mean deff and first-order Rao–Scott adjustments are possible, if cell and marginal design-effect estimates are provided but not the design-based co-variance-matrix estimate of proportion estimators.

7.4   TEST OF HOMOGENEITY

In survey analysis literature, a test of homogeneity is usually used to study the homogeneity of the distribution of a response variable over a set of non-overlapping regions where independent samples are drawn using multi-stage sampling designs (e.g. Rao and Thomas 1988). It is thus assumed that the regions are segregated classes so that all elements in a sample cluster fall in the same region (class of the predictor variable). The classes of the response variable are typically cross-classes which cut across the regions. More generally, the test of homogeneity can be taken as the simplest example of a logit model with a binary or polytomous response variable and one categorical predictor variable whose type in practice is not restricted to a segregated class.

For a homogeneity hypothesis, assuming that columns of the table are formed by the classes of the response variable and rows constitute the regions, it is assumed that each row-wise sum of cell proportions is equal to one. The population table is thus as follows:

Region	Response variable 1	2	...	k	...	c	Total
1	p_{11}	p_{12}	...	p_{1k}	...	p_{1c}	1
2	p_{21}	p_{22}	...	p_{2k}	...	p_{2c}	1
.		.					
.		.					
.		.					
j	p_{j1}	p_{j2}	...	p_{jk}	...	p_{jc}	1
.		.					
.		.					
.		.					
r	p_{r1}	p_{r2}	...	p_{rk}	...	p_{rc}	1

For simplicity we consider the case of only two regions, and assume that the regions are of segregated classes type. A hypothesis of homogeneity of a c category response variable for $r = 2$ regions was given in Section 7.3 as $H_0 : p_{1k} = p_{2k}$, where $p_{1k} = N_{1k}/N_1$ and $p_{2k} = N_{2k}/N_2$ are unknown population proportions in the first and second regions, respectively, and $k = 1, \ldots, c$. The hypothesis can be written, using vectors, as $H_0 : \mathbf{p}_1 = \mathbf{p}_2$, where $\mathbf{p}_j = (p_{j1}, \ldots, p_{j,c-1})'$ denotes the population vector of row proportions p_{jk} in region j. There are thus $c - 1$ elements in each regional proportion vector, because the proportions must sum up to one independently for each region. Further, we denote by $\mathbf{p} = (p_{+1}, \ldots, p_{+,c-1})'$ the unknown common proportion vector under H_0, where $p_{+k} = N_{+k}/N$ and $N_{+k} = N_{1k} + N_{2k}$.

The estimated regional proportion vectors, based on independent samples from the regions, are denoted by $\hat{\mathbf{p}}_j = (\hat{p}_{j1}, \ldots, \hat{p}_{j,c-1})'$, where $\hat{p}_{jk} = \hat{n}_{jk}/\hat{n}_j$ is a consistent estimator of the corresponding population proportion p_{jk}, and $\hat{n}_{jk}$ and $\hat{n}_j$ are scaled weighted-up cell and marginal frequencies accounting for unequal element inclusion probabilities and adjustment for nonresponse, so that $\sum_{k=1}^{c} \hat{n}_{jk} = \hat{n}_j$. The $\hat{p}_{jk}$ are ratio estimators when we work with subgroups of the regional samples whose sizes are not fixed in advance, as we assume here as in the goodness-of-fit case. This also holds, for example, for the demonstration data sets from the MFH and OHC Surveys.

Design-based Wald Statistic

Let us denote by $\hat{\mathbf{V}}_{des}(\hat{\mathbf{p}}_1)$ the consistent covariance-matrix estimator of the proportion estimator vector $\hat{\mathbf{p}}_1$ in the first region, and have $\hat{\mathbf{V}}_{des}(\hat{\mathbf{p}}_2)$ correspondingly for $\hat{\mathbf{p}}_2$ in the second region. The covariance-matrix estimators can be calculated for each region in a similar manner as for the goodness-of-fit case. Using $\hat{\mathbf{V}}_{des}(\hat{\mathbf{p}}_1)$ and $\hat{\mathbf{V}}_{des}(\hat{\mathbf{p}}_2)$, a design-based Wald statistic X_{des}^2 of a homogeneity hypothesis for two regions is given by

$$X_{des}^2 = (\hat{\mathbf{p}}_1 - \hat{\mathbf{p}}_2)'(\hat{\mathbf{V}}_{des}(\hat{\mathbf{p}}_1) + \hat{\mathbf{V}}_{des}(\hat{\mathbf{p}}_2))^{-1}(\hat{\mathbf{p}}_1 - \hat{\mathbf{p}}_2), \qquad (7.20)$$

because of segregated classes and $r = 2$. The Wald statistic is asymptotically chi-squared with $(2 - 1) \times (c - 1) = (c - 1)$ degrees of freedom. And also, if $c = 2$, then X_{des}^2 reduces to $X_{des}^2 = (\hat{p}_{11} - \hat{p}_{21})^2 / (\hat{v}_{des}(\hat{p}_{11}) + \hat{v}_{des}(\hat{p}_{21}))$. X_{des}^2 in (7.20) does not directly generalize to the case with more than two regions but is more complicated (see e.g. Rao and Thomas 1988).

The statistic X_{des}^2 can be expected to work reasonably if a large number of sample clusters are available in each region. But if this is not the case, an instability problem can be encountered. F-corrected Wald statistics may then be used instead. By using $f = m - H$ as the overall degrees of freedom for the estimate $(\hat{\mathbf{V}}_{des}(\hat{\mathbf{p}}_1) + \hat{\mathbf{V}}_{des}(\hat{\mathbf{p}}_2))$, where m and H are the total number of sample clusters and strata in the two regions, the corrections are given by

$$F_{1.des} = \frac{f - (c - 1) + 1}{f(c - 1)} X_{des}^2, \qquad (7.21)$$

which is referred to the F-distribution with $(c - 1)$ and $(f - (c - 1) + 1)$ degrees of freedom, and, further,

$$F_{2.des} = X_{des}^2 / (c - 1), \qquad (7.22)$$

which is referred to the F-distribution with $(c - 1)$ and f degrees of freedom. These test statistics can be effective in reducing the effect of instability if f is not large relative to the number c of classes in the response variable.

Adjustments to Pearson and Neyman Test Statistics

A Pearson test statistic for the homogeneity hypothesis in the case of $r = 2$ regions is

$$X_P^2 = \sum_{j=1}^{2} \sum_{k=1}^{c} \frac{\hat{n}_j(\hat{p}_{jk} - \hat{p}_{+k})^2}{\hat{p}_{+k}} = (\hat{\mathbf{p}}_1 - \hat{\mathbf{p}}_2)'(\hat{\mathbf{P}}/\hat{n}_1 + \hat{\mathbf{P}}/\hat{n}_2)^{-1}(\hat{\mathbf{p}}_1 - \hat{\mathbf{p}}_2), \quad (7.23)$$

where $\hat{p}_{+k} = (\hat{n}_1\hat{p}_{1k} + \hat{n}_2\hat{p}_{2k})/(\hat{n}_1 + \hat{n}_2)$ are marginal proportion estimators over the rows of the table, i.e. estimators of the elements p_{+k} of the hypothesized common proportion vector $\mathbf{p}$ under H_0, and $\hat{\mathbf{P}} = \text{diag}(\hat{\mathbf{p}}) - \hat{\mathbf{p}}\hat{\mathbf{p}}'$ such that $\hat{\mathbf{P}}/\hat{n}_1$ is the multinomial covariance-matrix estimator of the estimator vector $\hat{\mathbf{p}}$ for the first region, and $\hat{\mathbf{P}}/\hat{n}_2$ correspondingly for the second region. Also, if $c = 2$, then X_P^2 reduces to $\hat{n}_1\hat{n}_2(\hat{p}_{11} - \hat{p}_{21})^2 / ((\hat{n}_1 + \hat{n}_2)\hat{p}_{+1}(1 - \hat{p}_{+1}))$.

As an alternative, a Neyman test statistic can be used which can be derived from the design-based Wald statistic (7.20) by assuming independent multinomial sampling in both regions:

$$X_N^2 = \sum_{j=1}^{2} \sum_{k=1}^{c} \frac{\hat{n}_j(\hat{p}_{jk} - \hat{p}_{+k})^2}{\hat{p}_{jk}} = (\hat{\mathbf{p}}_1 - \hat{\mathbf{p}}_2)'(\hat{\mathbf{P}}_1/\hat{n}_1 + \hat{\mathbf{P}}_2/\hat{n}_2)^{-1}(\hat{\mathbf{p}}_1 - \hat{\mathbf{p}}_2), \quad (7.24)$$

where $\hat{\mathbf{P}}_1 = \text{diag}(\hat{\mathbf{p}}_1) - \hat{\mathbf{p}}_1\hat{\mathbf{p}}_1'$ and $\hat{\mathbf{P}}_1/\hat{n}_1$ is the multinomial covariance-matrix estimator for the first region, and $\hat{\mathbf{P}}_2/\hat{n}_2$ correspondingly for the second region. Also, if $c = 2$ then X_N^2 reduces to $(\hat{p}_{11} - \hat{p}_{21})^2/(\hat{p}_{11}(1 - \hat{p}_{11})/$ $\hat{n}_1 + \hat{p}_{21}(1 - \hat{p}_{21})/\hat{n}_2)$. Note that the matrix formulae of X_P^2 and X_N^2 resemble that of the design-based Wald statistic, the only difference being which covariance-matrix estimator is used.

The Pearson and Neyman test statistics are valid for a simple random sample, i.e. they are chi-squared with $(c - 1)$ degrees of freedom for two regions. But under more complex designs, the statistics require adjustments that account for clustering effects. The adjustments are basically similar to those for the goodness-of-fit test but, technically, they are obtained by different formulae.

For a mean deff adjustment and for a first-order Rao–Scott adjustment to X_P^2 and X_N^2, the cell design-effect estimates in both regions are needed, and for a second-order Rao–Scott adjustment, a generalized design-effects matrix estimate is required. The design-effect estimators in region j are of the form

$$\hat{d}_{jk} = \hat{d}(\hat{p}_{jk}) = \hat{n}_j\hat{v}_{jk}/(\hat{p}_{+k}(1 - \hat{p}_{+k})), \quad j = 1, 2 \quad \text{and} \quad k = 1, ..., c,$$

where $\hat{v}_{1k}$ is the kth diagonal element of the covariance-matrix estimate $\hat{\mathbf{V}}_{des}(\hat{\mathbf{p}}_1)$ in the first region, and $\hat{v}_{2k}$ is the corresponding element of $\hat{\mathbf{V}}_{des}(\hat{\mathbf{p}}_2)$. The generalized design-effects matrix estimate is

$$\hat{\mathbf{D}} = \frac{\hat{n}_1\hat{n}_2}{\hat{n}_1 + \hat{n}_2}\hat{\mathbf{P}}^{-1}(\hat{\mathbf{V}}_{des}(\hat{\mathbf{p}}_1) + \hat{\mathbf{V}}_{des}(\hat{\mathbf{p}}_2)). \quad (7.25)$$

Mean deff adjustments to the Pearson and Neyman test statistics are

$$X_P^2(\hat{d}_.) = X_P^2/\hat{d}_. \quad \text{and} \quad X_N^2(\hat{d}_.) = X_N^2/\hat{d}_. \quad (7.26)$$

where

$$\hat{d}_. = \sum_{j=1}^{2} \sum_{k=1}^{c} \hat{d}_{jk}/(2c)$$

is the mean of the design-effect estimates. By using the eigenvalues $\hat{\delta}_k$ of $\hat{\mathbf{D}}$, the first-order Rao–Scott adjustments to Pearson and Neyman test statistics

(7.23) and (7.24) are given by

$$X_P^2(\hat{\delta}_.) = X_P^2/\hat{\delta}_. \quad \text{and} \quad X_N^2(\hat{\delta}_.) = X_N^2/\hat{\delta}_. \quad (7.27)$$

where

$$\hat{\delta}_. = \text{tr}(\hat{\mathbf{D}})/(c-1) = \frac{1}{c-1}\sum_{j=1}^{2}\left(1 - \frac{\hat{n}_j}{\hat{n}_1 + \hat{n}_2}\right)\sum_{k=1}^{c}\frac{\hat{p}_{jk}}{\hat{p}_{+k}}(1 - \hat{p}_{jk})\hat{d}_{jk}$$

is an estimator of the mean $\bar{\delta}$ of the eigenvalues δ_k of the unknown generalized design-effects matrix $\mathbf{D}$. Note that an estimate $\hat{\delta}_.$ can also be computed directly from $\hat{\mathbf{D}}$ by first calculating the sum of its diagonal elements, i.e. the trace. Both adjustments are referred to the chi-squared distribution with $(c-1)$ degrees of freedom. The adjustments are approximative in the sense that they can be expected to work reasonably if the design-effect estimates, or the eigenvalues, do not vary considerably. A second-order adjustment to X_P^2 and X_N^2 is more appropriate if the variation in the eigenvalue estimates $\hat{\delta}_k$ is noticeable. For the Pearson statistic this adjustment is given by

$$X_P^2(\hat{\delta}_., \hat{a}^2) = X_P^2(\hat{\delta}_.)/(1 + \hat{a}^2), \quad (7.28)$$

where $\hat{a}^2$ is the squared coefficient of variation of the eigenvalue estimates $\hat{\delta}_k$. It is obtained by the formula

$$\hat{a}^2 = \sum_{k=1}^{c-1}\hat{\delta}_k^2/((c-1)\hat{\delta}_.^2) - 1,$$

where the sum of squared eigenvalues can be obtained as the trace of the generalized design-effects matrix estimate raised to the second power:

$$\sum_{k=1}^{c-1}\hat{\delta}_k^2 = \text{tr}(\hat{\mathbf{D}}^2).$$

The second-order Rao–Scott corrected Pearson test statistic is asymptotically chi-squared with Satterthwaite adjusted degrees of freedom $df_S = (c-1)/(1 + \hat{a}^2)$. Similar adjustment can be carried out to the first-order corrected Neyman statistic $X_N^2(\hat{\delta}_.)$ in (7.27).

If the regional covariance-matrix estimates $\hat{\mathbf{V}}_{des}(\hat{\mathbf{p}}_1)$ and $\hat{\mathbf{V}}_{des}(\hat{\mathbf{p}}_2)$ are based on a relatively small number of sample clusters, they might be unstable and therefore, F-corrected first-order test statistics can be used instead. The Pearson statistic in (7.27) with an F-correction for two regions is given by

$$FX_P^2(\hat{\delta}_.) = X_P^2(\hat{\delta}_.)/(c-1) \quad (7.29)$$

referred to the F-distribution with $(c - 1)$ and f degrees of freedom. This correction is analogous for the Neyman statistic.

Residual Analysis

Under rejection of the null hypothesis H_0 of homogeneity, the standardized residuals can be computed to detect cell deviations from the hypothesized proportions. Using the cell design-effect estimates $\hat{d}_{jk}$ we calculate the design-based standardized residuals

$$\hat{e}_{jk} = (\hat{p}_{jk} - \hat{p}_{+k})/\text{s.e.}_{des}(\hat{p}_{jk} - \hat{p}_{+k}), \quad j = 1, 2 \quad \text{and} \quad k = 1, \dots, c, \quad (7.30)$$

where a standard-error estimator s.e.$_{des}(\hat{p}_{jk} - \hat{p}_{+k})$ of a raw residual is obtained from the design-based variance estimator, given by

$$\hat{v}_{des}(\hat{p}_{1k} - \hat{p}_{+k}) = \frac{\hat{n}_2(\hat{n}_2\hat{d}_{1k} + \hat{n}_1\hat{d}_{2k})}{(\hat{n}_1 + \hat{n}_2)^2}\hat{p}_{+k}(1 - \hat{p}_{+k})/\hat{n}_1, \quad k = 1, \dots, c$$

for the first region, and

$$\hat{v}_{des}(\hat{p}_{2k} - \hat{p}_{+k}) = \frac{\hat{n}_1(\hat{n}_2\hat{d}_{1k} + \hat{n}_1\hat{d}_{2k})}{(\hat{n}_1 + \hat{n}_2)^2}\hat{p}_{+k}(1 - \hat{p}_{+k})/\hat{n}_2, \quad k = 1, \dots, c$$

for the second region. Note that under simple random sampling, when $\hat{d}_{1k} = \hat{d}_{2k} = 1$, these variance estimators reduce to

$$\hat{v}_{srs}(\hat{p}_{1k} - \hat{p}_{+k}) = \frac{\hat{n}_2}{\hat{n}_1 + \hat{n}_2}\hat{p}_{+k}(1 - \hat{p}_{+k})/\hat{n}_1, \quad k = 1, \dots, c$$

for the first region, and

$$\hat{v}_{srs}(\hat{p}_{2k} - \hat{p}_{+k}) = \frac{\hat{n}_1}{\hat{n}_1 + \hat{n}_2}\hat{p}_{+k}(1 - \hat{p}_{+k})/\hat{n}_2, \quad k = 1, \dots, c$$

for the second region. It can be inferred from these formulae that under positive intra-cluster correlation, smaller design-based standardized residuals are obtained than those from the equations based on an assumption of simple random sampling. The design-based standardized residuals can be referred to the critical values from the standard normal $N(0,1)$ distribution.

Example 7.2

Test of homogeneity for two populations in the OHC Survey. We consider the test of homogeneity of class proportions of the variable PSYCH, which is the

Table 7.2 Class proportions of PSYCH (psychic symptoms) in public services and other industries in the OHC Survey (design-effect estimates in parentheses).

Type of industry	PSYCH 1	PSYCH 2	3	Total	Sample size
Public services	0.2939 (2.02)	0.3345 (1.24)	0.3716 (1.74)	1.00	1184
Other industries	0.3526 (1.73)	0.3216 (1.23)	0.3258 (1.57)	1.00	6657
Over industries	0.3437	0.3236	0.3327	1.00	7841

first principal component of nine psychic symptoms measuring overall psychic strain, categorized into three nearly equally-sized classes. The two populations are formed by type of industry of establishment, constructed so that public services constitute the first class and all the other industries are put into the second class (Table 7.2). Note that the classes follow industrial stratification and thus are segregated classes, and independent samples can be assumed to be drawn from each population. Of the 250 sample clusters available, 49 are in the first class and 201 in the second, and the regional element data sets are taken to be self-weighting.

In public services, a larger proportion of serious psychic symptoms (class 3) is obtained than in other industries. A homogeneity hypothesis $H_0 : p_{1k} = p_{2k}, k = 1, 2, 3$, of the class proportions over the two populations is stated to examine the variation. Cell design-effect estimates, with an average 1.59, indicate a moderate clustering effect, which should be accounted for in a testing procedure. For the calculation of valid test statistics, we first obtain the two full covariance-matrix estimates $\hat{\mathbf{V}}_{des}(\hat{\mathbf{p}}_1)$ and $\hat{\mathbf{V}}_{des}(\hat{\mathbf{p}}_2)$. These are

$$\hat{\mathbf{V}}_{des}(\hat{\mathbf{p}}_1) = 10^{-5} \times \begin{bmatrix} 35.3394 & -12.1408 & -23.1986 \\ -12.1408 & 23.3570 & -11.2161 \\ 23.1986 & -11.2161 & 34.4148 \end{bmatrix},$$

and

$$\hat{\mathbf{V}}_{des}(\hat{\mathbf{p}}_2) = 10^{-5} \times \begin{bmatrix} 5.9177 & -2.3978 & -3.5200 \\ -2.3978 & 4.0417 & -1.6439 \\ -3.5200 & -1.6439 & 5.1639 \end{bmatrix}.$$

Because $c - 1 = 2$, we use the first two classes of PSYCH and the first 2×2 submatrices from the estimates $\hat{\mathbf{V}}_{des}(\hat{\mathbf{p}}_1)$ and $\hat{\mathbf{V}}_{des}(\hat{\mathbf{p}}_2)$ in the calculation of

Wald statistics and Rao–Scott adjustments. For a design-based Wald test (7.20) of homogeneity, we get $X^2_{des} = 8.62$ with 2 degrees of freedom and a p-value 0.0134, thus indicating non-homogeneity of the proportions over the populations. F-corrections (7.21) and (7.22) to X^2_{des} give $F_{1.des} = 4.29$, which referred to the F-distribution with 2 and 244 degrees of freedom attains a p-value 0.0147, and $F_{2.des} = 4.31$, which referred to the F distribution with 2 and 245 degrees of freedom attains a p-value 0.0144. These corrections do not have a large impact on X^2_{des} due to the relatively large total number of sample clusters, in which case $\hat{\mathbf{V}}_{des}(\hat{\mathbf{p}}_1)$ and $\hat{\mathbf{V}}_{des}(\hat{\mathbf{p}}_2)$ can be assumed stable.

As another valid testing procedure, we calculate the second-order Rao–Scott adjustments (7.28) to the Pearson and Neyman test statistics (7.23) and (7.24). The unadjusted statistics give observed values $X^2_P = 16.93$, with a p-value 0.0002, and $X^2_N = 17.77$, with a p-value 0.0001, both significant at the 0.001 level, so they are very liberal relative to X^2_{des} as expected. For the Rao–Scott adjustments, a generalized design-effects matrix estimate (7.25) is first obtained:

$$\hat{\mathbf{D}} = \begin{bmatrix} 2.013\,74 & -0.036\,63 \\ 0.355\,54 & 1.239\,77 \end{bmatrix}.$$

The mean of the diagonal elements of $\hat{\mathbf{D}}$ is $\hat{\delta}_{.} = \text{tr}(\hat{\mathbf{D}})/2 = 1.627$, and the sum of the squared eigenvalues is $\sum_{k=1}^{2} \hat{\delta}_k^2 = \text{tr}(\hat{\mathbf{D}}^2) = 5.566$. The second-order correction factor is thus $(1 + \hat{a}^2) = 1.052$, and this with Satterthwaite adjusted degrees of freedom $df_S = 1.902$ gives $X^2_P(\hat{\delta}_{.}, \hat{a}^2) = 9.89$, with a p-value 0.0063, and $X^2_N(\hat{\delta}_{.}, \hat{a}^2) = 10.38$, with a p-value 0.0049, both significant at the 0.01 level. The results are somewhat liberal relative to those from the Wald test. These test results indicate that the design-based Wald statistic works adequately in the OHC case, unlike the MFH case (see Example 7.1).

We finally calculate the first-order adjustments (7.26) and (7.27) to X^2_P and X^2_N under the assumption that the only information provided for a homogeneity test is that in Table 7.2. The estimated mean design effect is $\hat{d}_{.} = 1.59$, and the corresponding adjustments to X^2_P and X^2_N are $X^2_P(\hat{d}_{.}) = 10.66$, with a p-value 0.0048, and $X^2_N(\hat{d}_{.}) = 11.19$, with a p-value 0.0037, both significant at the 0.01 level. By using cell design-effect estimates and cell proportions, we obtain $\hat{\delta}_{.} = 1.627$, giving the first-order Rao–Scott adjustments $X^2_P(\hat{\delta}_{.}) = 10.41$, with a p-value 0.0055, and $X^2_N(\hat{\delta}_{.}) = 10.92$, with a p-value 0.0043, which are also significant at the 0.01 level. The F-corrections (7.29) to X^2_P and X^2_N give $FX^2_P = 5.20$, with a p-value 0.0061, and $FX^2_N = 5.46$, with a p-value 0.0048, indicating no obvious change to the results from the first-order corrected counterparts, again demonstrating stability of the testing situation.

Because all the tests suggest rejection of H_0 at least at the 0.05 level, we calculate the design-based standardized residuals $\hat{e}_{jk}$ for both classes. By using (7.30) these are:

PSYCH	Public services $\hat{e}_{1k}$	Other industries $\hat{e}_{2k}$
1	−2.79	2.79
2	0.78	−0.78
3	2.35	−2.35

The residuals sum up to zero across public services and other industries. Note that from absolute values of the standardized residuals the largest are in the first and third PSYCH classes. In the third PSYCH class, the direction of the difference favours those from public services, whereas in the first class the situation is opposite. The design-based standardized residuals also exceed the 1% critical value 2.33 from the standard normal $N(0,1)$ distribution in these classes.

In the case where all relevant information is available, we conclude that the design-based Wald statistic provides an adequate and usable testing procedure for the homogeneity hypothesis. And if only cell design effects are provided, but not the two regional covariance-matrix estimates, we would choose the Rao–Scott adjustment to a Pearson or Neyman test statistic. But inferential conclusions remain unchanged independently of the test statistic chosen in the case considered; the strength of the conclusion to reject the null hypothesis of homogeneity of PSYCH proportions over the two populations, however, varies somewhat.

Logit modelling provides a convenient general framework for the test of a homogeneity hypothesis. A test of homogeneity of PSYCH proportions in the INDU (type of industry) classes in a 2×3 table can be taken as a simple example of a logit model. A test of homogeneity is obtained by fitting the saturated logit model INTERCEPT + INDU, say, for PSYCH logits and then by testing by the Wald test the significance of the INDU term. This can be carried out, for example, with the PC CARP program by using the option for logistic regression. A valid test of the significance of the INDU term, executed as an F-correction to an adjusted design-based Wald test statistic, gives the observed value 4.03, with a p-value 0.0190, when referred to the F-distribution with 2 and 244 degrees of freedom. The result, although slightly more conservative, is compatible with the previous results from the F-corrections $F_{1.des}$ and $F_{2.des}$ to the Wald test statistic X^2_{des}.

The Case of More than Two Regions

We have considered a test of homogeneity for two regions, where the regions constitute segregated classes. Derivation of a design-based Wald statistic, and the Rao–Scott adjustments to the Pearson and Neyman test statistics, for the case of more than two segregated regions is straightforward, but involves more matrix algebra. We omit the derivations and refer the reader to Rao and Thomas (1988).

The test of homogeneity for segregated classes is a special case of a more general testing situation with any type of a categorical predictor variable. This case, with a binary response variable, is considered in Chapter 8 for logit modelling. There, the assumption of segregated-type regions is relaxed, and we work with cross-classes also for the predictor variable. Then, the design-based covariance matrices of the response variable proportions cannot be estimated separately in the predictor classes, as was done in the segregated regions case, but the between-region covariance must be estimated as well. This covariance was assumed zero for segregated regions.

7.5   TEST OF INDEPENDENCE

A test of independence is applied to study whether there is nonzero association between two categorical variables within a population. Organized in an $r \times c$ contingency table, the data are thus assumed to be drawn from a single population with no fixed margins. Therefore, it is assumed that the sum of all population proportions p_{jk} in the population table equals one. The population table is thus:

First variable	Second variable 1	2	...	k	...	c	Total
1	p_{11}	p_{12}	...	p_{1k}	...	p_{1c}	p_{1+}
2	p_{21}	p_{22}	...	p_{2k}	...	p_{2c}	p_{2+}
.	.	.					
.	.	.					
j	p_{j1}	p_{j2}	...	p_{jk}	...	p_{jc}	p_{j+}
.	.	.					
.	.	.					
r	p_{r1}	p_{r2}	...	p_{rk}	...	p_{rc}	p_{r+}
Total	p_{+1}	p_{+2}	...	p_{+k}	...	p_{+c}	1

For the formulation of the null hypothesis, and for the interpretation of test results, it is important to note that we are now working in a symmetrical case where neither of the classification variables is assumed to be a predictor. The two response variables with r and c categories are typically of cross-classes or mixed-classes type so that they cut across the strata and clusters. A hypothesis of independence of the response variables was formulated in Section 7.3 as $H_0 : p_{jk} = p_{j+}p_{+k}$, where $p_{jk} = N_{jk}/N$, and $p_{j+} = \sum_{k=1}^{c} p_{jk}$ and $p_{+k} = \sum_{j=1}^{r} p_{jk}$ are marginal proportions with $j = 1, \ldots, r$ and $k = 1, \ldots, c$. It is obvious that if the actual unknown cell proportions p_{jk} were close to the expected cell proportions $p_{j+}p_{+k}$ under the null hypothesis, then the two variables can be assumed independent. This fact is utilized in the construction of appropriate test statistics for the independence hypothesis.

For the derivation of the test statistics of independence, we write the null hypothesis in an equivalent form, $H_0 : F_{jk} = p_{jk} - p_{j+}p_{+k} = 0$, where $j = 1, \ldots, r - 1$ and $k = 1, \ldots, c - 1$ because of the constraint $\sum_{j=1}^{r} \sum_{k=1}^{c} p_{jk} = 1$. The F_{jk} are thus the residual differences between the unknown cell proportions and their expected values under the null hypothesis which states that the residual differences are all zero. The residuals can then be collected in a column vector $\mathbf{F} = (F_{11}, \ldots, F_{1,c-1}, \ldots, F_{r-1,1}, \ldots, F_{r-1,c-1})'$ with a total of $(r - 1)(c - 1)$ rows.

The estimated cell proportions $\hat{p}_{jk} = \hat{n}_{jk}/n$, obtained from a sample of n elements, provide consistent estimators of the corresponding unknown proportions p_{jk}, where $\hat{n}_{jk}$ are scaled weighted-up cell frequencies accounting for unequal element inclusion probabilities and nonresponse, such that $\sum_{j=1}^{r} \sum_{k=1}^{c} \hat{n}_{jk} = n$. The $\hat{p}_{jk}$ are ratio estimators when working with a subgroup of the total sample whose size is not fixed in advance, such as the demonstration data sets from the MFH and OHC Surveys. As for the goodness-of-fit and homogeneity hypotheses, we also make this assumption here.

Covariance-matrix Estimators

Let us first derive the covariance-matrix estimators of the estimated vector $\hat{\mathbf{F}}$ of the residual differences under various assumptions on the sampling design, to be used for a design-based Wald statistic and for Pearson and Neyman test statistics. The estimated vector of residual differences is

$$\hat{\mathbf{F}} = (\hat{F}_{11}, \ldots, \hat{F}_{1,c-1}, \ldots, \hat{F}_{r-1,1}, \ldots, \hat{F}_{r-1,c-1})', \qquad (7.31)$$

where $\hat{F}_{jk} = \hat{p}_{jk} - \hat{p}_{j+}\hat{p}_{+k}$, and $\hat{p}_{j+}$ and $\hat{p}_{+k}$ are estimators of the corresponding marginal proportions. For the design-based Wald statistic, we derive the consistent covariance-matrix estimator $\hat{\mathbf{V}}_F$ of $\hat{\mathbf{F}}$, accounting for complexities of the sampling design, given by

$$\hat{\mathbf{V}}_F = \hat{\mathbf{H}}'\hat{\mathbf{V}}_{des}\hat{\mathbf{H}}, \qquad (7.32)$$

where the $(r - 1)(c - 1) \times (r - 1)(c - 1)$ matrix $\hat{\mathbf{H}}$ is the matrix of partial derivatives of $\mathbf{F}$ with respect to p_{jk}, evaluated at $\hat{p}_{jk}$. The matrix $\hat{\mathbf{V}}_{des}$ is a consistent estimator of the asymptotic covariance matrix $\mathbf{V}/n$ of the vector of cell proportion estimators $\hat{\mathbf{p}} = (\hat{p}_{11}, \ldots, \hat{p}_{1,c-1}, \ldots, \hat{p}_{r-1,1}, \ldots, \hat{p}_{r-1,c-1})'$. An estimate $\hat{\mathbf{V}}_{des}$ is obtained by the linearization method as used previously for the goodness-of-fit and homogeneity hypotheses. In practice, $\hat{\mathbf{V}}_{des}$ can be calculated from the element-level data set, for example, with the SUDAAN procedure CATAN by fitting a full-interaction linear model without an intercept, with the categorical variables as the model terms. The estimated model coefficients then coincide with the observed proportions, and the covariance-matrix estimate of the coefficients provides an estimate $\hat{\mathbf{V}}_{des}$.

The two multinomial covariance matrix estimators of $\hat{\mathbf{F}}$ are as follows. For the Pearson test statistic we derive an expected multinomial covariance-matrix estimator $\hat{\mathbf{P}}_{0F}/n$ of $\hat{\mathbf{F}}$ under the null hypothesis such that

$$\hat{\mathbf{P}}_{0F} = \hat{\mathbf{H}}'\hat{\mathbf{P}}_0\hat{\mathbf{H}}, \tag{7.33}$$

where $\hat{\mathbf{P}}_0 = \text{diag}(\hat{\mathbf{p}}_0) - \hat{\mathbf{p}}_0\hat{\mathbf{p}}_0'$ with $\hat{\mathbf{p}}_0$ being the vector of expected proportions under the null hypothesis, i.e. a vector with elements $\hat{p}_{j+}\hat{p}_{+k}$. And for the Neyman test statistic, we derive an observed multinomial covariance-matrix estimator $\hat{\mathbf{P}}_F/n$ of $\hat{\mathbf{F}}$ given by

$$\hat{\mathbf{P}}_F = \hat{\mathbf{H}}'\hat{\mathbf{P}}\hat{\mathbf{H}}, \tag{7.34}$$

where $\hat{\mathbf{P}} = \text{diag}(\hat{\mathbf{p}}) - \hat{\mathbf{p}}\hat{\mathbf{p}}'$. Note that all the covariance-matrix estimators of $\hat{\mathbf{F}}$ are of a similar form and use the same matrix $\hat{\mathbf{H}}$ of partial derivatives.

Design-based Wald Statistic

By using the estimated vector $\hat{\mathbf{F}}$ of residual differences with its consistent covariance-matrix estimate $\hat{\mathbf{V}}_F$ from (7.32), we obtain for the independence hypothesis a design-based Wald statistic

$$X_{des}^2 = \hat{\mathbf{F}}'\hat{\mathbf{V}}_F^{-1}\hat{\mathbf{F}}, \tag{7.35}$$

which is asymptotically chi-squared with $(r-1)(c-1)$ degrees of freedom. As in the Wald tests for goodness-of-fit and homogeneity, this test statistic can suffer from instability problems in cases where only few degrees of freedom f are available for an estimate $\hat{\mathbf{V}}_F$. F-corrections to X_{des}^2 can then be used, where

$$F_{1.des} = \frac{f - (r-1)(c-1) - 1}{f(r-1)(c-1)} X_{des}^2, \tag{7.36}$$

which is referred to the F-distribution with $(r-1)(c-1)$ and $(f-(r-1)$

$(c - 1) - 1)$ degrees of freedom, and

$$F_{2.des} = \frac{X_{des}^2}{(r - 1)(c - 1)},$$ (7.37)

which in turn is referred to the F-distribution with $(r - 1)(c - 1)$ and f degrees of freedom.

Adjustments to Pearson and Neyman Test Statistics

A Pearson test statistic for an independence hypothesis in Section 7.3 was given as

$$X_P^2 = n \sum_{j=1}^{r} \sum_{k=1}^{c} \frac{(\hat{p}_{jk} - \hat{p}_{j+}\hat{p}_{+k})^2}{\hat{p}_{j+}\hat{p}_{+k}}.$$ (7.38)

A Neyman test statistic can be used as an alternative and is given by

$$X_N^2 = n \sum_{j=1}^{r} \sum_{k=1}^{c} \frac{(\hat{p}_{jk} - \hat{p}_{j+}\hat{p}_{+k})^2}{\hat{p}_{jk}}.$$ (7.39)

Observed values of these statistics can be obtained from the estimated cell and marginal proportions. And under simple random sampling, both test statistics are asymptotically chi-squared with $(r - 1)(c - 1)$ degrees of freedom.

For a convenient common framework, we write the Pearson and Neyman test statistics (7.38) and (7.39) using the corresponding matrix formulae,

$$X_P^2 = n\hat{\mathbf{F}}' \hat{\mathbf{P}}_{OF}^{-1} \hat{\mathbf{F}}$$ (7.40)

for the Pearson statistic, where the null multinomial covariance-matrix estimator $\hat{\mathbf{P}}_{OF}/n$ from (7.33) is used, and

$$X_N^2 = n\hat{\mathbf{F}}' \hat{\mathbf{P}}_F^{-1} \hat{\mathbf{F}}$$ (7.41)

for the Neyman statistic, where the empirical multinomial covariance-matrix estimator $\hat{\mathbf{P}}_F/n$ from (7.34) is used. Note that both statistics mimic the design-based Wald statistic X_{des}^2 in (7.35), the only difference being which covariance-matrix estimator of the residual differences is used. It should also be noted that in the calculation of X_{des}^2, X_P^2 and X_N^2 the vector $\hat{\mathbf{F}}$ is an $(r - 1)(c - 1)$ column vector, and the covariance-matrix estimates are $(r - 1)(c - 1) \times (r - 1)(c - 1)$ matrices. Thus, for example, in a 2×2 table, $\hat{\mathbf{F}}$ and the covariance-matrix estimates $\hat{\mathbf{P}}_{OF}$ and $\hat{\mathbf{P}}_F$ reduce to scalars.

In complex surveys there is a similar motivation to adjusting the statistics X_P^2 and X_N^2 for the clustering effect as in the corresponding tests of goodness of fit and homogeneity. Asymptotically valid adjusted test statistics are obtained using second-order Rao–Scott corrections given by

$$X_P^2(\hat{\delta}_., \hat{a}^2) = X_P^2/(\hat{\delta}_.(1 + \hat{a}^2)) \tag{7.42}$$

for the Pearson statistic (7.40), where

$$\hat{\delta}_. = \mathrm{tr}(\hat{\mathbf{D}})/((r-1)(c-1))$$

is the mean of the eigenvalues $\hat{\delta}_l$ of the generalized design-effects matrix estimate

$$\hat{\mathbf{D}} = n\hat{\mathbf{P}}_{OF}^{-1}\hat{\mathbf{V}}_F, \tag{7.43}$$

and

$$\hat{a}^2 = \sum_{l=1}^{(r-1)(c-1)} \hat{\delta}_l^2/((r-1)(c-1)\hat{\delta}_.^2) - 1$$

is again the squared coefficient of variation of the eigenvalue estimates $\hat{\delta}_l$, with the sum of squared eigenvalues given by

$$\sum_{l=1}^{(r-1)(c-1)} \hat{\delta}_l^2 = \mathrm{tr}(\hat{\mathbf{D}}^2).$$

The second-order adjusted statistic (7.42) is asymptotically chi-squared with Satterthwaite adjusted degrees of freedom

$$\mathrm{df}_S = \frac{(r-1)(c-1)}{(1 + \hat{a}^2)}.$$

A similar second-order correction can also be made to X_N^2. There, a design-effects matrix estimate $\hat{\mathbf{D}} = n\hat{\mathbf{P}}_{OF}^{-1}\hat{\mathbf{V}}_F$ can alternatively be used.

Both the design-based Wald statistic X_{des}^2 and the second-order Rao–Scott adjustments to X_P^2 and X_N^2 require availability of the full covariance-matrix estimate $\hat{\mathbf{V}}_{des}$ of the cell proportion estimators $\hat{p}_{jk}$. In secondary analysis situations this estimate is not necessarily provided, but cell design-effect estimates $\hat{d}_{jk}$, possibly with marginal design-effect estimates $\hat{d}_{j+}$ and $\hat{d}_{+k}$, might be reported. By using these design-effect estimates, approximative first-order corrections can then be obtained. The simplest mean deff adjustment to the Pearson statistic X_P^2 is calculated using the mean of the estimated cell design effects given by

$$X_P^2(\hat{d}_.) = X_P^2/\hat{d}_., \tag{7.44}$$

where $\hat{d}_. = \sum_{j=1}^{r} \sum_{k=1}^{c} \hat{d}_{jk}/(rc)$ is the average cell design effect. And the first-order Rao–Scott adjustment to X_P^2 is given by

$$X_P^2(\hat{\delta}_.) = X_P^2/\hat{\delta}_. \tag{7.45}$$

where $\hat{\delta}_.$ can be calculated from the cell and marginal design effects by

$$\hat{\delta}_. = \frac{1}{(r-1)(c-1)} \sum_{j=1}^{r} \sum_{k=1}^{c} \frac{\hat{p}_{jk}(1-\hat{p}_{jk})}{\hat{p}_{j+}\hat{p}_{+k}} \hat{d}_{jk} - \sum_{j=1}^{r}(1-\hat{p}_{j+})\hat{d}_{j+} - \sum_{k=1}^{c}(1-\hat{p}_{+k})\hat{d}_{+k}$$

without calculating the generalized design-effects matrix itself. Similar corrections can again be made to X_N^2. The statistics $X_P^2(\hat{d}_.)$ and $X_P^2(\hat{\delta}_.)$ are referred to the chi-squared distribution with $(r-1)(c-1)$ degrees of freedom. $X_P^2(\hat{\delta}_.)$ is usually superior to $X_P^2(\hat{d}_.)$, and the statistic $X_P^2(\hat{\delta}_.)$ can be expected to work adequately if the variation in the eigenvalue estimates $\hat{\delta}_l$ is small.

If instability problems due to a relatively small f are expected, an F-correction to $X_P^2(\hat{\delta}_.)$ can be obtained given by

$$FX_P^2(\hat{\delta}_.) = X_P^2(\hat{\delta}_.)/((r-1)(c-1)), \tag{7.46}$$

which is referred to the F-distribution with $(r-1)(c-1)$ and f degrees of freedom. A similar correction is also available for the first-order adjusted Neyman statistic $X_N^2(\hat{\delta}_.)$.

Residual Analysis

If the null hypothesis of independence is rejected, then the standardized design-based cell residuals can be obtained for a closer examination of deviations from H_0. These residuals are given by

$$\hat{e}_{jk} = \frac{\hat{F}_{jk}}{\text{s.e.}(\hat{F}_{jk})}, \tag{7.47}$$

where s.e.$(\hat{F}_{jk})$ is the design-based standard-error estimate of $\hat{F}_{jk}$, i.e. square root of the corresponding variance estimate from (7.32). Under positive intra-cluster correlation, these design-based residuals tend to be smaller than the corresponding residuals calculated assuming simple random sampling. These would be obtained by inserting $\text{s.e.}_0(\hat{F}_{jk})$ in place of $\text{s.e.}(\hat{F}_{jk})$, where $\text{s.e.}_0(\hat{F}_{jk})$ is the multinomial standard-error estimate of $\hat{F}_{jk}$, i.e. the square root of the corresponding variance estimate from (7.33).

Example 7.3

Test of independence of health hazards of work and psychic strain in the OHC Survey. Let us study whether the variables PHYS (physical health hazards of work: none or some) and PSYCH (overall psychic strain classified into three nearly equally sized classes) are associated or not. Note that both classification variables constitute cross-classes. The appropriate cross-tabulation is displayed in Table 7.3.

A hypothesis of independence is stated as $H_0 : p_{jk} = p_{j+}p_{+k}$ with $j = 1,2$ and $k = 1,2,3$, or, analogously, $H_0 : p_{11} - p_{1+}p_{+1} = 0$ and $p_{12} - p_{1+}p_{+2} = 0$. The design-effect estimates of the cell proportions indicate a noticeable clustering effect, which is due to strong intra-cluster correlation for the variable PHYS, as can be seen from the corresponding marginal design-effect estimate which is 7.17. There is a natural interpretation for this unusually large design-effect estimate: separate establishments tend to be internally homogeneous with respect to physical working conditions, but sites from different industries can differ noticeably each from another in their working conditions. For the variable PSYCH, on the other hand, marginal design effects are only moderate, which is also understandable because experiencing psychic symptoms cannot be expected to be a strongly workplace-specific phenomenon. The mean of cell design-effect estimates is also quite large, 2.51. It is therefore important that a valid testing procedure should account for the clustering effect.

For the test statistics (7.35), (7.38) and (7.39), the corresponding covariance-matrix estimates $\hat{\mathbf{V}}_F$, $\hat{\mathbf{P}}_{0F}$ and $\hat{\mathbf{P}}_F$ of residual differences $\hat{F}_{jk}$ are required.

Table 7.3   Cell and marginal proportions of variables PHYS (physical health hazards) and PSYCH (overall psychic strain) in the OHC Survey (design-effect estimates in parentheses).

PHYS	PSYCH 1	2	3	Total	n
None	0.2276	0.2188	0.2078	0.6543	5130
	(2.09)	(2.26)	(2.63)	(7.17)	
Some	0.1161	0.1047	0.1250	0.3457	2711
	(2.82)	(2.37)	(2.87)	(7.17)	
Total	0.3437	0.3236	0.3327	1.00	
	(1.77)	(1.23)	(1.61)		
n	2695	2537	2609		7841

Technically, in the calculation of these estimates, the full $(rc) \times (rc)$ estimate $\hat{\mathbf{H}}$ of the partial derivatives and the corresponding full covariance-matrix estimates $\hat{\mathbf{V}}_{des}$, $\hat{\mathbf{P}}_0$ and $\hat{\mathbf{P}}$ are used, but in the construction of the test statistics, only the $(r-1)(c-1) \times (r-1)(c-1)$ submatrices of these matrices are used. For the 2×3 table, we thus calculate the 6×6 full matrices, but use only the 2×2 submatrices of these. A full 6×6 covariance-matrix estimate $\hat{\mathbf{V}}_{des}$ is first obtained using the linearization method. It is

$$\hat{\mathbf{V}}_{des} = 10^{-5} \begin{bmatrix} 4.6922 & 0.3207 & 0.6599 & -1.6442 & -1.6965 & -2.3321 \\ 0.3207 & 4.9264 & 1.7922 & -2.5751 & -2.1611 & -2.3030 \\ 0.6599 & 1.7922 & 5.5279 & -2.8972 & -2.5938 & -2.4890 \\ -1.6442 & -2.5751 & -2.8972 & 3.6938 & 1.9619 & 1.4608 \\ -1.6965 & -2.1611 & -2.5938 & 1.9619 & 2.8332 & 1.6562 \\ -2.3321 & -2.3030 & -2.4890 & 1.4608 & 1.6562 & 4.0072 \end{bmatrix}.$$

In addition to $\hat{\mathbf{V}}_{des}$, the matrix $\hat{\mathbf{H}}$ of partial derivatives is calculated to obtain the covariance-matrix estimate $\hat{\mathbf{V}}_F = \hat{\mathbf{H}}'\hat{\mathbf{V}}_{des}\hat{\mathbf{H}}$ of the vector of the residual differences, $\hat{\mathbf{F}}$. In the construction of the Wald statistic we use the 2×1 vector of residual differences,

$$\hat{\mathbf{F}} = \begin{bmatrix} \hat{F}_{11} \\ \hat{F}_{12} \end{bmatrix} = \begin{bmatrix} \hat{p}_{11} - \hat{p}_{1+}\hat{p}_{+1} \\ \hat{p}_{12} - \hat{p}_{1+}\hat{p}_{+2} \end{bmatrix} = 10^{-3} \begin{bmatrix} 2.778 \\ 7.162 \end{bmatrix},$$

and the corresponding 2×2 submatrix from the full $\hat{\mathbf{V}}_F$, calculated as

$$\hat{\mathbf{V}}_F = 10^{-6} \begin{bmatrix} 7.8147 & -2.8281 \\ -2.8281 & 6.3930 \end{bmatrix}.$$

For the design-based Wald statistic $X_{des}^2 = \hat{\mathbf{F}}'\hat{\mathbf{V}}_F^{-1}\hat{\mathbf{F}}$, we obtain an observed value $X_{des}^2 = 13.41$, which, referred to the chi-squared distribution with 2 degrees of freedom, attains a p-value 0.0012, significant at the 0.01 level. The F-corrections (7.36) and (7.37) to X_{des}^2 give observed values $F_{1.des} = 6.68$, which, referred to the F-distribution with 2 and 244 degrees of freedom, attains a p-value 0.0015, and $F_{2.des} = 6.71$, which with 2 and 245 degrees of freedom attains the same p-value. The F-corrections do not contribute noticeably to the uncorrected X_{des}^2.

For the alternative asymptotically valid tests based on the second-order adjustment to the Pearson test statistic X_P^2, or the Neyman statistic X_N^2, we first calculate the estimated generalized design-effects matrix (7.43):

$$\hat{\mathbf{D}} = n\hat{\mathbf{P}}_{0F}^{-1}\hat{\mathbf{V}}_F = \begin{bmatrix} 1.30761 & 0.21651 \\ 0.08616 & 1.05628 \end{bmatrix}.$$

The first-order adjustment factor is $\hat{\delta}_. = \text{tr}(\hat{\mathbf{D}})/2 = 1.182$, and the sum of squared eigenvalues is $\sum_{l=1}^{2} \hat{\delta}_l = \text{tr}(\hat{\mathbf{D}}^2) = 2.863$, giving a second-order correction factor, $(1 + \hat{a}^2) = 1.025$. These figures indicate that the eigenvalues are close to one on average, and their variation is negligible.

For the unadjusted test statistics (7.38) and (7.39), the observed values $X_P^2 = 16.40$ and $X_N^2 = 16.59$ are obtained, both of which, referred to the chi-squared, distribution with 2 degrees of freedom, attain a p-value 0.0003, which is significant at the 0.001 level. Note that X_P^2 and X_N^2 are considerably liberal relative to X_{des}^2. For the second-order Rao–Scott adjusted Pearson statistic (7.42), an observed value $X_P^2(\hat{\delta}_., \hat{a}^2) = 13.68$ is obtained, which, referred to the chi-squared distribution with Satterthwaite adjusted degrees of freedom $\text{df}_s = 1.952$, attains a p-value 0.0010. This test appears somewhat liberal relative to the design-based Wald statistic, which seems also to work reasonably in this OHC Survey testing situation (see Example 7.2).

With availability of only limited information, we calculate the first-order adjustments (7.44), (7.45) and (7.46) to the Pearson statistic by using the design-effect estimates in Table 7.3. The mean deff adjustment, with an observed value $X_P^2(\hat{d}_.) = 6.60$ and a p-value 0.0369, is overly conservative relative to the first-order Rao–Scott adjustment $X_P^2(\hat{\delta}_.) = 14.02$, with a p-value 0.0009, and its F-correction $FX_P^2(\hat{\delta}_.) = 7.01$, which attains a p-value 0.0011. Conservativity of the mean deff adjustment arises because $\hat{d}_. = 2.51$ considerably overestimates the mean $\bar{\delta}$ of the true eigenvalues, and the estimate $\hat{\delta}_. = 1.182$, calculated using cell and marginal design-effect estimates, provides a much better estimate. This suggests a warning against the use of the mean deff adjustment if either of the classification variables is strongly intra-cluster correlated. The F-corrected first-order Rao–Scott adjustment works very reasonably when compared to the design-based Wald statistic and the second-order Rao–Scott adjustment.

The tests suggest rejection of the null hypothesis of independence of PHYS and PSYCH. We finally calculate the design-based standardized cell residuals by using (7.47):

	PHYS	
	None	Some
PSYCH	$\hat{e}_{1k}$	$\hat{e}_{2k}$
1	0.99	−0.99
2	2.83	−2.83
3	−3.40	3.40

The residual analysis shows that the largest deviations are in the last PSYCH class so that the direction of the difference favours those suffering from physical health hazards of work. Standardized residuals in these classes exceed the 0.1% critical value 2.58 from the $N(0,1)$ distribution. Note that the sum of residuals is zero across the two PHYS classes.

Also in this testing situation, as in Example 7.2, the design-based Wald statistic behaves adequately due to the relatively large number of sample clusters (250), and we may conclude that the Wald test provides a reasonable testing procedure for the independence hypothesis of PHYS and PSYCH. And if only the cell and marginal design effects are provided, we would choose the F-corrected first-order Rao–Scott adjustment to the Pearson (or Neyman) statistic. But if only the cell design effects are provided and not the marginal design effects, difficulties would arise in obtaining an approximately valid testing procedure because of the apparent over-conservativity of the mean deff adjustment in such a case.

The test of independence in a two-way table can also be executed as a test of no interaction for an appropriate log-linear model with two categorical variables. The independence test is obtained by fitting the saturated log-linear model INTERCEPT + PHYS + PSYCH + PHYS*PSYCH, say, and then by testing with the Wald test the significance of the interaction of PHYS and PSYCH, i.e. the item PHYS*PSYCH. This can be done, for example, with the SUDAAN procedures CROSSTAB or CATAN. The design-based Wald statistic implemented in CROSSTAB gives an observed value 13.83, with a p-value 0.0012, and is compatible with the previous results. The following section of SUDAAN CROSSTAB statements for the independence test, applied to the individual-level data set, and the accompanying output, close this example.

```
1   PROC CROSSTAB DATA=<dataset>  DESIGN=WR;
2   NEST     STRATUM CLUSTER;
3   WEIGHT   _ONE_;
4   SUBGROUP PHYS PSYCH;
5   LEVELS   2 3;
6   TABLES   PHYS*PSYCH;
7   TEST     LLCHISQ;
```

```
Number of observations read    : 7841 Weighted count: 7841
Number of observations skipped : 0
(WEIGHT variable nonpositive)
Denominator degrees of freedom : 245
```

```
Chi Square Test of Independence for PHYS and PSYCH
----------------------------------------------------------
                       P-value  Degrees of freedom
            LLChiSq    LLChiSq  LLChiSq
----------------------------------------------------------
            13.83      0.0012      2
----------------------------------------------------------
```

7.6   CHAPTER SUMMARY AND FURTHER READING

Summary

For a goodness-of-fit test and tests of homogeneity and independence on tables from complex surveys, testing procedures are available that properly account for the complexities of the sampling design. These complexities include the weighting of observations for obtaining consistently estimated proportions, and intra-cluster correlations, which arise due to the clustering and are usually positive. Generally valid testing procedures include the design-based Wald test and the second-order adjustment to the Pearson and Neyman test statistics.

The design-based Wald test can be expected to work adequately when working with large samples where a large number of sample clusters are also available. This was the case in the OHC Survey. A drawback to the Wald test is its sensitivity to such small-sample situations where only a small number of sample clusters are present, leading to unexpectedly liberal test results. The MFH Survey appeared to be an example of such a design. The degrees-of-freedom corrections to the Wald statistic, leading to F-type test statistics, can be used to account for possible instability. The second-order Rao–Scott adjustment to the Pearson and Neyman test statistics can be expected not to be seriously sensitive to instability problems. This adjustment appeared to work reasonably in both the OHC and MFH Surveys.

A full design-based covariance-matrix estimate is required for the design-based Wald test and for the second-order Rao–Scott adjustments. In secondary analyses on published tables, where such a covariance-matrix estimate is not supplied, only approximately valid first-order testing procedures are available. The mean deff adjustment to the standard test statistics can be used if only the cell design-effect estimates are provided. But this adjustment can be overly conservative, as appeared in the example from the OHC Survey. The first-order Rao–Scott adjustment is superior to the mean deff adjustment, and, using an F-correction, the first-order adjustment is also able to account for possible instability problems, as appeared in the MFH Survey example.

The Wald test of independence with the corresponding F-corrections is available, for example, in the SUDAAN software and in the PC CARP and WesVarPC programs. An element-level survey data set must be available for these software, which, in addition to the analysis variables, should include the appropriate design variables for stratum and cluster identification, and a weight variable. Because a test of homogeneity can also be taken as a simple application of logit modelling, and a test of independence in turn as an application of log-linear modelling, programs for fitting these models, implemented in the above-mentioned software products, can also be used.

In hypothesis testing, a vector of finite-population cell proportions was considered. But if the finite population is large, these proportions are close to the corresponding cell probabilities of the infinite superpopulation from which the finite population can be regarded as a single realization. Thus, the design-based inferences considered here also constitute inference on the parameters of the appropriate infinite superpopulation.

Further Reading

The analysis of one-way and two-way frequency tables has received considerable attention in the survey analysis literature in recent years. The articles of Holt *et al.* (1980), and Rao and Scott (1981, 1984, 1987), cover important theoretical developments of the 1980s. More applied sources include Binder *et al.* (1984), Hidiroglou and Rao (1987a, 1987b), and Rao and Thomas (1988, 1989).

There are also overviews and more specialized materials available on this topic, such as the articles by Freeman and Nathan in the *Handbook of Statistics* (vol. 6, 1988), the article by Rao in Puri *et al.* (eds, 1987), and the articles by Binder *et al.* and by Hidiroglou and Paton, in MacNeill and Umphrey (eds, 1987), and a section in Särndal *et al.* (1992). The duality between design-based and model-based inference is discussed, e.g. in Rao and Thomas (1988) and in Skinner *et al.* (1989).

8

Multivariate Survey Analysis

Multivariate methods provide powerful tools for the analysis of complex survey data. Multivariate analysis is discussed in this chapter in the case of one response variable and a set of predictor or explanatory variables. For this kind of analysis situation, logit models and linear models are widely used. Proper methods are available for fitting these models for intra-cluster correlated response variables from complex sampling designs. These methods have also been implemented in software products for survey analysis. With logit and linear modelling in complex surveys, as with the analysis of two-way tables, it is important to eliminate the effects of clustering from the estimation and test results. Examination of recent methodology for this task, supplemented with numerical examples, is the main focus in this chapter.

8.1   RANGE OF METHODS

The aim in fitting multivariate models is to find a scientifically interesting but parsimonious explanation of the systematic variation of the response variable. This is achieved by modelling the variation with a reasonable set of predictor variables using the available survey data. For example, in a health survey based on a cluster sample of households, variation of health status and use of health services is to be studied in order to find possible high-risk population subgroups to target in developing a health promotion programme. Certain socioeconomic determinants of the sample households and demographic and behavioural characteristics of household members are used as predictor variables. In an educational survey based on cluster sampling of teaching groups, one may wish to study the effect of the teacher, and that of the students, on the differences in learning. Further, in a survey on health-related working conditions, the association of perceived psychic (psychological or mental) strain with certain physical and other working conditions can be studied, based again on data from cluster sampling with industrial establishments as the clusters. In all these surveys, the data would be collected with cluster sampling, but inferences concern mainly a person-level population or, more generally, relationships of the person-level variables under a superpopulation framework.

Response variables in the example surveys were binary (chronic sickness is present or not present; psychic strain is low or high), polytomous (learning outcomes are poor, medium, or good), or quantitative or continuous (the number of physician visits; principal component score of psychic strain). Logit modelling on a binary or polytomous response and linear modelling on continuous measurements provide two popular approaches to these cases. If cluster sampling is used, as in the example surveys, the response variables are exposed to intra-cluster correlations. The consequences of the intra-cluster correlation are discussed briefly in the following introductory example.

Introductory Example

Let us consider more closely the cases of a binary and a continuous response variable. With categorical predictors, the data for a binary response can be arranged in a table of proportions, and for a continuous response, in a table of means. From the OHC Survey we have the following table of perceived overall psychic strain (PSYCH), which is originally a continuous variable of scores of the first principal component from a set of psychic symptoms. For a binary response, the variable PSYCH is recoded so that the value zero indicates strain below the mean (low-strain group), and the value one indicates strain above the mean (high-strain group). In the table, we have three categorical predictors, each with two classes: sex and age of respondent, and the variable PHYS (physical health hazards) which measures physical working conditions coded so that the value one indicates more hazardous work. The domains are formed by cross-classifying the predictors, and they cut across the sample clusters. The main interests are in the relation of psychic strain to physical working conditions. In Example 7.3, statistically significant dependence was noted for these variables, although in a slightly different setting where PSYCH was recoded as a three-class variable.

The percentage of persons experiencing above average psychic strain in the whole sample is of course 50%, and in the risk group (PHYS $=1$) this percentage was noted to be 52.2%, i.e. only slightly larger than in the other group. But, when inspecting the variation of percentage estimates in Table 8.1, it appears that there are certain subgroups with a large proportion of persons suffering from psychic strain. For both sexes, the proportions tend to increase with increasing age and, in a given age group, the proportions are higher for those involved in physically more hazardous work. There might also exist an interaction between age and physical working conditions.

Thus, the variation in the proportions of the binary response is quite

Table 8.1 Proportion (%) of persons in the upper psychic strain group, and mean of the continuously measured psychic strain, in domains formed by sex, age and physical working conditions of respondent, and design-effect estimates of the proportions and means (the OHC Survey; $n = 7841$ employees).

Domain	SEX	AGE	PHYS	PSYCH %	deff	PSYCH Mean	deff
1	Males	−44	0	41.9	1.16	−0.193	1.14
2			1	47.2	1.33	−0.084	1.36
3		45−	0	46.1	0.87	−0.075	1.05
4			1	52.0	1.18	0.139	1.25
5	Females	−44	0	54.1	1.23	0.065	1.61
6			1	62.0	1.38	0.264	1.46
7		45−	0	53.2	1.65	0.098	1.74
8			1	70.0	1.47	0.656	1.44
Total sample				50.0	1.69	0.000	1.97

logical. Obviously, the variation in the means of the corresponding continuously measured psychic strain follows a similar pattern. A logit analysis would be chosen for the analysis of the domain proportions, and linear modelling is appropriate for the domain means. Because the predictors are categorical, an analysis-of-variance-type model would be selected in both cases. If the data were obtained with simple random sampling, the analysis would be technically a standard one: take a procedure for binomial logit analysis and for linear analysis of variance from any commercial program package (SAS, SPSS, BMDP, GLIM), search for well-fitting and parsimonious logit and linear models, and draw conclusions.

But in the OHC Survey, cluster sampling was used with establishments as the clusters. Positive intra-cluster correlation can thus be expected for the response variable PSYCH, as in Example 7.3. This correlation can disturb the analysis in such a way that if it is ignored, erroneous conclusions might be drawn. From Table 8.1 it can be seen that design-effect estimates of proportions are larger than one on average, with an overall design-effect estimate 1.7, indicating a noticeable clustering effect. For a proper analysis, this clustering effect should be taken into account, and a simpler model for the variation of PSYCH proportions can be obtained than by ignoring the clustering effects, as will be seen in Example 8.1.

Two Main Approaches

There are two main approaches available for proper multivariate analysis of an intra-cluster correlated response variable such as PSYCH. If intra-cluster correlation is taken as a nuisance, one may make efforts to eliminate this disturbance effect from the estimation and test results, as was done in Chapter 7. The *nuisance approach*, covering a variety of methods for logit and linear modelling, has been developed over a long period, mainly within the context of traditional survey sampling. This approach is sometimes referred to as the *aggregated* approach.

If, on the other hand, clustering is interesting as a structural property of the population, it can be examined with appropriate models. This approach has been mainly developed under a general framework of *multi-level modelling* for hierarchically structured data sets, and for the analysis of longitudinal data under a certain version of the so-called *generalized estimating equations* (GEE) methodology. Both methodologies can also be applied to multivariate analysis of correlated responses from clustered designs. However, for complex surveys, the nuisance approach has had a dominant role, and it is the main approach used here. The alternative approach, which can also be called *disaggregated*, will be briefly discussed in this chapter and demonstrated in Chapter 9.

Estimation Methods

In the nuisance approach there are alternative asymptotically valid estimation methods for modelling intra-cluster correlated response variables. For a binary or polytomous response variable, we apply the method of *weighted least squares estimation* of model parameters in cases where the data are arranged in a multidimensional table such as Table 8.1. Intra-cluster correlation is accounted for by using a design-based covariance-matrix estimate of proportion estimators. The method, called henceforth the *WLS method*, will be discussed in Section 8.4 for logit and linear modelling of categorical data. The WLS method, introduced in Grizzle *et al.* (1969) and Koch *et al.* (1975), is applicable to a combination of linear, logarithmic and exponential functions on proportions. Thus, in addition to logit and linear models, log-linear models are also covered. In complex surveys, these models can be fitted with the WLS method by using, for example, the SUDAAN procedure CATAN.

A widely used method for fitting models for binary, polytomous and continuous response variables in complex surveys is based on a modification of maximum likelihood (ML) estimation. The traditional ML estimation of model parameters is for response variables from simple random sampling, for which appropriate likelihood functions can be derived under standard distributional assumptions. But for more complex designs no convenient likelihood functions are available, and therefore, a method called *pseudolikelihood estimation* is used instead. Intra-cluster correlation is accounted for in this method in a similar way to the WLS method. The method is henceforth called the *PML method*, and it will be considered in Section 8.5 for logit analysis on a binary response. In linear modelling on a continuous response, the estimation reduces to solving weighted normal equations. These are special cases of a broad methodology for fitting *generalized linear models* following Nelder and Wedderburn (1972) and McCullagh and Nelder (1989), covering, for example linear, logit and log-linear models. Programs are available for logit modelling with the PML method under clustered designs, such as the SUDAAN procedure LOGISTIC and the logistic option in the PC CARP and WesVarPC programs. For linear modelling in complex surveys, the SUDAAN procedure REGRESS, the OSIRIS program REPERR, or the regression option of the PC CARP and WesVarPC programs can be used.

The third method is based on the simplest version of the GEE methodology of generalized estimating equations (Liang and Zeger 1986). The model parameters are estimated using the so-called *multivariate quasilikelihood* method and, for example, for a binary response, intra-cluster correlations are parametrized using pair-wise odds ratios. We will briefly discuss this method in Section 8.5, because the method, like the PML method, has its roots in generalized linear models methodology.

In WLS and PML methods, design-based Wald test statistics and second-order Rao–Scott adjusted test statistics can be used, providing asymptotically valid testing procedures. However, the test statistics may suffer from instability problems, especially when the number of sample clusters is small. Instability can disturb the behaviour of a design-based Wald statistic, resulting in overly liberal test results relative to the nominal levels and leading to unnecessarily complex models. This property is similar to that noted for Wald tests on two-way tables, and it holds for Wald statistics in both WLS and PML methods. To protect against the effects of instability, certain degrees-of-freedom corrections such as *F*-corrections are available.

Although there are many similarities in the WLS and PML methods, their applicability and properties differ in certain important respects. For further discussion, we next define the main types of linear and logit models, and more formally introduce the corresponding models.

8.2 TYPES OF LINEAR AND LOGIT MODELS

Three Types of Models

In linear models the expectation of a continuous response variable is related to a linear expression on the predictors. In logit models, a nonlinear function of the expectation of a binary response variable, called a *logit* or *logistic* function, is related to a linear expression on the predictors. Note that both models share the property that the expression on the predictors is a linear one. But the essential difference is that in a linear model this predictor part is linearly related to the response variable, and in a logit model, a nonlinear relationship is postulated.

For introducing the types of linear and logit models, it is instructive to consider separately the case of multidimensional tables with categorical predictors and the case where the predictors are purely continuous (or at least one of them is). In both instances, the response variable can be binary, polytomous, or a quantitative or continuous one.

In multidimensional tables such as Table 8.1, the predictors are categorical qualitative or categorized quantitative variables, and depending on additional assumptions on their types, special cases of linear and logit models are obtained. In models of analysis-of-variance (ANOVA) type, the classes of each predictor are taken to be qualitative. Sex, occupation, social class, and type of industry, are examples of commonly used predictors. For categorized quantitative predictors, monotonic ordering can be assumed on the classes of each predictor, and desired scores can be assigned to the classes. The predictors can then be taken to be continuous, leading to regression-type models. Age, systolic blood pressure, monthly income of a household, and first principal component of psychic symptoms, are examples of such predictors, each categorized into a small number of classes. Note that the classes of an originally quantitative variable can also be taken to be qualitative, as in Example 7.3. If both qualitative and quantitative categorical predictors are present, we may call the model an analysis of covariance or ANCOVA type model. For ANOVA and ANCOVA models, it is common to include interaction terms in the model and test their significance which often constitutes an essential part of model building.

Sometimes it is desirable to work with quantitative predictors without categorizing them and arranging the data in a multidimensional table. Thus we have at least one continuous predictor, and depending on the types of the other predictors, the corresponding models are obtained. If all the predictors are continuous measurements, we have a regression-type model, and additional qualitative predictors result in an ANCOVA-type model. It should be noted that in this case we actually model individual-level differences, whereas in the former case we are modelling differences between subgroups of the population.

In the analysis of a continuous response variable, the traditional ANOVA, regression analysis and ANCOVA models constitute the commonly used special cases of a linear model. We use analogous terminology for logit models with a binary or polytomous response variable. For these, we hence have the corresponding logit ANOVA, logit (or logistic) regression, and logit (or logistic) ANCOVA types of models.

Logit and Linear Models for Proportions

The following examples often deal with logit and linear modelling on domain proportions of a binary response variable, because of the simplicity and popularity of this analysis situation in practice. Let us thus introduce the logit and linear models in the case where the data are organized in a multidimensional table such that there are u domains which are formed by cross-classifying the categorical predictors, and the response variable is binary. A logit or a linear model can then be postulated for examining the systematic variation of the estimated domain proportions of the response variable across the domains. The situation is thus essentially similar to that of Table 8.1.

Under a logit model we deal with logarithms of *ratios of proportions* p_{j1} and p_{j2}, where the former is the proportion of 'success'. We denote this proportion by p_j; thus the other is $p_{j2} = 1 - p_j$. The variation is modelled by relating the functions of the form $\log(p_j/(1 - p_j))$ of the unknown proportions p_j to linear functions of the form $b_1 x_{j1} + b_2 x_{j2} + \cdots + b_s x_{js}$, where 'log' refers to natural logarithm. A function $\log(p_j/(1 - p_j))$ is called the *logit* or *log odds* of success. In the linear functions, b_k are the model coefficients to be estimated, of which the first coefficient b_1 is an intercept term. The values x_{jk} are for the fixed variables x_k, with a constant value of one assigned to the first variable x_1. Other variables depend on the model type. In logit ANOVA, x_k are indicator variables for the classes of the predictors. In logistic regression they are continuous-valued scores assigned to the classes, or the original continuous measurements. And in logit ANCOVA the fixed variables constitute a mixture of indicator variables and continuous variables. Interpretation of the coefficients b_k depends on the model type and on the parametrization used under a specific model. An advantage of the logit model is that *odds-ratio*-type statistics are readily available, and in special cases, interpretations with the concepts of independence and conditional independence are also possible.

Under linear modelling on proportions, on the other hand, we deal directly with *differences of proportions*. Thus, the population proportions p_j are related linearly to the linear functions $b_1 x_{j1} + b_2 x_{j2} + \cdots + b_s x_{js}$. This model formulation can be equally appropriate as a logit formulation, and it involves certain convenient interpretations. But interpretations by independence or related natural terminology are excluded.

The logit and linear models can be compactly written in a matrix form. Let $\mathbf{p} = (p_1, \ldots, p_u)'$ be the vector of unknown domain proportions, $\mathbf{b} = (b_1, \ldots, b_s)'$ be the vector of model coefficients, and let $\mathbf{X}$ be the $u \times s$ matrix of x_{jk} such that the columns of the matrix represent the values of the fixed variables x_j. Usually $\mathbf{X}$ is called the *model matrix*. A hypothesized model can be written in the form

$$F(\mathbf{p}) = \mathbf{Xb}, \tag{8.1}$$

where, in the case of a logit model, the function vector $F(\mathbf{p})$ of the unknown proportion vector $\mathbf{p}$ is formulated as

$$F(\mathbf{p}) = F(\mathbf{f}(\mathbf{b})) = \log\left(\frac{\mathbf{f}(\mathbf{b})}{1 - \mathbf{f}(\mathbf{b})}\right), \tag{8.2}$$

and, in the case of a linear model, the function vector $F(\mathbf{p})$ equals $\mathbf{p}$ because F is simply an identity function. Further, for a logit model, the function vector $\mathbf{f}(\mathbf{b})$ is derived using the inverse of the logit function:

$$\mathbf{f}(\mathbf{b}) = F^{-1}(\mathbf{Xb}) = \frac{\exp(\mathbf{Xb})}{1 + \exp(\mathbf{Xb})}, \tag{8.3}$$

where 'exp' refers to the exponential function. For a linear model this function vector is obviously $\mathbf{f}(\mathbf{b}) = \mathbf{Xb}$. An important motivation for the logit function is that the values of the function vary between zero and one, i.e. in the same range as the proportions p_j themselves. Therefore, predicted proportions from a fitted logit model always fall in the range $(0,1)$. This property does not necessarily hold for the linear model formulation.

As an illustration of the matrix expressions (8.1)–(8.3), let us consider the case with two dichotomous predictors A and B for logit and linear ANOVA models for proportions p_j of a binary response variable. There are thus four domains $(u = 4)$ and the table of the unknown proportions p_j is:

Domain	A	B	p_j
1	1	1	p_1
2	1	2	p_2
3	2	1	p_3
4	2	2	p_4

We have three sources of variation in the table: that due to the effect of A, that due to the effect of B, and that due to the effect of the interaction of A and B. In order to cover all these sources of variation, a total of four coefficients b_k are included in the model $F(\mathbf{p}) = \mathbf{Xb}$. The coefficient b_1 is the intercept, b_2 is assigned to A, b_3 is assigned to B, and b_4 is assigned to the interaction of A and B. This model is called a *saturated* model, and by choosing a specific model matrix $\mathbf{X}$ it can be expressed as

$$\begin{bmatrix} F(p_1) \\ F(p_2) \\ F(p_3) \\ F(p_4) \end{bmatrix} = \begin{bmatrix} 1 & 1 & 1 & 1 \\ 1 & 1 & -1 & -1 \\ 1 & -1 & 1 & -1 \\ 1 & -1 & -1 & 1 \end{bmatrix} \begin{bmatrix} b_1 \\ b_2 \\ b_3 \\ b_4 \end{bmatrix}, \tag{8.4}$$

where for a logit model the functions $F(p_j)$ are the logits

$$F(p_j) = \text{logit}(p_j) = \log\left(\frac{p_j}{1 - p_j}\right), \quad j = 1, 2, 3, 4,$$

and for a linear model the functions are $F(p_j) = p_j$. In the model matrix $\mathbf{X}$ of (8.4) we first have a column of ones for the indicator variable x_1. Then, there are three columns of contrasts with values 1 or -1, of which the first is for the predictor A, i.e. for the indicator variable x_2, the second is for the predictor B, i.e. for the indicator variable x_3, and the last one is for the interaction of A and B, i.e. for the indicator variable x_4. Note that each indicator variable sums to zero in this parametrization, and there is one indicator variable for each predictor and their interaction, because the predictors are two-class variables. Generally, there are $t - 1$ columns in the model matrix for a t-class variable, and $(t - 1) \times (v - 1)$ columns for an interaction of a t-class variable and a v-class variable, corresponding to the degrees of freedom for a model term. The sum of these degrees of freedom is the number s of model coefficients.

The parametrization just applied is sometimes called a *marginal* or full-rank centre-point parametrization. It is used as a default in many programs for logit analysis such as the SAS procedure CATMOD and the BMDP program PLR. Under this parametrization, for categorical predictors with more than two classes, each indicator variable is used with the others to contrast a given class with the average of all classes. For example, in a logit ANOVA model, the coefficients b_k indicate differential effects on a logit scale, i.e. with respect to the average of all the fitted logits, and in a linear ANOVA model, they indicate differential effects on the untransformed scale, i.e. with respect to the average of all the fitted proportions.

It is important for proper inferences that we are fully aware of the specific parametrization applied, because there are also other commonly used parametrizations. For example, in the SUDAAN procedures CATAN and LOGISTIC, a parametrization called *partial* or reference-cell is applied as a default. There, a specific reference class is assumed, and each indicator variable is used with the others to compare a given class with the reference class. Under this parametrization, we put zeros in place of -1 in the previous model matrix **X**. This parametrization is especially useful when a definite reference group can be stated. In a logit model, the coefficients now indicate differential effects with respect to the fitted logit in the reference class, and in a linear model, differential effects with respect to the fitted proportion in the reference class. An odds ratio $\exp(b_k)$ interpretation is readily available for logit models under the partial parametrization.

Under these parametrizations we have for the functions $F(p_j)$:

Marginal	Partial
$F(p_1) = b_1 + b_2 + b_3 + b_4$	$F(p_1) = b_1 + b_2 + b_3 + b_4$
$F(p_2) = b_1 + b_2 - b_3 - b_4$	$F(p_2) = b_1 + b_2$
$F(p_3) = b_1 - b_2 + b_3 - b_4$	$F(p_3) = b_1 + b_3$
$F(p_4) = b_1 - b_2 - b_3 + b_4$	$F(p_4) = b_1 + b_4$

Note that, because the functions $F(p_j)$ must be equal for both parametrizations, the corresponding coefficients b_k from these parametrizations cannot coincide. So, for example, the coefficient b_1 in the marginal parametrization is not equal to the b_1 in the partial parametrization.

Our discussion so far has been on logit and linear ANOVA models on domain proportions. A similar discussion applies for linear ANOVA models on domain means of a continuous response. For logit and linear regression and ANCOVA models on binary responses, and for the corresponding linear models on continuous responses, the model matrices, however, are different, involving different interpretation of the model parameters.

Model Building in Practice

When fitting a specified logit or linear model, the primary task is to estimate the model coefficients b_k and the variances of the estimated coefficients by using an appropriate computer program on the available sample data set. Using the resulting estimates, adequacy of the model is assessed by examining the goodness of fit of the model, and tests of linear hypotheses on model coefficients are executed. Model building in practice often involves repetition of this procedure several times for alternative models.

Let us consider further the logit and linear models on proportions. In a model-fitting procedure using a standard notation, the previous ANOVA-type models can be written as $F(P) = \log(P/(1-P)) = A + B + A*B$ for a logit model, and $F(P) = P = A + B + A*B$ for a linear model with a binary response variable. There are three model terms corresponding to the predictors: two main effects and an interaction term. The model is saturated because it includes all the terms possible in this situation; the intercept term is included as a default in all the models. This kind of notation is commonly used for requesting a specified model structure, i.e. the terms desired in the linear part of the model, in many programs for linear and logit analysis.

A saturated model, including all possible main effects and interaction terms, is seldom interesting because the model includes as many parameters as there are degrees of freedom available. Also, the saturated model fits perfectly to the data. In a model-building procedure, the aim is to reduce the saturated model in order to find a well-fitting model which is parsimonious, so that as few model terms as possible are included.

By using the above notation, the possible models in these logit and linear ANOVA cases are as follows:

F (P) = A + B + A * B   (saturated model),
F (P) = A + B   (main effects model),
F (P) = A   (model for the predictor A only),
F (P) = B   (model for the predictor B only), and
F (P) = INTERCEPT   (null model).

Reduced models are obtained by hierarchically removing statistically nonsignificant terms from a model. This procedure corresponds to removing columns (or sets of columns) from the model matrix. Usually, a well-fitting model for further use and for interpretation is found between the saturated and null models.

A model-building procedure in linear ANOVA on domain means resembles that of logit and linear ANOVA on domain proportions. In logistic and linear regression or ANCOVA-type models involving continuous predictors, an appropriate model is usually searched for by consecutively entering statistically significant or scientifically interesting terms, beginning from the null model. In these models it should be noted that interactions are not allowed between the continuous predictors.

In complex surveys, estimation of the model coefficients of a logit ANOVA, ANCOVA or regression model on domain proportions can be executed by either the WLS or the PML method. For a linear model on such proportions, the PML method reduces to the WLS method. For logistic regression and ANCOVA models on a binary or polytomous response with strictly continuous predictors, the PML method is used. A special case of the PML method, reducing to OLS-type estimation, is applied to linear models on continuous responses. In practice, all these models can be conveniently fitted with special software for survey analysis, such as SUDAAN and PC CARP.

Before entering into the details of modelling by WLS and PML methods, we discuss in greater depth the special features of multivariate analysis when working with complex surveys. Several options will be introduced for proper analysis under different sampling-design assumptions.

8.3 OPTIONS FOR ANALYSIS

Here we introduce a set of options for multivariate analysis of complex survey data involving clustering, stratification, multi-stage sampling, and nonignorable nonresponse. In the presence of such complexities, consistent estimators of model coefficients and their variances, and valid test results, can be obtained by appropriately weighting the observations due to unequal inclusion probabilities and nonresponse, and by appropriately accounting for the intracluster correlations.

Three specific analysis options are presented: the *full design-based* option DES, the *extravariation-oriented* option EFF using effective sample sizes, and the *SRS-based* option assuming simple random sampling. With the DES and EFF options, the design complexities can be accounted for, and the SRS-based option is used as a reference for the other options when quantifying the effects of the design complexities on analysis results. All the options will be used in logit and linear ANOVA, regression and ANCOVA modelling on multidimensional tables of domain proportions. The full design-based and the SRS-based options will be used in logistic and linear regression and ANCOVA-type analyses on binary or continuous responses, with continuous measurements as the predictors.

Full Design-based DES Option

Under the full design-based option, intra-cluster correlations, unequal element inclusion probabilities, and adjustment for nonresponse can be properly accounted for. This option is evidently the most appropriate for multivariate analysis generally in complex surveys. Therefore, the DES option is widely used in survey analysis, and it will be adopted in this chapter as the main analysis option.

The DES option can in practice be applied in various ways, depending on special features of the sampling design and on software available for the analysis. Sampling designs involving weighting due to stratification or post-stratification and several stages of sampling, often require approximations to conveniently fit the DES option. For data from two-stage stratified cluster-sampling with a large population of clusters, a simple solution for this option is to reduce the design to one-stage stratified sampling where the primary sampling units are assumed to be drawn with replacement. The approximation is common in complex analytical surveys and is available in SUDAAN and PC CARP. It is the only method for the DES option available to OSIRIS users. Use of this approximation requires access to an element-level data set, which includes variables for stratum and cluster identification and for weighting. The approximation was used in the design-based covariance-matrix estimation in Section 6.2, and in the analysis of frequency tables in Chapter 7, by using the appropriate design option (WR) in the SUDAAN applications.

In a more advanced use of the DES option, additional features of the sampling design can be accounted for if necessary for proper estimation. Examples are when the variation is due to several stages of sampling or sampling of clusters is with unequal probabilities without replacement. This presupposes the availability of population counts at each stage of sampling, and the calculation of single and joint selection probabilities of each primary sampling unit and each pair of PSUs in each first-stage stratum. Thus, more information must be supplied for an analysis program. For SUDAAN users, these complexities can be accounted for by using the design options WOR (without replacement) and UNEQWOR (WOR with unequal selection probabilities of the PSUs). Similar facilities are available in PC CARP.

In addition to the above refinements, analysis under the DES option can involve reorganization of the sample clusters into strata using the collapsed stratum technique, if only one primary sampling unit was originally drawn from each stratum, as was the case in the MFH Survey. In some cases, additional weighting for poststratification is desirable. All these features have been implemented in the above-mentioned software.

In multivariate analysis of domain proportions of a binary response, it is assumed for the DES option that an appropriate design-based covariance-matrix estimate of proportions is available. In Section 6.2, we introduced a technique for obtaining a consistent covariance-matrix estimate based on the linearization method. This estimate is allowed to be nondiagonal, because the correlations of the proportions from separate domains can be nonzero, which is the case when working with cross-classes or mixed classes.

The other two analysis options assume a diagonal covariance matrix of the domain proportions. The validity of this assumption depends on the domain structure, and the assumption applies strictly to segregated-classes-type-domains. The strength of the clustering effect determines the type of the corresponding diagonal covariance-matrix estimator, leading to two different options, one assuming extravariation due to the clustering effect and the other where such extravariation is not present.

Extravariation-oriented EFF Option

This option uses the effective domain sample sizes introduced in Section 6.3. The option is usable for multivariate analysis on domain proportions in cases where the binary response variable is intra-cluster correlated and the domains constitute segregated classes. In that case, it can be assumed that correlations of the proportions from separate domains are zero, because all elements in a given cluster fall in the same domain. The clustering effect can cause extra-binomial variation because of dependencies of the elements in a cluster. If the clustering effect is negligible, these correlations would be near zero, and variances of the estimators could be approximated by binomial variance estimators. But if the clustering effect causes positive intra-cluster correlation, extravariation to the binomial variance is involved. True variances of the domain proportions are thus larger than their binomial counterparts.

Extra-binomial variation can be accounted for by using the covariance-matrix estimators of domain proportions introduced in Section 6.3, based on either of the two methods of effective sample sizes. Therefore, an analysis under the EFF option is parallel to the first-order techniques using the mean deff adjustment considered in Chapter 7. Approximatively valid analysis can be expected also for mixed classes or also cross-classes-type domains, if the corresponding design-based covariance-matrix estimate appears nearly diagonal. In practice, analysis under the EFF option can be executed with standard analysis programs by using element weights. The weighting is based on rescaling the original relative element weights so that they, instead of the actual sample size, sum up to an effective sample size.

SRS-based Option

This option assumes simple random sampling with replacement. Two alternative versions of the SRS-based option can be derived. Under the *weighted* version of the SRS-based option, it is assumed that the domain proportions are consistently estimated by using the relative element weights, and a binomial covariance matrix is assumed for these proportions. Because the weighting is the only complexity accounted for by this option, the analysis can be carried out by standard programs by using the original relative element weights.

Under the *unweighted* version of the SRS-based option, called the *IID-based* option (independent identically distributed), simple random sampling with replacement is assumed, and the data set is assumed self-weighting. Thus, all the complexities of the sampling design are ignored. Analysis under the IID option therefore corresponds to the standard *model-based* analysis. And analysis under the weighted SRS option can be called *model-based using auxiliary information*, because auxiliary information is incorporated in the model-based analysis in the form of relative element weights.

Because the two versions of the SRS-based option are not valid for complex surveys involving clustering, they will be used as reference options to the full design-based DES option and in the construction of appropriate generalized design-effect matrices. The weighted SRS-based option is used when assessing the magnitude of the clustering effects on results from multivariate analyses, and the IID-based option can be used as a reference option to the DES option when examining the effects of all the complexities of the sampling design on analysis results.

The analysis options with respect to the sampling design are summarized below:

Option	Allowing weights	Allowing extravariation	Allowing full design
Full design-based			
DES	yes	yes	yes
Extravariation-oriented			
EFF	yes	yes	no
Simple random sampling			
SRS	yes	no	no
IID	no	no	no

It should be noticed that in multivariate survey analysis, as in the analysis of two-way tables, the design-based approach to inference also constitutes inference on the parameters of the corresponding superpopulation model, provided that the finite population is large (see Rao and Thomas 1988).

8.4   ANALYSIS OF CATEGORICAL DATA

The WLS method of weighted least squares estimation provides a flexible technique for the analysis of categorical data covering ANOVA, regression and ANCOVA logit and linear models on domain proportions. Allowing all the complexities of a sampling design including stratification, clustering and weighting, the full design-based DES option provides a generally valid WLS analysis. Analysis under the SRS option assuming simple random sampling serves as a reference when studying the effects of clustering and weighting on results. The EFF option of effective sample sizes is useful if only extra-binomial variation appears.

The WLS method is computationally simple because it is non-iterative for both logit and linear models on proportions. The alternative PML method of pseudolikelihood estimation for logit models is, as an iterative method, computationally more demanding, but for linear models, PML reduces to WLS. For logit regression with continuous predictors which are not categorized, the PML method can be used but the WLS method is inappropriate. The application area of the WLS method is thus more limited than that of PML.

In surveys with large samples, closely related results are usually attained by the WLS and PML methods. But in fitting ANOVA-type models there can be many multi-class predictors included in the model and, therefore, the number of domains can be large, and a large element-level sample size is required to obtain a reasonably large number of observations falling in each domain. This is especially important for the WLS method which is mainly used in large-scale surveys where the sample sizes can be in thousands of persons, as is the case in the OHC and MFH Surveys. For proper behaviour of both WLS and PML methods, a large number of sample clusters is beneficial. Recall that this property holds for the OHC Survey.

We consider the WLS method for a binary response variable and a set of categorical predictors. The data can thus be arranged into a multidimensional table, such as Table 8.1, where the u domains are formed by cross-classifying the categorical predictors and the proportions p_j of the binary response are estimated in each domain. The consistent estimates $\hat{p}_j$, used under the DES, EFF and SRS options, are weighted ratio-type estimators of the form $\hat{p}_j = \hat{n}_{j1}/\hat{n}_j$, where $\hat{n}_{j1}$ is the weighted sample sum of the binary response in domain j, and $\hat{n}_j$ are weighted domain sample sizes. The unweighted proportion estimates $\hat{p}_j^U$, used under the IID option, are obtained by using the unweighted counterparts n_{j1} and n_j.

When applying the WLS method for logit and linear modelling under an analysis option, the starting point is the calculation of the corresponding proportion estimate vector and its covariance-matrix estimate. By using these estimates, the model coefficients are estimated, together with a covariance matrix of the estimated coefficients, and using these, fitted proportions and their covariance-matrix estimates are obtained. Further, the Wald test of good-ness of fit of the model, and desired Wald tests of linear hypotheses on the model coefficients, are executed. Finally, residual analysis is carried out to more closely examine the fit of the selected model.

Design-based WLS Estimation

Under the DES option, a consistent *WLS estimator* $\hat{\mathbf{b}}_{des}$, denoted $\hat{\mathbf{b}}$ for short in this section, of the $s \times 1$ model coefficient vector $\mathbf{b}$ for a model $F(\mathbf{p}) = \mathbf{Xb}$ is given by

$$\hat{\mathbf{b}} = (\mathbf{X}'(\mathbf{H}\hat{\mathbf{V}}_{des}\mathbf{H})^{-1}\mathbf{X})^{-1}\mathbf{X}'(\mathbf{H}\hat{\mathbf{V}}_{des}\mathbf{H})^{-1}F(\hat{\mathbf{p}}), \qquad (8.5)$$

where $\hat{\mathbf{V}}_{des}$ is a consistent estimator of the covariance matrix of the domain proportion estimator vector $\hat{\mathbf{p}}$, and $\mathbf{H}\hat{\mathbf{V}}_{des}\mathbf{H}$ is a covariance-matrix estimator of the function vector $F(\hat{\mathbf{p}})$. An estimate $\hat{\mathbf{V}}_{des}$ is obtained using the linearization method as was described in Section 6.2. The WLS estimating equations (8.5) are thus based on the consistently estimated functions $F(\hat{p}_j)$ and their design-based covariance-matrix estimate. The equations also indicate that no iterations are needed to obtain the estimates $\hat{b}_k$.

The WLS estimator $\hat{\mathbf{b}}$ from (8.5) applies for both logit and linear models on domain proportions. But the matrix $\mathbf{H}$ in the covariance-matrix estimator of the function vector differs. In the logit model, the diagonal $u \times u$ matrix $\mathbf{H}$ of partial derivatives of the functions $F(\hat{p}_j)$ has diagonal elements of the form $h_j = 1/(\hat{p}_j(1 - \hat{p}_j))$. And in the linear model, the matrix $\mathbf{H}$ is an identity matrix with ones on the main diagonal and zeros elsewhere.

Under a partial parametrization of a logit ANOVA model (see Section 8.2), where the columns of the model matrix $\mathbf{X}$ corresponding to the classes of the predictors are binary variables, a log odds ratio interpretation can be given to the estimates $\hat{b}_k$. Thus, an estimate $\exp(\hat{b}_k)$ is the *odds ratio* for the corresponding class with respect to the reference class adjusted for the effects of the other terms in the model. This interpretation of the estimated model coefficients is common in epidemiology and also in social sciences.

A covariance-matrix estimate $\hat{\mathbf{V}}_{des}(\hat{\mathbf{b}})$ of the estimated model coefficients $\hat{b}_k$ from (8.5) is used in obtaining Wald test statistics for the coefficients. This $s \times s$ covariance matrix is given by

$$\hat{\mathbf{V}}_{des}(\hat{\mathbf{b}}) = (\mathbf{X}'(\mathbf{H}\hat{\mathbf{V}}_{des}\mathbf{H})^{-1}\mathbf{X})^{-1}. \qquad (8.6)$$

With proper choice of $\mathbf{H}$ this estimator applies again for both logit and linear models. Diagonal elements of $\hat{\mathbf{V}}_{des}(\hat{\mathbf{b}})$ provide the design-based variance estimates $\hat{v}_{des}(\hat{b}_k)$ of the estimated coefficients $\hat{b}_k$ to be used in obtaining the corresponding standard-error estimates $\text{s.e}_{des}(\hat{b}_k) = \hat{v}_{des}^{1/2}(\hat{b}_k)$. Under a logit model, using these standard-error estimates, for example, an approximative 95% confidence interval for an odds ratio $\exp(b_k)$ can be calculated:

$$\exp(\hat{b}_k \pm 1.96 \times \text{s.e.}_{des}(\hat{b}_k)). \tag{8.7}$$

Two additional covariance-matrix estimators are useful in practice. These are the $u \times u$ covariance-matrix estimator $\hat{\mathbf{V}}_{des}(\hat{\mathbf{F}})$ of the vector $\hat{\mathbf{F}} = \mathbf{X}\hat{\mathbf{b}}$ of the fitted functions, and the covariance-matrix estimator $\hat{\mathbf{V}}_{des}(\hat{\mathbf{f}})$ of the vector $\hat{\mathbf{f}} = F^{-1}(\mathbf{X}\hat{\mathbf{b}})$ of the fitted proportions. These are

$$\hat{\mathbf{V}}_{des}(\hat{\mathbf{F}}) = \mathbf{X}\hat{\mathbf{V}}_{\mathbf{des}}(\hat{\mathbf{b}})\mathbf{X}' \tag{8.8}$$

and

$$\hat{\mathbf{V}}_{des}(\hat{\mathbf{f}}) = \hat{\mathbf{H}}^{-1}\hat{\mathbf{V}}_{des}(\hat{\mathbf{F}})\hat{\mathbf{H}}^{-1}. \tag{8.9}$$

For a linear model these covariance matrices obviously coincide, because the fitted functions are equal to the fitted proportions. For a logit model, the diagonal matrix $\hat{\mathbf{H}}$ has diagonal elements of the form $\hat{h}_j = 1/(\hat{f}_j(1 - \hat{f}_j))$, and the terms $\hat{f}_j = f_j(\hat{\mathbf{b}})$ are elements of the vector $\hat{\mathbf{f}}$ of fitted proportions calculated using the equation

$$\hat{\mathbf{f}} = \mathbf{f}(\hat{\mathbf{b}}) = \exp(\mathbf{X}\hat{\mathbf{b}})/(1 + \exp(\mathbf{X}\hat{\mathbf{b}})). \tag{8.10}$$

The diagonal elements of the covariance-matrix estimates (8.8) and (8.9) are needed to obtain the design-based standard errors of the fitted functions and of the fitted proportions.

Goodness of Fit and Related Tests

Examining goodness of fit of the model is an essential part of a logit and linear modelling procedure on domain proportions. Various goodness-of-fit statistics can be obtained by first partitioning the total variation *(total chi-square)* in the table into the variation due to the model *(model chi-square)* and into the residual variation *(residual chi-square)*. Hence we have

total chi-square = model chi-square + residual chi-square

similarly to the partition of the total sum of squares for usual linear regression and ANOVA. A design-based Wald test statistic X^2_{des} measuring the residual variation is commonly used as an indicator of goodness of fit of the model. This statistic is given by

$$X^2_{des} = (F(\hat{\mathbf{p}}) - \mathbf{X}\hat{\mathbf{b}})'(\mathbf{H}\hat{\mathbf{V}}_{des}\mathbf{H})^{-1}(F(\hat{\mathbf{p}}) - \mathbf{X}\hat{\mathbf{b}}), \tag{8.11}$$

which is asymptotically chi-squared with $u - s$ degrees of freedom under the DES option. A small value of this statistic, relative to the residual degrees of freedom, indicates good fit of the model, and obviously, the fit is perfect for a saturated model. A Wald statistic denoted by $X^2_{des}(overall)$, measuring the variation due to the overall model, is used to test the hypothesis that all the model coefficients are zero. It is given by

$$X^2_{des}(overall) = F(\hat{\mathbf{p}})'(\mathbf{H}\hat{\mathbf{V}}_{des}\mathbf{H})^{-1}F(\hat{\mathbf{p}}) - X^2_{des}, \qquad (8.12)$$

where the first quadratic form measures the total variation and the second is the residual chi-square (8.11) for the model under consideration. This statistic is asymptotically chi-squared with s degrees of freedom. Also, a Wald statistic denoted by $X^2_{des}(gof)$ can be constructed for the hypothesis that all the model parameters, except the intercept, are zero. This statistic is defined as the difference of the observed values of the residual chi-square statistic (8.11) for the model where only the intercept is included and for the model including all the terms of the current model, and therefore, it is asymptotically chi-squared with $s - 1$ degrees of freedom. The statistic $X^2_{des}(overall)$ is sometimes called a test for the overall model, and $X^2_{des}(gof)$ a test of goodness of fit. They are implemented in the SUDAAN modelling procedures, for example, in CATAN for WLS analysis. Note that all these test statistics apply for both logit and linear models on domain proportions.

Linear hypotheses H_0: $\mathbf{Cb} = \mathbf{0}$ on the model coefficient vector $\mathbf{b}$ can be tested using the Wald statistic

$$X^2_{des}(\mathbf{b}) = (\mathbf{C}\hat{\mathbf{b}})'(\mathbf{C}\hat{\mathbf{V}}_{des}(\hat{\mathbf{b}})\mathbf{C}')^{-1}(\mathbf{C}\hat{\mathbf{b}}), \qquad (8.13)$$

where $\mathbf{C}$ is the desired $c \times s$ $(c \leq s)$ matrix of contrasts. The statistic is asymptotically chi-squared with c degrees of freedom under the DES option. This statistic is used, for example, in the testing of hypotheses H_0: $b_k = 0$ on single model parameters using the Wald statistics

$$X^2_{des}(b_k) = \hat{b}^2_k/\hat{v}_{des}(\hat{b}_k), \quad k = 1, \ldots, s,$$

which are asymptotically chi-squared with one degree of freedom. Note that for the corresponding *t-test statistic* the equation $t^2_{des}(\hat{b}_k) = X^2_{des}(\hat{b}_k)$ holds.

Another asymptotically valid testing procedure for linear hypotheses on model parameters is based on a second-order Rao–Scott adjustment to a binomial-based Wald test statistic using the Satterthwaite method. This technique is similar to that used in Chapter 7 on the Pearson and Neyman test statistics. We first calculate the WLS estimate $\hat{\mathbf{b}} = \hat{\mathbf{b}}_{bin}$ by using in (8.5) the binomial covariance-matrix estimate $\hat{\mathbf{V}}_{bin}$ of $\hat{\mathbf{p}}$ in place of $\hat{\mathbf{V}}_{des}$, and construct the corresponding Wald test statistic $X^2_{bin}(\mathbf{b})$:

$$X^2_{bin}(\mathbf{b}) = (\mathbf{C}\hat{\mathbf{b}})'(\mathbf{C}\hat{\mathbf{V}}_{bin}(\hat{\mathbf{b}})\mathbf{C}')^{-1}(\mathbf{C}\hat{\mathbf{b}}),$$

where $\hat{\mathbf{V}}_{bin}(\hat{\mathbf{b}})$ is the covariance-matrix estimate of the binomial WLS estimates obtained by using in (8.6) the estimate $\hat{\mathbf{V}}_{bin}$ in place of $\hat{\mathbf{V}}_{des}$. The second-order corrected Wald statistic is given by

$$X^2_{bin}(\mathbf{b}; \hat{\delta}_{\cdot}, \hat{a}^2) = \frac{X^2_{bin}(\mathbf{b})}{\hat{\delta}_{\cdot}(1 + \hat{a}^2)}, \tag{8.14}$$

where the first-order and second-order adjustment factors $\hat{\delta}_{\cdot}$ and $(1 + \hat{a}^2)$ are calculated from the $c \times c$ generalized design-effects matrix estimate

$$\hat{\mathbf{D}} = (\mathbf{C}\hat{\mathbf{V}}_{bin}(\hat{\mathbf{b}})\mathbf{C}')^{-1}(\mathbf{C}\hat{\mathbf{V}}_{des}(\hat{\mathbf{b}})\mathbf{C}') \tag{8.15}$$

so that

$$\hat{\delta}_{\cdot} = \mathrm{tr}(\hat{\mathbf{D}})/c$$

is the mean of the eigenvalues $\hat{\delta}_k$ of the generalized design-effects matrix estimate, and

$$(1 + \hat{a}^2) = \sum_{k=1}^{c} \hat{\delta}_k^2/(c\hat{\delta}_{\cdot}^2),$$

where the sum of squared eigenvalues is calculated by the formula

$$\sum_{k=1}^{c} \hat{\delta}_k^2 = \mathrm{tr}(\hat{\mathbf{D}}^2).$$

The second-order adjusted statistic $X^2_{bin}(\mathbf{b}; \hat{\delta}_{\cdot}, \hat{a}^2)$ is asymptotically chi-squared under the DES option with Satterthwaite adjusted degrees of freedom $\mathrm{df}_S = c/(1 + \hat{a}^2)$. If $c = 1$, as in tests on separate parameters of a model, we have $(1 + \hat{a}^2) = 1$ because the generalized design-effects matrix reduces to a scalar and the adjustment reduces to a first-order adjustment. The test statistic (8.13) and the corresponding second-order adjustment (8.14) are available in the modelling procedures of the SUDAAN software, in which, in addition to linear hypotheses on model parameters, Satterthwaite adjustments to the goodness-of-fit statistics $X^2_{des}(overall)$ and $X^2_{des}(gof)$ are also included.

Unstable Situations

Because the Wald statistics X^2_{des}, $X^2_{des}(overall)$ and $X^2_{des}(gof)$ of goodness of fit, and the statistic $X^2_{des}(\mathbf{b})$ of linear hypotheses on model parameters, are asymptotically chi-squared under the DES option, they can be expected to work reasonably well if the number m of sample clusters is large relative to the number u of domains. But the test statistics can become overly liberal relative to the nominal significance levels if the covariance-matrix estimate $\hat{\mathbf{V}}_{des}$ appears unstable (see Section 6.2). This can happen if the degrees of freedom $f = m - H$ are small for an estimate $\hat{\mathbf{V}}_{des}$, relative to the residual or model degrees of freedom.

There are certain F-corrected Wald test statistics available to protect against the effects of instability similar to those used in Chapter 7 for hypotheses of homogeneity and independence. For the goodness-of-fit test statistic (8.11), these degrees-of-freedom corrections are

$$F_{1.des} = \frac{f - (u - s) + 1}{f(u - s)} X^2_{des},\tag{8.16}$$

referred to the F-distribution with $(u - s)$ and $(f - (u - s) + 1)$ degrees of freedom, and

$$F_{2.des} = X^2_{des}/(u - s),\tag{8.17}$$

referred in turn to the F-distribution with $(u - s)$ and f degrees of freedom. These F-corrections can also be derived for the Wald statistics $X^2_{des}(overall)$ and $X^2_{des}(gof)$, by using the corresponding degrees of freedom s or $(s - 1)$ in place of $(u - s)$.

Similar F-corrections can be derived for the Wald test statistics of linear hypotheses on model parameters. For the statistic (8.13) these are:

$$F_{1.des}(\mathbf{b}) = \frac{f - c + 1}{fc} X^2_{des}(\mathbf{b})\tag{8.18}$$

and

$$F_{2.des}(\mathbf{b}) = X^2_{des}(\mathbf{b})/c,\tag{8.19}$$

referred to the F-distributions with c and $(f - c + 1)$, and c and f degrees of freedom, respectively.

Second-order Rao–Scott adjustments can be expected to be robust to instability problems. However, for the statistic (8.14), an F-correction can be derived. It is given by

$$F_{bin}(\mathbf{b}; \hat{\delta}_{\cdot}) = (1 + \hat{a}^2) X^2_{bin}(\mathbf{b}; \hat{\delta}_{\cdot}, \hat{a}^2)/c = X^2_{bin}(\mathbf{b})/(c\hat{\delta}_{\cdot}),\tag{8.20}$$

which is referred to the F-distribution with df_S and f degrees of freedom.

The impact of these F-corrections on p-values of the tests is small if f is large. However, if f is relatively small, and especially if f and the residual degrees of freedom are close, the corrections can be effective. Under serious instability, the statistics $F_{1.des,}$ and $F_{1.des}(\mathbf{b})$ or $F_{bin}(\mathbf{b}; \hat{\delta}.)$, are preferable. These corrections have been implemented as testing options in software products for logit and linear WLS modelling in complex surveys, such as the SUDAAN procedure CATAN. Alternatively, one can use certain more advanced corrections, based on dimensionality reduction of the original design-based covariance-matrix estimate $\hat{\mathbf{V}}_{des}$, as proposed by Singh (1985), or on modified Wald statistics which use various smoothing techniques on an estimate $\hat{\mathbf{V}}_{des}$ introduced in Lehtonen (1990).

Residual Analysis

It is desirable to examine more closely the fit of the selected model by calculating the raw and standardized residuals. These can be used in detecting possible outlying domain proportions. The raw residuals are simple differences $(\hat{p}_j - \hat{f}_j)$ of the fitted proportions $\hat{f}_j$ from the corresponding observed proportions $\hat{p}_j$. Under the DES option, the standardized residuals are calculated by first obtaining a covariance-matrix estimate $\hat{\mathbf{V}}_{res}$ of the raw residuals given by

$$\hat{\mathbf{V}}_{res} = \mathbf{H}^{-1}(\mathbf{H}\hat{\mathbf{V}}_{des}\mathbf{H} - \hat{\mathbf{V}}_{des}(\hat{\mathbf{F}}))\mathbf{H}^{-1}, \qquad (8.21)$$

where $\mathbf{H}\hat{\mathbf{V}}_{des}\mathbf{H}$ and $\hat{\mathbf{V}}_{des}(\hat{\mathbf{F}})$ are the design-based covariance matrix estimates of the vector $F(\hat{\mathbf{p}})$ of the observed functions and the vector $\hat{\mathbf{F}} = \mathbf{X}\hat{\mathbf{b}}$ of the fitted functions, respectively, and the matrix $\mathbf{H}$ depends on which model type, logit or linear, is fitted. Using (8.21), the standardized residuals are calculated as

$$\hat{e}_j = (\hat{p}_j - \hat{f}_j)/\sqrt{\hat{v}_j}, \quad j = 1, ..., u, \qquad (8.22)$$

where $\hat{v}_j$ are the diagonal elements of the residual covariance matrix $\hat{\mathbf{V}}_{res}$. A large standardized residual indicates that the corresponding domain is poorly accounted for by the model. Because the standardized residuals are approximate standard normal variates, they can be referred to critical values from the $N(0,1)$ distribution.

WLS Method for the SRS Option

A principal property of the WLS method is its flexibility, not only for various model formulations, but also for alternative sampling designs. The design-based WLS method appeared valid under the DES option involving a complex multi-stage design with clustering and stratification. But the WLS method can be used also for simpler designs with the choice of an appropriate proportion estimator and its covariance-matrix estimator reflecting the complexities of the sampling design.

Under the SRS option, the consistent proportion estimate $\hat{\mathbf{p}}$ and its binomial covariance-matrix estimate $\hat{\mathbf{V}}_{bin}(\hat{\mathbf{p}})$ are used in equations (8.5)–(8.10) to obtain the corresponding WLS estimate $\hat{\mathbf{b}}$ of model coefficients and the covariance-matrix estimates $\hat{\mathbf{V}}_{bin}(\hat{\mathbf{b}})$, $\hat{\mathbf{V}}_{bin}(\hat{\mathbf{F}})$ and $\hat{\mathbf{V}}_{bin}(\hat{\mathbf{f}})$. By using these estimates, observed values of the Wald test statistics (8.11)–(8.13) can be calculated. The same holds for the IID option, where the unweighted counterparts $\hat{\mathbf{p}}^{U}$ and $\hat{\mathbf{V}}_{bin}(\hat{\mathbf{p}}^{U})$ are used. Recall that the binomial covariance-matrix estimators were derived in Section 6.3. The WLS estimating equations indicate that the estimates $\hat{b}_k$ obtained under the SRS or IID option would not numerically coincide with those from the DES option.

The SRS and IID options are restrictive in the sense that the effect of clustering on standard-error estimates of estimated model coefficients and on observed values of the corresponding Wald test statistics cannot be accounted for. This effect is indicated in design-effect estimates of model coefficient estimates. For the SRS option, the design-effect estimates are calculated by using the diagonal elements of the covariance-matrix estimates $\hat{\mathbf{V}}_{des}(\hat{\mathbf{b}})$ and $\hat{\mathbf{V}}_{bin}(\hat{\mathbf{b}})$ of the model coefficients. Hence we have:

$$\hat{d}(\hat{b}_k) = \hat{v}_{des}(\hat{b}_k)/\hat{v}_{bin}(\hat{b}_k), \quad k = 1, ..., s. \qquad (8.23)$$

These design-effect estimates are provided when using, for example, the SUDAAN procedure CATAN.

WLS Method for the EFF Option

As noted earlier, design-effect estimates $\hat{d}(\hat{p}_j)$ of proportion estimators $\hat{p}_j$ larger than one indicate that positive intra-cluster correlation is present, involving larger variance estimates of the proportion estimators than those expected under simple random sampling. And possibly nonzero covariances of separate domain proportions can be expected when working with mixed or cross-classes-type domains. If these covariances can be assumed zero, which is the case with segregated domains, only extra-binomial variation needs to be accounted for. This can be accomplished by using the EFF option based on effective domain sample sizes.

Under the EFF option, the WLS method uses binomial-type covariance-matrix estimators, where variances of the domain proportions are estimated using rescaled binomial variance estimators. In these variance estimators, effective sample sizes $\bar{n}_j$, based on the constant deff adjustment, or $\tilde{n}_j$, based on the domain-specific deff adjustment, are used instead of the original domain sample sizes $\hat{n}_j$, leading to the corresponding covariance-matrix estimators

$\hat{\mathbf{V}}_{1.eff}$ and $\hat{\mathbf{V}}_{2.eff}$ introduced in Section 6.3. Recall that the diagonal elements of $\hat{\mathbf{V}}_{2.eff}$ are those of the design-based estimate $\hat{\mathbf{V}}_{des}$, and $\hat{\mathbf{V}}_{1.eff} = \hat{d}.\hat{\mathbf{V}}_{bin}$, where $\hat{d}.$ is the mean of the design-effect estimates $\hat{d}(\hat{p}_j)$.

As with the SRS option, conducting a WLS analysis under the EFF option requires that in addition to $\hat{\mathbf{p}}$, a covariance-matrix estimate $\hat{\mathbf{V}}_{1.eff}$ or $\hat{\mathbf{V}}_{2.eff}$ is used in all the equations (8.5)–(8.13) in place of the design-based estimate $\hat{\mathbf{V}}_{des}$. Because the estimate $\hat{\mathbf{V}}_{1.eff}$ can be taken as a smoothed version of $\hat{\mathbf{V}}_{2.eff}$, using $\hat{\mathbf{V}}_{1.eff}$ provides a preferable WLS analysis under the EFF option for the cases where variance estimates of domain proportion estimators are expected to be unstable. The EFF option with $\hat{\mathbf{V}}_{1.eff}$ also provides numerically equal WLS estimates $\hat{b}_k$ as the SRS option. In practice, logit and linear modelling under the EFF option can be executed with standard programs for WLS analysis, such as the SAS procedure CATMOD, by first deriving appropriately rescaled relative element weights. In this, only the domain design-effect estimates $\hat{d}(\hat{p}_j)$ are required.

The EFF option provides valid WLS logit and linear modelling if separate domain proportions can be assumed uncorrelated as is the case with segregated classes. In practice, the option is found to work reasonably well also when working with mixed classes or cross-classes if the design-based covariance-matrix estimate $\hat{\mathbf{V}}_{des}$ appears nearly diagonal. Because this option provides first-order adjustments, overly conservative test results can be attained if $\hat{\mathbf{V}}_{des}$ is apparently nondiagonal and covariances are positive. Therefore, although technically simple and executable with standard software, the EFF option should be used cautiously as an alternative to the full design-based option.

Criteria for Choosing a Model Formulation

Which one of the model formulations for proportions, logit or linear, should be chosen? In certain sciences, one type is more standard than the other, but taking an explicit position in favour of either of the types generally is not possible. It appears that there are gains with the logit formulation, such as possibilities for interpretation with odds ratios, and in certain cases with standard independence concepts. Moreover, being a member of the broad category of so-called exponential family models, a logit model for binomial proportions involves convenient statistical properties which are not shared with linear models for binomial proportions. Although these properties do not necessarily apply to logit models in complex surveys, attention has also been directed to the use of logit models for this kind of surveys.

The linear model formulation on proportions, on the other hand, provides a simple modelling approach that is especially convenient for those familiar with linear ANOVA on continuous measurements. Being additive on a linear scale, the coefficients of a linear model describe differences of the proportions themselves, not their logits. In practice, however, logit and linear WLS estimation results on model coefficients do not markedly differ if proportions are in the range 0.2–0.8, say. In the following example, we compare the logit and linear model formulations in a typical health sciences analysis.

Example 8.1

Logit and linear ANOVA with the WLS method. Let us apply the WLS method for logit and linear modelling on domain proportions in the simple OHC Survey setting displayed in Table 8.1. Our aim is to model the variation of domain proportions of the binary response variable PSYCH, measuring overall psychic strain, across the $u = 8$ domains formed by sex and age of respondent, and the variable PHYS describing the respondent's physical working conditions. Table 8.2 provides a more complete description of the analysis situation. The original and effective domain sample sizes $\hat{n}_j$, $\tilde{n}_j$ and $\bar{n}_j$, and the number m_j of sample clusters covered by each domain, are included in addition to the domain proportions $\hat{p}_j$ and design effects $\hat{d}_j$. Note that the domain proportions vary around the value 0.5.

The full design-based DES option provides valid WLS logit and linear modelling in this analysis. The sampling design involves clustering effects, as indicated by design-effect estimates of proportions being on average greater than one. The average design-effect estimate is 1.28. Further, the domains constitute cross-classes, which is indicated by the fact that each domain covers a reasonably large number of sample clusters. More apparently, this property can be seen from the design-based covariance-matrix estimate $\hat{\mathbf{V}}_{des}$ of domain proportions displayed in Figure 8.1. It can be noted that there exist nonzero covariance terms in the off-diagonal part of the covariance-matrix estimate. The estimate also seems relatively stable, because covariance estimates are

Table 8.2 Proportion of persons in the upper psychic strain group, with design-effect estimates of the proportions, and orginal and effective domain sample sizes and the number of sample clusters (the OHC Survey).

Domain	SEX	AGE	PHYS	$\hat{p}_j$	$\hat{d}_j$	$\hat{n}_j$	$\tilde{n}_j$	$\bar{n}_j$	m_j
1	Males	−44	0	0.419	1.16	1734	1491	1350	230
2			1	0.472	1.33	1578	1188	1228	198
3		45−	0	0.461	0.87	690	790	537	186
4			1	0.520	1.18	483	409	376	138
5	Females	−44	0	0.541	1.23	1966	1599	1530	240
6			1	0.620	1.38	447	324	348	152
7		45−	0	0.532	1.65	740	448	576	185
8			1	0.700	1.47	203	138	158	101
Total sample				0.500	1.69	7841	6387	6104	250

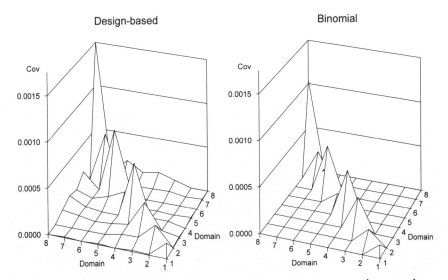

Figure 8.1 Design-based and binomial covariance-matrix estimates $\hat{\mathbf{V}}_{des}$ and $\hat{\mathbf{V}}_{bin}$ of domain proportion estimates $\hat{p}_j$.

much smaller than the corresponding variance estimates. The condition number of $\hat{\mathbf{V}}_{des}$ is 12.1 which also indicates stability. The corresponding binomial covariance-matrix estimate $\hat{\mathbf{V}}_{bin}$ is displayed for comparison.

We consider the model-building process under the DES option, and use the other options as a reference. There are three predictors, and together with their main effects, an intercept, and four interaction terms, a total of eight model terms appear in the saturated logit and linear ANOVA models, which can be written in the form:

```
F(P) = INTERCEPT + SEX + AGE + PHYS + SEX*AGE +
       SEX*PHYS + AGE*PHYS + SEX*AGE*PHYS
```

where the function is $F(P)=\log(P/(1-P))$ for the logit model and $F(P)=P$ for the linear model, and P stands for proportions of the upper PSYCH group.

In the model-building process, we first fit the saturated logit and linear models and test the significance of the interaction term of all the three predictors. If it appears nonsignificant, we remove the term, and study the two-variable interactions, each in turn, for further reduction of the model. Model building is completed when a reasonably well-fitting reduced model is attained. This stepwise process is an example of the so-called *backward elimination* common in fitting of log-linear and logit ANOVA models. Models were fitted using the SUDAAN procedure CATAN, which covers both logit and linear ANOVA modelling on proportions with the WLS method. For this procedure, we recode PSYCH and PHYS so that the original values of zero are recoded to the value 2 to obtain estimates for the desired groups.

Let us consider more closely the results on logit model fitting. Under the DES option, the main effects model appeared reasonably well-fitting and could not be further reduced. Results for the model reduction are in Table 8.3. There, the values of X^2_{des} for a difference Wald statistic are obtained, for example, in the comparison of the saturated model 5 and the model 4. The difference statistic is calculated as $X^2_{des}(overall; 5) - X^2_{des}(overall; 4) = 78.84 - 76.90 = 1.94$, and compared to the chi-squared distribution with one degree of freedom attains a nonsignificant p-value 0.1635, and thus, the interaction term can be removed from the model 5. The observed value of the Wald statistic of goodness of fit of the main effects model (Model 1) is $X^2_{des} = 78.84 - 72.39 = 6.45$, which with 4 degrees of freedom attains a p-value 0.1681, indicating reasonably good fit.

Substantial reduction of the saturated logit model was possible, and the model-building procedure produced a quite simple structure including the main effects terms only. So, the suspected interaction of SEX and PHYS appeared nonsignificant. We turn to this conclusion later when fitting logit models under the other analysis options.

In the partial parametrization of CATAN, for each predictor the model coefficient for the last class is set to zero. The last class of the last domain is the reference domain, here the domain 7 in Table 8.2 (i.e. recoded PHYS = 2). There are four coefficients b_k to be estimated in the main effects models. WLS estimates $\hat{b}_k$ are actually obtained under the following model matrix:

$$\mathbf{X} = \begin{bmatrix} 1 & 1 & 1 & 1 \\ 1 & 1 & 1 & 0 \\ 1 & 1 & 0 & 1 \\ 1 & 1 & 0 & 0 \\ 1 & 0 & 1 & 1 \\ 1 & 0 & 1 & 0 \\ 1 & 0 & 0 & 1 \\ 1 & 0 & 0 & 0 \end{bmatrix}.$$

The fitted models can be written with $\hat{b}_k$ and the model matrix as

$$F(\hat{f}_j) = \hat{b}_1 + \hat{b}_2(\text{SEX})_j + \hat{b}_3(\text{AGE})_j + \hat{b}_4(\text{PHYS})_j, \quad j = 1, \ldots, 8,$$

where $F(\hat{f}_j) = \log(\hat{f}_j/(1 - \hat{f}_j))$ for the logit model, and $F(\hat{f}_j) = \hat{f}_j$ for the linear model, and the indicator variable values for SEX, AGE and PHYS are in the second, third and fourth columns of the model matrix $\mathbf{X}$.

Let us consider more closely the estimation and test results for the main

Table 8.3   Observed values of the Wald statistics $X^2_{des}(overall)$ for overall models, and the difference statistics X^2_{des} when compared with reduced logit ANOVA models, under the DES option.

Model	df	Overall X^2_{des}	p-value	Model comparison	df	Difference X^2_{des}	p-value
5	8	78.84	0.0000	–	1	–	–
4	7	76.90	0.0000	5–4	1	1.94	0.1635
3	6	76.09	0.0000	4–3	1	0.81	0.3693
2	5	74.78	0.0000	3–2	1	1.31	0.2533
1	4	72.39	0.0000	2–1	1	2.39	0.1218

Model 5: SEX + AGE + PHYS + SEX*AGE + SEX*PHYS + AGE*PHYS + SEX*AGE*PHYS
Model 4: SEX + AGE + PHYS + SEX*AGE + SEX*PHYS + AGE*PHYS
Model 3: SEX + AGE + PHYS + SEX*PHYS + AGE*PHYS
Model 2: SEX + AGE + PHYS + SEX*PHYS
Model 1: SEX + AGE + PHYS

effects logit model. The estimation results for the model coefficients are displayed in Part A of the output page below. In Part B, the test results for the model adequacy and for the model terms are displayed. The relevant statements for fitting the main effects logit model are also provided.

In Part A of the output page, a negative value of the estimated coefficients $\hat{b}_2$ and $\hat{b}_3$ for males and for the younger group is obtained as expected, and the corresponding t-tests attain significant p-values. The sex-age adjusted estimate $\hat{b}_4$ for the PHYS class of more hazardous work is positive, involving a clearly significant t-test. It should be noticed that the absolute value of the t-test statistic used here corresponds to the square root of the F-corrected Wald statistic (8.19). The design-effect estimates $\hat{d}(\hat{b}_k)$ of the estimated model coefficients are larger than one due to the clustering effect. Thus, binomial standard-error estimates of the model coefficients would be smaller than the corresponding design-based estimates.

Using the estimate $\hat{b}_4 = 0.2568$ for the interesting parameter of the PHYS class of more hazardous work, the corresponding age–sex adjusted odds ratio estimate with its 95% confidence interval can be obtained by (8.7). The odds ratio estimate is $\exp(\hat{b}_4) = 1.29$, and its 95% confidence interval is

$$\exp(0.2568 \pm 1.96 \times 0.0574) = (1.16, 1.45).$$

The age–sex adjusted odds of experiencing a higher level of psychic strain is thus 1.3 times higher for persons under more hazardous working conditions than for those in the group of less hazardous work. This result is consistent with the t-test results, because the confidence interval does not include the value one, which is the odds ratio for the reference group.

Input statements for SUDAAN procedure CATAN on WLS logit ANOVA:

```
1 PROC CATAN  DATA=<dataset> DESIGN=WR;
2 NEST        STRATUM CLUSTER;
3 WEIGHT      _ONE_;
4 SUBGROUP    PSYCH SEX AGE PHYS;
5 LEVELS      2 2 2 2;
6 MODEL       PSYCH=SEX AGE PHYS / LOGIT;
7 TEST        WALDCHI WALDF ADJWALDF SATADJCHI SATADJF;
```

Output for the fitted main effects logit model (Model 1):

```
Number of observations read        : 7841   Weighted count: 7841
Observations used in the analysis  : 7841   Weighted count: 7841
Observations with missing values   :    0   Weighted count:    0
Denominator degrees of freedom     :  245
```

A. Estimates of model coefficients, design effects and standard errors, and t-test results:

by: Independent Variables and Effects.

Independent Variables and Effects	Beta Coeff	DEFF Beta	SE Beta	T-Test B=0	PVal for Test B=0
Intercept	0.2766	1.43	0.0635	4.36	0.0000
Sex					
Males	-0.4663	1.32	0.0579	-8.06	0.0000
Females	0.0000	.	0.0000	.	.
Age					
-44	-0.1385	1.17	0.0570	-2.43	0.0159
45-	0.0000	.	0.0000	.	.
Physical health hazards					
1	0.2568	1.24	0.0574	4.48	0.0000
0	0.0000	.	0.0000	.	.

B. ANOVA table of test results on model adequacy and model terms.

by: Contrast.

Contrast	Degrees of Freedom	S_waite Adj DF	Wald ChiSq	PVal Wald ChiSq	Wald F	PVal Wald F
OVERALL MODEL	4	3.84	72.39	0.0000	18.10	0.0000
GOODNESS OF FIT	3	2.92	70.61	0.0000	23.54	0.0000
INTERCEPT	.	.	.	.	.	.
SEX	1	1.00	64.92	0.0000	64.92	0.0000
AGE	1	1.00	5.90	0.0151	5.90	0.0159
PHYS	1	1.00	20.04	0.0000	20.04	0.0000

Contrast	Adj Wald F	PVal Adj Wald F	S_waite Adj ChiSq	PVal S_waite Adj ChiSq	S_waite Adj F	PVal S_waite Adj F
OVERALL MODEL	17.88	0.0000	71.74	0.0000	18.70	0.0000
GOODNESS OF FIT	23.35	0.0000	77.60	0.0000	26.56	0.0000
INTERCEPT	.	.	.	.	.	.
SEX	64.92	0.0000	64.92	0.0000	64.92	0.0000
AGE	5.90	0.0159	5.90	0.153	5.90	0.0159
PHYS	20.04	0.0000	20.04	0.0000	20.04	0.0000

We next turn to the ANOVA table where the test results are displayed. There is a set of observed values from different Wald test statistics and their *F*-corrections. Let us consider more closely the tests for the model terms. The test statistic Wald ChiSq corresponds to the original design-based Wald statistic (8.13), the statistic Wald F is the *F*-corrected statistic (8.19), and the statistic Adj Wald F is the *F*-corrected statistic (8.18). The statistic S_waite Adj ChiSq is the Satterthwaite corrected binomial statistic (8.14), and finally, the statistic S_waite Adj F is the *F*-corrected statistic (8.20). The design-based Wald statistic $X^2_{des}(\mathbf{b})$ and the second-order corrected binomial statistic $X^2_{bin}(\mathbf{b}; \hat{\delta}_., \hat{a}^2)$ provide similar results. The design-based Wald statistic thus works adequately in this case, which is primarily due to the stability of the covariance-matrix estimate $\hat{\mathbf{V}}_{des}(\hat{\mathbf{b}})$. Because there is a large number of degrees of freedom f for an estimate $\hat{\mathbf{V}}_{des}(\hat{\mathbf{b}})$, the *F*-corrected tests do not contribute substantially to the *p*-values of the original tests.

Although there is no controversy about the results from the alternative test statistics in this analysis situation, there can be situations where the choice of an adequate statistic is crucial. This is especially so if the number *m* of sample clusters is small and the number of domains *u* is close to *m*. Then, some of the *F*-corrected statistics can be chosen to protect against the effects of instability.

For a more detailed examination of the model fit, let us now calculate the fitted proportions and the raw and standardized residuals for a residual analysis. These are displayed in Table 8.4.

The observed and fitted proportions are close, except in the last three domains where the largest raw residuals can be obtained. The standardized residuals in the last two groups exceed the 5% critical value 1.96 from the $N(0,1)$ distribution; so the model fit is somewhat questionable for these domains. It should be noticed that the fitted proportions and the residuals are independent of the parametrization of the model.

It would be useful to consider briefly the logit analysis under the other analysis options as a reference to the results from the full design-based DES

Table 8.4 Observed and fitted proportions $\hat{p}_j$ and $\hat{f}_j$ with their standard errors, and raw and standardized residuals $(\hat{p}_j - \hat{f}_j)$ and $\hat{e}_j$ for the logit ANOVA Model 1 under the DES option.

Domain	SEX	AGE	PHYS	$\hat{p}_j$	s.e.($\hat{p}_j$)	$\hat{f}_j$	s.e. ($\hat{f}_j$)	$(\hat{p}_j - \hat{f}_j)$	$\hat{e}_j$
1	Males	−44	0	0.419	0.0128	0.419	0.0114	0.0000	0.000
2			1	0.472	0.0145	0.482	0.0122	−0.0100	−1.270
3		45−	0	0.461	0.0178	0.453	0.0142	0.0082	0.771
4			1	0.520	0.0247	0.517	0.0167	0.0029	0.160
5	Females	−44	0	0.541	0.0125	0.534	0.0115	0.0062	1.306
6			1	0.620	0.0270	0.597	0.0160	0.0222	2.012
7		45−	0	0.532	0.0236	0.569	0.0156	−0.0363	−2.073
8			1	0.700	0.0391	0.630	0.0199	0.0692	1.993

option. In this, we are especially interested in the importance of the term SEX*PHYS, describing the interaction of SEX and PHYS, which appeared nonsignificant under the DES option. The results from the Wald tests are in Table 8.5.

The interaction of SEX and PHYS appears significant when ignoring the clustering effect by using the SRS option. A more complex model is thus obtained than under the DES option. Under the EFF option, the results are intermediate but both tests indicate nonsignificance of the interaction term. These results suggest further warnings on ignoring the clustering effect even if it is not very serious as indicated in the medium-sized domain design-effect estimates. Also, the EFF method provides reasonable test results, which are in this case somewhat liberal compared to the DES option.

Let us turn to the corresponding design-based analysis with a linear model for the proportions of Table 8.2. In this situation, logit and linear formulations of an ANOVA model lead to similar results because proportions do not deviate much from the value 0.5. The main effects model (Model 1) is chosen, and results on model fit, residuals, and on significance of the model terms, are close to those for the logit model. But the estimates of the model coefficients differ and are subject to different interpretations. For the logit model with the partial parametrization, an estimated coefficient indicates differential effect on a logit scale of the corresponding class from the estimated intercept being the fitted logit for the reference domain. And for the linear model, an estimated coefficient indicates differential effect on a linear scale of the corresponding class from the estimated intercept which is now the fitted proportion for the reference domain.

The linear model formulation thus involves a more straightforward interpretation of the estimates of the model coefficients. Under Model 1 these estimates are:

$$\hat{b}_1 = 0.5705 \quad \text{(Intercept)}$$
$$\hat{b}_2 = -0.1172 \quad \text{(Differential effect of SEX = Males)}$$
$$\hat{b}_3 = -0.0355 \quad \text{(Differential effect of AGE = -44)}$$
$$\hat{b}_4 = 0.0650 \quad \text{(Differential effect of PHYS = 1)}$$

The fitted proportion for falling into the upper psychic strain group is thus 0.57 for females in the older age group whose working conditions are less

Table 8.5 Wald tests $X^2(\mathbf{b})$ for the significance of the interaction term SEX * PHYS in Model 2 under the analysis options DES, EFF and SRS.

Term	df	DES X^2_{des}	p-value	EFF $X^2_{2.eff}$	p-value	EFF $X^2_{1.eff}$	p-value	SRS X^2_{bin}	p-value
SEX*PHYS	1	2.39	0.1218	2.68	0.1016	3.09	0.0788	3.97	0.0463

hazardous, and for males in the same age group, $0.57 - 0.12 = 0.45$. The highest fitted proportion, $0.57 + 0.07 = 0.64$, is for the older age-group females under more hazardous work. Also, the fitted proportions are close to those obtained with the corresponding logit ANOVA model.

Under the EFF, SRS and IID analysis options, the models can be fitted with standard software for logit and linear WLS analysis, such as the SAS procedure CATMOD. The default parametrization in CATMOD, however, differs from that of the SUDAAN procedure CATAN. In order to obtain similar parametrization, CATAN can be used by requesting an analysis assuming simple random sampling. Generally, an appropriate weight variable is required for the SRS and EFF options. In the OHC Survey data set, however, the relative element weights are equal to one. Under the EFF option, the rescaled element weights can thus be calculated using the domain design-effect estimates as follows:

$w_{jk} = 1/\hat{d}_j, j = 1, \ldots, u, k = 1, \ldots, n_j$, for the domain-specific deff adjustment,

$w_l = 1/\hat{d}_., l = 1, \ldots, 7841$, for the constant deff adjustment.

The weights sum up to 6387 and 6104, respectively, as can be seen from Table 8.2. Effective sample sizes are thus noticeably smaller than the original sample size, 7841. In other words, with an about 20% smaller simple random sample, equal precision for the proportion and other estimates would have been attained as it was with the original sample based on the clustered design. This was the price to be paid to fulfil the scientific goals of the OHC Survey.

8.5 LOGISTIC AND LINEAR REGRESSION

The PML method of pseudolikelihood is often used on complex survey data for logit analysis in similar analysis situations to the WLS method. But the applicability of the PML method is wider, covering not only models on domain proportions of a binary or polytomous response but also the usual regression-type settings with continuous measurements as the predictors. We consider in this section first a PML analysis on domain proportions, and then, a more general situation of logit modelling of a binary response with a mixture of continuous measurements and categorical variables as predictors. Finally, an example is given on linear modelling for a continuous response variable in an ANCOVA setting.

In PML estimation of model coefficients and their asymptotic covariance matrix, we use a modification of the maximum likelihood (ML) method. In the ML estimation for simple random samples, we work with unweighted observations and appropriate likelihood equations can be constructed, based on standard distributional assumptions, to obtain the maximum likelihood estimates of the model coefficients and the corresponding covariance-matrix estimate. Using these estimates, standard likelihood ratio and binomial-based

Wald test statistics can be used for testing the model adequacy and linear hypotheses on the model coefficients.

Under more complex designs involving element weighting and clustering, a maximum likelihood estimator of the model coefficients and the corresponding covariance-matrix estimator are not consistent and, moreover, the standard test statistics are not asymptotically chi-squared with appropriate degrees of freedom. For consistent estimation of model coefficients, the standard likelihood equations are modified to cover the case of weighted observations. In addition to this, a consistent covariance-matrix estimator of the PML estimators is constructed such that the clustering effects are properly accounted for. Using these consistent estimators, appropriate asymptotically chi-squared test statics are derived.

The PML method can be conveniently introduced in a similar setting to the WLS method, assuming again a binary response variable and a set of categorical predictors. The data set is arranged in a multidimensional table, such as Table 8.1, with u domains, and our aim is to model the variation of the domain proportion estimates $\hat{p}_j$ across the domains. The variation is modelled by a logit model of the type given in (8.1) and (8.2). A PML logit analysis for domain proportions, covering logit ANOVA, ANCOVA and regression models with categorical predictors, can be carried out under any of the analysis options previously introduced, by using the corresponding domain proportion estimator vector and its covariance-matrix estimate, and the steps in model-building are equivalent to those in the WLS method. The design-based DES option provides a generally valid PML logit analysis for complex surveys, and the EFF option can be used to account for extra-binomial variation. In practice, a PML logit analysis under the DES option requires access to specialized software, such as the SUDAAN procedure LOGISTIC and the logistic regression option of PC CARP.

Design-based and Binomial PML Methods

Under both DES and SRS options, a consistent *PML estimator* $\hat{\mathbf{b}}_{pml}$ for the vector $\mathbf{b}$ of the s model coefficients b_k in a logit model $F(\mathbf{p}) = \mathbf{Xb}$ is obtained by iteratively solving the PML estimating equations

$$\mathbf{X}'\mathbf{Wf}(\hat{\mathbf{b}}_{pml}) = \mathbf{X}'\mathbf{W}\hat{\mathbf{p}}, \qquad (8.24)$$

where $\mathbf{W}$ is a $u \times u$ diagonal weight matrix with weights $w_j = \hat{n}_j$ on the main diagonal, and $\mathbf{f} = \exp(\mathbf{Xb})/(1 + \exp(\mathbf{Xb}))$ is the inverse function of the logit function. It is essential in (8.24) that the weighted domain sample sizes $\hat{n}_j$ and the weighted proportion estimates $\hat{p}_j$ be used, not their unweighted counterparts n_j and $\hat{p}_j^U$ as in the ML method i.e. under the IID option. This is for consistency of the PML estimators. The corresponding vector (8.5) of the WLS estimates can be used as an initial value for the PML iterations. Note that under the linear formulation of the ANOVA model, the function vector $\mathbf{f}(\hat{\mathbf{b}}_{pml})$ would be linear in $\hat{b}_k$ and, thus, no iterations are needed and the PML estimation reduces to WLS estimation. Henceforth in this section we denote the vector of PML estimates of logit model coefficients by $\hat{\mathbf{b}}$ for short.

Because the vector $\hat{\mathbf{b}}$ of PML estimates is equal under the DES and SRS options, so also are the vectors $\hat{\mathbf{F}} = \mathbf{X}\hat{\mathbf{b}}$ and $\hat{\mathbf{f}} = F^{-1}(\mathbf{X}\hat{\mathbf{b}})$ of fitted logits and fitted proportions. The equality also holds for estimated odds ratios, which can be obtained as $\exp(\hat{b}_k)$ under the partial parametrization of the model. Fitted proportions $\hat{f}_j = f_j(\hat{\mathbf{b}})$ are estimated under both options by the formula

$$\hat{\mathbf{f}} = \mathbf{f}(\hat{\mathbf{b}}) = \exp(\mathbf{X}\hat{\mathbf{b}})/(1 + \exp(\mathbf{X}\hat{\mathbf{b}})). \tag{8.25}$$

Let us derive under the SRS and DES options the $s \times s$ covariance-matrix estimators of the PML estimator vector $\hat{\mathbf{b}}$ calculated by (8.24). Assuming simple random sampling, the covariance-matrix estimator is given by

$$\hat{\mathbf{V}}_{bin}(\hat{\mathbf{b}}) = (\mathbf{X}'\mathbf{W}\hat{\Delta}\mathbf{W}\mathbf{X})^{-1}, \tag{8.26}$$

where the diagonal elements of the diagonal $u \times u$ matrix $\hat{\Delta}$ are binomial-type variances $\hat{f}_j(1 - \hat{f}_j)/\hat{n}_j$. The binomial covariance-matrix estimator (8.26) is not consistent for complex sampling designs involving clustering. For these designs, we derive a more complicated consistent covariance-matrix estimator which is valid under the DES option:

$$\hat{\mathbf{V}}_{des}(\hat{\mathbf{b}}) = \hat{\mathbf{V}}_{bin}(\hat{\mathbf{b}})\mathbf{X}'\mathbf{W}\hat{\mathbf{V}}_{des}\mathbf{W}\mathbf{X}\hat{\mathbf{V}}_{bin}(\hat{\mathbf{b}}). \tag{8.27}$$

This estimator is of a 'sandwich' form such that the design-based covariance-matrix estimator $\hat{\mathbf{V}}_{des}$ of the proportion vector $\hat{\mathbf{p}}$ acts as the 'filling'.

Approximate confidence intervals for odds ratio estimates $\exp(b_k)$ under the DES and SRS options can be calculated by (8.7) using the corresponding variance estimates $\hat{v}_{des}(\hat{b}_k)$ and $\hat{v}_{bin}(\hat{b}_k)$ of the PML estimates $\hat{b}_k$, as in the WLS method. Also, the design-effect estimates $\hat{d}(\hat{b}_k)$ of the model coefficients $\hat{b}_k$ can be obtained by (8.23), again analogously to the WLS method.

Expressions for the consistent covariance-matrix estimators $\hat{\mathbf{V}}_{des}(\hat{\mathbf{F}})$ and $\hat{\mathbf{V}}_{des}(\hat{\mathbf{f}})$ of the vector $\hat{\mathbf{F}}$ of fitted logits and the vector $\hat{\mathbf{f}}$ of fitted proportions are similar under the DES option to those of the WLS method, as given in equations (8.8) and (8.9). The PML analogue $\hat{\mathbf{V}}_{des}(\hat{\mathbf{b}})$ from (8.27) and the corresponding matrix $\hat{\mathbf{H}}$ must of course be used in the equations. And under the SRS option, the covariance-matrix estimators $\hat{\mathbf{V}}_{bin}(\hat{\mathbf{F}})$ and $\hat{\mathbf{V}}_{bin}(\hat{\mathbf{f}})$ are derived similarly by using the binomial estimator (8.26) in the equations in place of its design-based counterpart.

A residual covariance-matrix estimator is needed for conducting a proper residual analysis under the DES option. This $u \times u$ estimator is given by

$$\hat{\mathbf{V}}_{res} = \mathbf{A}\hat{\mathbf{V}}_{des}\mathbf{A}', \qquad (8.28)$$

where the matrix $\mathbf{A}$ is obtained by the formula

$$\mathbf{A} = \mathbf{I} - \hat{\Delta}\mathbf{WX}(\mathbf{X}'\mathbf{W}\hat{\Delta}\mathbf{WX})^{-1}\mathbf{X}'\mathbf{W}$$

with $\mathbf{I}$ being a $u \times u$ identity matrix. Using this estimate, design-based standardized residuals of the form (8.22) can then be calculated.

There are thus many similarities between the PML formulae and those derived for the WLS method. The main differences lie in the way the estimates of model coefficients and their covariance-matrix estimate are calculated. More similarities are evident in the testing procedures. All the test statistics derived for the WLS method are also applicable to the PML method.

Under the DES option, goodness of fit of the model can be tested with the design-based Wald statistic X_{des}^2 given by (8.11). When examining the model fit more closely, PML analogues to the Wald statistics $X_{des}^2(overall)$ and $X_{des}^2(gof)$ can be used. The Wald statistics (8.13) and (8.14) for linear hypotheses on model parameters are applicable as well. Finally, in unstable situations, the F-corrected Wald and Rao–Scott statistics (8.16)–(8.20) can be used. It should be noted that the PML estimates from (8.24) and the corresponding covariance-matrix estimate (8.27) must be used in the calculation of these test statistics under the DES option. All these test statistics are available, for example, in the SUDAAN procedure LOGISTIC for logit analysis for complex survey data.

In testing procedures for the SRS option, the corresponding binomial covariance-matrix estimates are used in the test statistics in place of those from the DES option. As an alternative to the Wald statistics, likelihood ratio (LR) test statistics can be used, which for the DES option should be adjusted using the Rao–Scott methodology. A second-order adjustment to LR test statistics similar to (8.14) for the binomial-based Wald statistic provides asymptotically chi-squared test statistics. The residual covariance-matrix estimate (8.28) can be used in deriving an appropriate generalized design-effects matrix estimate for the adjustments.

PML Method for Other Analysis Options

The main application area of the PML method for complex surveys is under the DES option, and the SRS or IID options are used as the reference when examining the effects of weighting and intra-cluster correlation on standard-error estimates of model coefficients and on p-values of Wald test statistics. The EFF option, aimed at adjusting for extra-binomial variation, can also be used in fitting logit models on domain proportions in special situations similar to those in the WLS method. The EFF option is executable with standard programs for ML logit analysis, by supplying appropriately rescaled relative element weights derived using the effective domain sample sizes.

Logistic Regression

The PML method can also be used in strictly regression-type logit analyses on a binary response variable from a complex survey, where the predictors are continuous measurements. Recall that the WLS method is not applicable in this case. In logistic regression we work with an element-level data set without aggregating these data into a multidimensional table. So, the measured values of the continuous predictor variables constitute the columns in an $n \times s$ model matrix X for a logistic regression model. But all the other elements of the PML estimation remain unchanged, and consistent PML estimates with their consistent covariance-matrix estimate are obtained in a similar way to that described for the DES option. Moreover, a logistic ANCOVA can be performed, by incorporating categorical predictors into the logistic regression model. Then, interaction terms of the continuous and categorical predictors can also be included.

A logistic regression model is usually built by entering predictors into the model by using subject-matter criteria or significance measures of potential predictors. In this, t-tests $t_{des}(b_k)$, or the corresponding Wald tests $X^2_{des}(b_k)$, on model coefficients can be used as previously and, under the DES option, asymptotic properties of these test statistics remain unchanged.

Instability of an estimate $\hat{V}_{des}(\hat{b})$ from (8.27) can destroy the distributional properties of the test statistics on model coefficients in such small-sample situations where the number of sample clusters is small. Usual degrees-of-freedom F-corrections to the Wald and t-test statistics can then be used. There are also special methods developed for unstable situations in logistic regression. In the method proposed by Morel (1989), a stabilizing adjustment is performed to the estimate $\hat{V}_{des}(\hat{b})$. This method is available in the logistic regression option of the PC CARP program.

A special version of the GEE methodology of generalized estimating equations can also be used for logistic modelling on complex survey data. In this method, the model coefficients are estimated using the quasilikelihood technique, and intra-cluster correlations, which are taken as nuisances, are parametrized by pair-wise odds ratios estimated independently of the model coefficients. Using an estimated intra-cluster correlation structure, a 'robust' estimator of the covariance matrix of the model coefficients can be obtained, basically similar to the 'sandwich' form in the PML method. The GEE estimators possess certain optimality properties not necessarily shared by the PML estimators.

Various assumptions can be stated on a 'working' intra-cluster correlation structure, leading to GEE models with different estimates of model coefficients and their covariance matrix. In the simplest case, assuming an independent correlation structure, the estimated model coefficients are identical to the PML estimates, and the GEE covariance matrix estimate is close to that from the PML method. Assuming more complex structures, however, these estimates differ from those obtained by the PML method.

Example 8.2

Logistic ANCOVA with the PML method. Let us consider in a slightly more general setting the analysis situation of Example 8.1, where a logit ANOVA model was fitted by the WLS method to proportions in a multidimensional table. We now fit a logistic ANCOVA model using the PML method, by entering some of the predictors as continuous measurements in the model. Note that the WLS method is inappropriate in this analysis situation. The DES option is applied, providing valid PML analysis, and the SRS option is used for examining the effects of clustering on analysis results.

The binary response variable PSYCH measures high psychic strain, and we take the variables AGE, PHYS (physical working conditions) and CHRON (chronic morbidity) as continuous predictors such that AGE is measured in years and PHYS and CHRON are binary. Thus there are four predictors, of which SEX is taken as a qualitative predictor. So, the interaction of SEX with AGE, PHYS and CHRON can also be examined.

A model with SEX, AGE, PHYS and CHRON as the main effects and an interaction term of SEX and AGE was taken as the final model, because the other interactions appeared nonsignificant at the 5% level. Results on the model coefficients are displayed below as a piece of output from the SUDAAN procedure LOGISTIC (the procedure RLOGIST in the SAS-callable version):

Independent Variables and Effects	Beta Coeff.	DEF Beta	SE Beta	T-Test B=0	PVal for Test B=0
Intercept	0.1964	1.56	0.1572	1.25	0.2127
Sex					
Males	−0.9926	1.43	0.2033	−4.88	0.0000
Females	0.0000	.	0.0000	.	.
Age	−0.0046	1.55	0.0041	−1.12	0.2624
Physical health					
hazards	0.2765	1.39	0.0596	4.64	0.0000
Chronic morbidity	0.5641	1.17	0.0575	9.82	0.0000
Sex, Age					
Males	0.0131	1.41	0.0051	2.56	0.0111
Females	0.0000	.	0.0000	.	.

The fitted logit ANCOVA model can be written using the estimated coefficients $\hat{b}_k$ and the corresponding model matrix $\mathbf{X}$ similarly to the ANOVA modelling in Example 8.1:

$$F(\hat{f}_l) = \hat{b}_1 + \hat{b}_2(\text{SEX})_l + \hat{b}_3(\text{AGE})_l + \hat{b}_4(\text{PHYS})_l$$

$$+ \hat{b}_5(\text{CHRON})_l + \hat{b}_6(\text{SEX}*\text{AGE})_l,$$

where $l = 1, \ldots, 7841$, and $F(\hat{f}_l) = \log(\hat{f}_l/(1 - \hat{f}_l))$. The values for the model terms are obtained from the corresponding columns of the 7841×6 model matrix $\mathbf{X}$. There, SEX, PHYS and CHRON are binary, and AGE has its original values (age in years). Note the difference in the ANCOVA model matrix when compared with that for the ANOVA model.

The t-tests on model coefficients indicate that the coefficients for the interesting predictors, physical working conditions and chronic morbidity, are strongly associated with experiencing psychic strain. Persons in hazardous work, and chronically ill persons, more likely suffer from psychic strain than healthy persons and persons whose working conditions are less hazardous. Note that the sex–age adjusted coefficient $\hat{b}_5$ for CHRON is larger than $\hat{b}_4$ for PHYS. Thus, in the model, chronic morbidity is more important as a predictor of psychic strain. This can also be seen in the odds ratio estimates.

By using the model coefficients $\hat{b}_4$ and $\hat{b}_5$ we calculate the odds ratios, which is easily done since both predictors are binary. Odds ratios with their approximative 95% confidence intervals (in parenthesis) are:

PHYS : Odds ratio $= \exp(0.2765) = 1.32$ $(1.17, 1.48)$,

CHRON : Odds ratio $= \exp(0.5641) = 1.76$ $(1.57, 1.97)$.

We may thus conclude that odds for experiencing a higher level of psychic strain, adjusted for sex, age and chronic morbidity, is about 1.3 times higher for those in more hazardous work than for those in less hazardous work. This conclusion was similar in Example 8.1, where a closely related odds ratio and confidence interval were obtained. Furthermore, the odds of experiencing much psychic strain, adjusted for sex, age and working conditions, are about 1.8 times higher for chronically ill persons than for healthier persons. Because neither of the 95% confidence intervals covers the value one, the corresponding odds ratios differ significantly (at the 5% level) from one. It should be noted that the binomial-based confidence intervals would be narrower especially for the predictor PHYS, for which the design-effect estimate is larger than for CHRON.

An analysis under the SRS option yields the same final model as the DES analysis, but the observed values of the test statistics are somewhat larger and thus, more liberal test results are attained. For example, the p-value of the

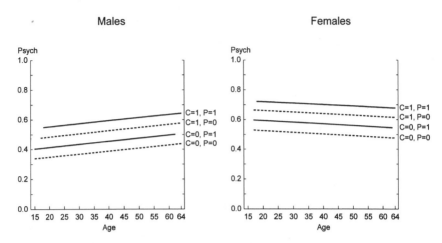

Figure 8.2 Fitted proportions of falling into the high psychic strain group for the final logistic ANCOVA model (C: CHRON; P: PHYS).

interaction term of SEX and AGE is 0.0041 under the SRS option and 0.0111 under the DES option.

Finally let us examine more closely the fitted proportions $\hat{f}_l$ for the upper psychic strain group under the present model. Note that these proportions are equal under the DES and SRS options. The results are summarized in Figure 8.2 by plotting the proportions against the predictors included in the model. Fitted proportions increase with increasing age for males, and decrease for females. At a given age, the proportions are larger for chronically ill and for those in more hazardous work than in the reference groups. Also, in females the fitted proportions tend to be larger than in males in all the corresponding domains, although the differences decline with increasing age.

Linear Modelling on Continuous Responses

We have extensively considered the modelling of binary response variables from complex surveys. The WLS and PML methods were used, covering logit and linear modelling on categorical data and logit modelling with continuous predictors. These types of multivariate models are most frequently found in analytical surveys, for example, in social and health sciences. But in some instances it is appropriate to model a quantitative or continuous response variable, such as the number of physician visits or blood pressure. We discuss briefly the special features of multivariate analysis in such cases, and give an illustrative example of a special case of linear ANCOVA.

Usual linear modelling provides a convenient analysis methodology for analysis situations with a continuous response variable and a set of predictors.

This situation was present in Table 8.1, where domain means of a continuously measured response variable PSYCH (the first standardized principal component of a set of psychic symptoms) were displayed. There, linear ANOVA could be used for the analysis of variation in these domain means. Also in Example 8.2, where the dichotomized PSYCH was analysed with a logistic ANCOVA model, the original continuous variable could be taken as the response as well, leading to linear ANCOVA modelling. For a simple random sample, the analysis would be based on ordinary least squares (OLS) estimation with a standard program such as the SAS procedure GLM. For the OHC Survey data set, which is based on cluster sampling, the design-based approach under the DES option provides proper linear modelling.

Under the DES option, similar complexities to those of the previous modelling techniques enter also into linear modelling. In the estimation technique and testing procedures, however, no novel elements are involved compared to those already introduced for modelling with the WLS and PML methods. So, we first aim at consistent estimation of the model coefficients and consistent estimation of the covariance matrix of the estimated coefficients. These require weighting with relative element weights, and the construction of a covariance-matrix estimator of the model coefficient estimates properly accounting for the clustering effects. Also, desired tests of model adequacy and of linear hypotheses on model coefficients can be executed using test statistics similar to the Wald and F-statistics used in the WLS and PML methods for logit and linear modelling on proportions.

Linear modelling under the DES option can in practice be carried out most conveniently with appropriate software, such as the SUDAAN procedure REGRESS and the regression option of the PC CARP program. An OSIRIS program, REPERR, can also be used for this purpose. In the next example we use the SUDAAN software for a linear ANCOVA model on psychic strain, which is taken as a continuous response variable.

Example 8.3

Linear ANCOVA modelling on perceived psychic strain. In Example 8.2, a logistic ANCOVA model was fitted on the dichotomized variable PSYCH of psychic strain. A linear ANCOVA model is now fitted on the original variable PSYCH, whose values are scores of the first standardized principal component of nine psychic symptoms. Thus, the average of PSYCH is zero and the variance is one. The distribution of PSYCH is, however, somewhat skewed; there are numerous persons in the data set not experiencing any of the psychic symptoms in question. The range of the values of PSYCH is $(-1, 4.7)$, and the median of the distribution is -0.4.

We include the same variables as in Example 8.2 as the potential predictors in the linear ANCOVA model. The predictor SEX is taken to be qualitative, and AGE, PHYS and CHRON are taken to be continuous, and we also study the pair-wise interactions of SEX and the continuous predictors. The model-building produces a similar ANCOVA model as in Example 8.2. Thus, all the main effects and the interaction of SEX and AGE appear significant.

The fitted linear ANCOVA model on PSYCH can be written using the estimated coefficients $\hat{b}_k$ and the corresponding model matrix $\mathbf{X}$, as in the logistic model on a binary PSYCH in Example 8.2:

$$\hat{f}_1 = \hat{b}_1 + \hat{b}_2(\text{SEX})_l + \hat{b}_3(\text{AGE}) + \hat{b}_4(\text{PHYS})_l + \hat{b}_5(\text{CHRON})_l + \hat{b}_6(\text{SEX} * \text{AGE})_l,$$

where $l = 1, \ldots, 7841$, and the values for the model terms are obtained from the model matrix $\mathbf{X}$ of Example 8.2. Results on the ANCOVA model coefficients with the continuously measured psychic strain as the response variable are displayed below as a piece of output from the SUDAAN procedure REGRESS:

Independent Variables and Effects	Beta Coeff.	DEF Beta	SE Beta	T-Test B=0	PVal for Test B=0
Intercept	−0.0121	1.70	0.0831	−0.15	0.8846
Sex					
Males	−0.4975	1.48	0.0997	−4.99	0.0000
Females	0.0000	.	0.0000	.	.
Age	−0.0001	1.60	0.0021	−1.02	0.9804
Physical health hazards	0.1772	1.37	0.0290	6.11	0.0000
Chronic morbidity	0.3922	1.17	0.0294	13.33	0.0000
Sex, Age					
Males	0.0057	1.39	0.0025	2.25	0.0252
Females	0.0000	.	0.0000	.	.

The signs of model coefficients and the t-test results follow a similar pattern to those in the corresponding logit ANCOVA model in Example 8.2. The model coefficients, however, have different interpretations from those in the logit model. In a logit ANCOVA we were working on a logit scale on the binary response, whereas we are now dealing with continuous measurements on a linear scale. Thus, the coefficients of the linear ANCOVA model can be interpreted in the usual linear regression context.

Under the SRS analysis option, the same ANCOVA model would result in, and the results on model coefficients would be equal. But the standard errors of the model coefficients would be smaller, because the design-effect estimates $\hat{d}(\hat{b}_k)$ are greater than one. However, this does not affect the inferences from the t-test results.

Table 8.7 Design-based linear ANCOVA on overall psychic strain (SUDAAN procedure REGRESS).

Model term	Beta coefficient	Design effect	Standard error	t-test	p-value
Intercept	−0.0121	1.70	0.0831	−0.15	0.8846
Sex					
Males	−0.4975	1.48	0.0997	−4.99	0.0000
Females	0.0000	.	0.0000	.	.
Age	−0.0001	1.60	0.0021	−1.02	0.9804
Physical health					
hazards	0.1772	1.37	0.0290	6.11	0.0000
Chronic					
morbidity	0.3922	1.17	0.0294	13.33	0.0000
Sex, Age					
Males	0.0057	1.39	0.0025	2.25	0.0252
Females	0.0000	.	0.0000	.	.

$\hat{d}(\hat{b}_k)$ are greater than one. However, this does not affect the inferences from the t-test results.

The continuous response variable PSYCH offered good possibilities for demonstration of linear modelling due to the continuity of the response variable, although the distribution was somewhat skewed. Count variables, such as the number of physician visits in a given time interval or related variables whose distribution can be very skewed, are often met with in practice. Modelling of such quantitative response variables can involve such symmetrizing transformations as logarithmic, often used in econometrics, or Box–Cox transformations, prior to the fitting of a linear model. Moreover, a linear model formulation can even be inappropriate for such variables. Then, other regression modelling techniques should be used: for example, Poisson regression and a negative binomial model to account for extra-Poisson variation. These methods belong to a class of generalized linear models for correlated response variables. For these models, for example, the pseudolikelihood and generalized estimating equations methods can be successfully used under the nuisance approach.

Methods for the Disaggregated Approach

Methods for multivariate analysis considered so far fall under the nuisance or aggregated approach, where the aim is to clean out the possibly disturbing clustering effects from the analysis results in order to attain consistent estimation and asymptotically valid testing. Under the disaggregated approach, on the other hand, intra-cluster correlation structures are intrinsically interesting, and the estimation of these correlations constitutes an essential part of the analysis. This often occurs in social and educational surveys when working with hierarchically structured data sets. Clustering with villages, establishments or schools constitute common examples of sources of such a hierarchical structure.

There are advanced methods available for multivariate analysis of intra-cluster correlated response variables from hierarchically structured data sets. The methodology of multi-level modelling is based on *generalized linear mixed models,* where certain random effects are incorporated in the model. These constitute a new class of models not yet considered in this book; in all the previous models the model parameters have been taken as fixed. Applications of multi-level modelling have been mainly in linear modelling of continuous response variables from educational surveys, where schools or teaching groups are used as the clusters (Goldstein 1987). For example, computer program MLn can be used for this kind of modelling. Multi-level models have also been developed for binary and polytomous responses, and appropriate computing algorithms are available by using, for example MLn. We will use multilevel logit modelling in Section 9.3 for a binary response variable from clustered educational data. There, a brief introduction to the method will be given.

Another method for multivariate analysis of correlated responses from hierarchically structured data is based on the GEE methodology of generalized estimating equations (Liang and Zeger 1986; Diggle *et al.* 1994). For example, in logit regression, the GEE method can be used for simultaneous estimation of regression coefficients and correlation parameters. Simultaneous estimation makes the method computationally more demanding than the previous GEE method for the nuisance approach where separate estimation of regression coefficients and correlation parameters is executed and the main scientific interest is in regression coefficients.

8.6   CHAPTER SUMMARY AND FURTHER READING

Summary

Linear and logit modelling of an intra-cluster correlated response variable were considered in this chapter mainly under the nuisance approach. The principal aim was to successfully remove the effects of intra-cluster correlations from the estimation and test results. Severity of these effects, however, varies under different sampling designs and therefore various analysis options were introduced for proper analysis in practice.

A full design-based DES option provides a generally valid analysis option for multivariate analysis in complex surveys. Under this option, the complexities of the sampling design can be properly accounted for, including clustering, stratification and weighting. Analysis under the DES option requires access to the element-level data set, and availability of proper software for survey analysis. A simple EFF option of effective sample sizes can be used for valid analysis in complex surveys if only extravariation such as extra-binomial variation is present. A practical value of the EFF method is in its applicability with standard SRS-based analysis programs by using appropriately rescaled weights. Also, under stratified element STR sampling and SRS sampling, the weighted SRS and IID options can be used for valid analysis. Under the SRS option, only the weighting is covered, and the IID option ignores all the sampling complexities. These options are thus inappropriate for clustered designs of complex surveys.

Under any of the analysis options, logit and linear ANOVA, ANCOVA and regression analysis on domain proportions of a binary or polytomous response variable can be carried out by the WLS method of weighted least squares estimation in a data set arranged into a multidimensional table. The WLS method, applied under the DES option, provides valid analysis for such tables from complex surveys. For reliable results, a large element sample and a large number of sample clusters are required; these conditions are usually met in large-scale analytical surveys such as the OHC Survey based on a stratified cluster-sampling design. With a small number of sample clusters, instability problems can arise, making the estimation and test results unreliable. This problem can be successfully handled using appropriate correction techniques. The SUDAAN procedure CATAN was used in an OHC Survey case study on logit and linear ANOVA modelling by the WLS method on domain proportions of a binary response variable.

The PML method of pseudolikelihood estimation can be used in similar analysis situations to the WLS method, but its main applications are in logistic regression with continuous predictors where the WLS method fails. Under the DES option, the PML method provides valid logit analysis for complex surveys. It is also beneficial for the PML method that the number of sample clusters is large, and similar corrections are available for unstable cases, as for the WLS method. We applied the PML method for logistic ANCOVA modelling in an OHC Survey case study on a binary response variable, by using the SUDAAN procedure LOGISTIC. The option for logistic regression of the PC CARP and WesVarPC programs could have been used instead.

The PML method covers not only logistic regression models but also other model types from the class of generalized linear models. So, linear models on continuous responses are also covered. We applied linear ANCOVA modelling in an OHC Survey case study with the SUDAAN procedure REGRESS. The regression option of the PC CARP and WesVarPC programs provide similar analysis facilities.

In the case studies on selected multivariate analysis situations from the OHC Survey, it appeared that accounting for sampling complexities, especially for the clustering effects, can be crucial for valid inferences. We shall demonstrate this important conclusion further in Chapter 9, where additional case studies from other complex survey data sets will be given.

The nuisance, or aggregated, approach provides a reasonable and manageable analysis strategy for different kinds of multivariate analysis situations on an intra-cluster correlated response variable. In the alternative disaggregated approach, the intra-cluster correlations are taken as intrinsically interesting parameters to be estimated as well as the model coefficients. We discussed briefly a version of the method of generalized estimating equations and multi-level modelling, applicable for hierarchically structured data sets. The method of multi-level modelling will be demonstrated in the next chapter.

Further Reading

Multivariate analysis of complex surveys has received considerable attention in the literature. Advances in the methodology can be found in Binder (1983), Rao and Scott (1984, 1987), Roberts *et al.* (1987), Rao *et al.* (1989) and Scott *et al.* (1990), covering, for example, the weighted least squares, pseudolikelihood and quasilikelihood methods for logit and related analysis of categorical data from complex surveys. The book edited by Skinner *et al.* (1989) covers many of the important advances in multivariate analysis under both the aggregated and disaggregated approaches. A review on the topic is in Binder *et al.* (1987). Hidiroglou and Paton (1987), Rao and Thomas (1988) and Lee *et al.* (1989) provide more applied sources on the methodology.

Rao *et al.* (1993) discuss regression analysis with two-stage cluster samples. Skinner *et al.* (1986) consider principal component analysis, and Skinner (1986) and Fuller (1987) address factor analysis for complex surveys. Binder (1992) addresses the fitting of proportional hazards models to complex survey data. The analysis of categorical data with nonresponse is considered in Binder (1991), and Glynn *et al.* (1993) consider multiple imputation in linear models.

Pfeffermann and LaVange (1989) consider linear regression models and Holt and Ewings (1989) propose a method for logistic regression in modelling structured populations under the disaggregated approach. Multi-level modelling, including the ML3 software and its more recent version MLn, is introduced in Goldstein (1987, 1991), Prosser *et al.* (1991), and Rasbash and Woodhouse (1995), and is further developed in Goldstein and Silver (1989), Goldstein and Rasbash (1992), and Goldstein (1995). Modelling by the generalized estimating equations is introduced in Liang and Zeger (1986), and is further developed in Liang *et al.* (1992). Breslow and Clayton (1993) give general results on approximate inference in the framework of generalized linear mixed models.

9

More Detailed Case Studies

Three additional case studies are selected to provide a more subject-matter-oriented demonstration of the methodology for survey estimation and analysis discussed in this book. The first case study (Section 9.1) is from a *business survey* and is an example of the estimation of a mean in a descriptive-type survey, where business firms are used as the units of data collection. There, different assumptions on the sampling design lead different estimates of the population mean and different standard-error estimates of the mean estimates. The other two case studies are from analytical surveys with stratified cluster sampling, and they concentrate on issues of analysis when using various modelling techniques.

In the case study from a *socioeconomic survey* (Section 9.2), a logit model is fitted to categorical data from a cluster sampling design with households as the clusters. The main emphasis is not only on pointing out the importance of accounting for the clustering effects, but also on the importance of adequate selection of the model type for the analysis. Here, analysis of variance and regression-type logit models are used which lead to different conclusions.

In the final case study (Section 9.3), we introduce and demonstrate an approach of modelling hierarchically structured data sets using multi-level logit models, applied to clustered survey data from an *educational survey*. These models differ from the methods of the nuisance approach, as used in the preceding case study, in the sense that in multi-level modelling, the hierarchical structure is emphasized as an essential phenomenon of the population to be taken into account in the model fitting. The results from multi-level modelling are compared with those from the nuisance approach.

9.1 ESTIMATION OF MEAN SALARY IN A BUSINESS SURVEY

Each year in August, Statistics Finland collects data from business firms in the commercial sector to estimate the average salary of employees in different occupations within this sector. The main concern in this case study is with appropriate estimation of average salaries and their accompanying sampling errors from this material by using methods which take the sample design into account. In this sampling design the primary sampling unit is the individual firm, which implies that data on salaries at the employee level are clustered by firms and so accordingly this design should be taken into account in the estimation. The actual sampling design is stratified one-stage cluster sampling. In the estimation of the average salaries in the business sector as a whole, as well as in certain occupational groups within this sector, three other sampling-design assumptions are also used for comparison. One of these designs is that used in the production of official statistics.

Sampling Design

The sampling frame used is the Business Register of Statistics Finland, in which the business firms in the commercial sector are divided into two subpopulations. The first comprises all the firms which are members of the Confederation of Commerce Employers (for short, CCE firms). From this subpopulation, the Confederation collects census data on salaries in different commercial occupations. The total number of employees in this subpopulation is 190 217. The average salaries calculated on the basis of the complete data set will be used as a point of reference in subsequent comparisons.

The other subpopulation comprises firms that are not members of the Confederation of Commerce Employers. From this subpopulation, Statistics Finland has selected a stratified simple random sample by using the individual firm as the primary sampling unit. The subpopulation consists of a total of 57 762 commercial employees, of whom 13 987 are included in the present sample of 744 firms. Our aim is to estimate the average salaries for different occupations from this sample.

The sampling frame for the present sample is the 1988 Business Register, from which the smallest companies (those employing 1–2 people) have been excluded. This leaves a population of 25 345 companies, which are stratified into five categories by number of employees and also into five categories by branch of business, giving a total of 25 strata. Sampling fractions vary by stratum; in some strata, all firms are included, and in others, only some of them. The order in which individual firms appear in the Business Register is then stratum-wise randomized. Next, starting from the top, the required number of units are sampled from each stratum. Insofar as the sampling takes place at the firm level, the sampling design may be described as *stratified simple random sampling without replacement*. If conclusions were to be drawn for the firm level, then the analysis would be carried out within a stratified simple random sampling design. For example, this sort of sample design is well-suited to the analysis of turnover and similar firm-level data.

However, the purpose here is to estimate the average salaries of employees in different occupations. This implies a different interpretation of the sampling design in that the individual employee who is the unit of analysis is not the primary sampling unit. The selection of a certain firm into the sample implies that all its employees are also included. Each selected firm should therefore be interpreted as a cluster, the elements of which are all the firm's employees. This sample design is described as *stratified one-stage cluster sampling*. There is only one single stage in the sampling procedure; namely, the sampling of the firms. Within each selected firm, then, data are collected on the salaries of all employees.

The specific concern here is with the regular monthly salaries of commercial occupations at the time of measurement in August 1991. These occupations are grouped according to the classification used by Statistics Finland. The average salaries of 22 occupational groups are regularly published, but some of these categories are so small that for reasons of confidentiality only the job title can be indicated. The focus here is restricted to the occupational groups which occur in at least 50 sampling units or firms. One item obviously of special interest is the average salary for the whole commercial sector, which in the present sample design comprises 744 firms or clusters with a total of 13 987 employees. When weighted by the inverse of the sampling rate, the size of the corresponding population is estimated to be $\hat{N} = 57\,762$ employees.

Weighting and Estimators of the Mean

For the present kind of sample data it is possible to construct different types of mean estimators depending on the assumptions made in the sampling design. In the following, four alternative sampling designs are presented with the corresponding mean and design-effect estimates. Appropriate variance estimators have been considered in Chapters 2, 3 and 5 and we omit them here.

Simple random sampling   The firm level is omitted and the sample at the employee level is interpreted as a simple random sample taken directly from the employee population. Thus the corresponding estimator of average salary is

$$\bar{y} = \frac{\hat{N}}{n} \sum_{k=1}^{n} y_k / \hat{N}, \tag{9.1}$$

where y_k is the salary of the kth employee in the sample and the joint sample size is $n = 13\,987$. The same weight $\hat{N}/n$ is used for all employees; this is the inverse of the approximate sampling rate. The weight is $\hat{N}/n = 57\,762/ 13\,987 = 4.13$. This coefficient could only be justified if the sampling had been carried out at the employee level and if neither stratification nor clustering had been done. In the present case, neither of these conditions hold. The variance of the mean estimator is useful in determining the estimate of the design effect, a measure which summarizes the effects of design complexities on variance estimation. As defined in Chapter 2, the design-effect estimator for the mean is a ratio of two variance estimators:

$$\text{deff}(\bar{y}^\star) = \frac{\hat{v}_{p(s)}(\bar{y}^\star)}{\hat{v}_{srs}(\bar{y})}, \tag{9.2}$$

where $\bar{y}^\star$ is an estimator of the mean under the actual sampling design $p(s)$ with a variance estimator $\hat{v}_{p(s)}(\bar{y}^\star)$ and $\hat{v}_{srs}(\bar{y})$ is the variance estimator of $\bar{y}$ under SRS. If the design effect is close to one, the actual sample design can be interpreted as an SRS design. In this case, the analysis does not require sampling-design identifiers and can be carried out using standard software packages such as SAS, SPSS or BMDP. In situations where cluster sampling is used, the design effect can be larger than one. Then, to obtain a proper analysis it is necessary to use specialized software such as SUDAAN or PC CARP with the appropriate design identifiers. Under the SRS design the design effect is by definition equal to one.

Stratified simple random sampling   Element-level sampling is assumed and each stratum is assigned its own weight. The estimator of the average salary is

$$\bar{y}_{str} = \sum_{h=1}^{H} \sum_{\beta=1}^{n_h} \frac{\hat{N}_h}{n_h} y_{h\beta} / \hat{N}. \tag{9.3}$$

The stratum-specific weights are $\hat{N}_h/n_h$, or the inverse of the sampling rate in stratum h where $\sum_{h=1}^{H} \hat{N}_h = \hat{N}$, and $\sum_{h=1}^{H} n_h = n$. It is worth noting that the weight remains constant for all employees in the same stratum even if (as indeed is the case in practice) they work at different companies.

Stratified cluster sampling with stratum-wise varying weights   The estimator for the mean is equal to that of stratified simple random sampling. However, the designs involve different estimators for the standard error, which can be used to determine confidence intervals, for instance. In stratified cluster sampling the design effect is usually larger than one (deff ≥ 1), depending on the internal homogeneity of the clusters with respect to the study variable.

Stratified cluster sampling with cluster-wise varying weights   This is a very realistic assumption in samples of business firms. The size of firms (i.e. the size of the cluster), measured in terms of the number of employees, usually varies considerably. In this case the design can be taken into account by estimating the mean using the *Horvitz–Thompson* estimator and regarding the relative size of a cluster as the sampling weight. Here, the relative size of a cluster is measured by the number of employees $n_{h\alpha}$ in a firm divided by the total number of employees N_h in the corresponding stratum. This will yield a cluster weight for a certain firm, and the inverse of this figure is accordingly the sampling weight for that particular firm. To match the sum of the weights with the total number of employees within the frame population, this figure must still be divided by the number m_h of sample firms in the stratum. Thus the mean estimator is

$$\bar{y}_{clu} = \sum_{h=1}^{H} \sum_{\alpha=1}^{m_h} \sum_{\beta=1}^{n_{h\alpha}} \frac{\hat{N}_h}{m_h \times n_{h\alpha}} y_{h\alpha\beta}/\hat{N}. \tag{9.4}$$

The estimator incorporates all the information concerning the sampling design: sampling weights which vary firm-wise, and stratification.

Results

The sample data have been analysed by the SUDAAN software so that the appropriate sample-design can be taken properly into account. Estimation under the four sampling design assumptions differ in their weighting schemes and they take the same sampling design into account to varying extents. The most realistic of these design assumptions is quite obviously stratified cluster sampling with cluster-wise varying weights, which incorporates all the information concerning the sampling design, whilst the SRS design is the simplest one. The results on these sampling designs can also be compared with the statistics on average salaries obtained by the Confederation of Commerce Employers from its census. In Table 9.1, these data are shown on the last line. The Statistics Finland sample specifies the estimated number of employees at 57 762, which means that the figure for the whole sector in August 1991 would have been $57\,762 + 190\,217 = 247\,979$ full-time employees.

Table 9.1 Average salary (FIM) of commercial sector employees in 1991 based on different sampling design assumptions and census data.

Sample design	Weighted sample size	Average salary	Standard error	deff
SRS	57 762	10 458	44	1.00
STR (stratified)	57 762	9 528	55	1.72
CLU (stratum weights)	57 762	9 528	60	2.10
CLU (cluster weights)	57 762	9 402	66	2.58
Census (CCE register)	190 217	9 098	. . .	. . .

The estimates from the SRS design gives the largest average salary at FIM 10 458. On the other hand, it also has the smallest standard error estimate of FIM 44. In other designs the average salary approximates the reference figure obtained from a census, which is FIM 9 098. Since this is the exact figure for the corresponding subpopulation, it obviously contains no standard error. The design which estimates closest to the reference figure is stratified cluster sampling with cluster-wise weights. The estimated average salary from this design is FIM 9 402. There, the primary sampling unit has been the firm, but the weighting is done at the employee level.

Comparison of the Results

Moving on to look at average salaries in selected commercial occupational groups, the following compares the figures from three sources: the Confederation of Commerce Employers register data, the Statistics Finland estimates based on the stratified one-stage simple random sampling, and finally the estimates obtained from the stratified cluster-sampling design with cluster-wise varying weights. The comparison covers the biggest occupational categories on which data have been obtained from at least 50 companies.

There are certain differences between the figures based on the census data and the sample compiled by Statistics Finland. However, since these differ-

Table 9.2 Average salaries in different occupational groups in August 1991: census of CCE member companies and the Statistics Finland sample.

| | Average salary in August 1991 | | |
| | CCE | STATFIN sample | |
Occupational group	census	CLU design	STR design
Shop managers	9 582	8 835	8 504
Service station workers	6 893	6 977	6 905
Shop assistants	. . .	. . .	. . .
Cleaners	6 840	5 417	5 386
Warehouse workers	7 106	7 112	7 082
Warehouse supervisors	. . .	. . .	. . .
Van/Lorry drivers	7 804	7 138	7 231
Forwarders	8 944	12 866	13 635
Other branches	8 407	7 656	7 748
Upper white-collar	15 131	14 432	14 395
Office management	19 212	19 659	19 774
Office supervisors	13 969	15 000	15 117
Clerical staff	8 881	10 157	10 151
Motor-transport workers	9 593	7 917	7 871
All occupational groups	9 098	9 402	9 528

ences only occur in a small number of occupational groups, it would seem useful to look more closely at the internal compatibility of occupational classifications used in different statistical sources. On average, the estimates from stratified cluster sampling with cluster-wise weights come closer to the census figures than those of Statistics Finland which are based on an assumption of stratified simple random sampling.

The use of complete design information significantly increases the standard errors of average salary estimates. One possible reason for this is that during the time lag between the compilation of the sampling frame and the sampling date, firms have moved up or down from their original size category but retained the weight of that stratum. This was evident in the design effects in the sample design employed by Statistics Finland (deff $= 1.72$). Firm-specific weights have two kinds of effects. Firstly, they lessen the above-mentioned frame-ageing problems by taking the actual size measure into account. Secondly, they introduce a clustering effect which results in positive intra-class correlation. Therefore the use of stratified sampling design with cluster-wise varying weights increases the standard errors of average salaries and, accordingly, the design effects.

Conclusions

This case study dealt with estimation under four different sampling-design assumptions on the very same data. The design assumptions varied in the degree to which the properties of the sample design were taken into account in the estimation. It was discovered that in the estimation of average salaries for different occupational groups, the sample design used by Statistics Finland should be interpreted as stratified one-stage cluster sampling in which business firms are the primary sampling clusters. Since the size of these clusters varies, the weighting should be done at firm level. The estimate of average salary produced by this sort of design came closest to the reference figure based on the CCE census data.

The relatively high design-effect estimates of the clustered designs ($2.10 \leq$ deff < 2.58) lend further support to the argument that there is a considerable clustering effect that should be to taken into account in the calculation of average salaries in business firms. Clustering effect here means that employees working in a certain occupation within the same firm (say, shop assistants) have more or less the same salary, whereas their salary is clearly different from the average pay for their occupation in other firms. This observation also supports the view that the calculation of average salaries should use weights at the cluster level. Another factor that speaks in favour of cluster-level weights is the wide range of variation in firm (cluster) size. The most natural way to do this is to apply *Horvitz-Thompson* estimators. Recent developments in business survey methodology are summarized in Cox *et al.* (1995).

9.2   PRIVATE SICKNESS INSURANCE IN SOCIOECONOMIC SURVEY

We demonstrate in this case study not only that accounting for the clustering effect is crucial, but also that the model formulation and assumptions on the predictors can be important. For this, we use the weighted least squares (WLS) and pseudolikelihood (PML) methods for logit ANOVA and ANCOVA modelling on domain proportions. Analysis results from the full design-based DES option are compared to those from SRS and IID options assuming simple random sampling. The study problem evaluates a sickness insurance scheme. The data make up a single selected regional stratum from the Finnish Health Security Survey sampling design, which involves clustering with households as clusters and weighting for nonresponse adjustment.

The Study Problem and the Data

An important aim of sickness insurance is to reduce differences between population subgroups in the utilization of health services, and to reduce the financial burden of illness on individuals and families. In Finland, a public sickness insurance scheme, covering the entire population, has been in force since 1964. In 1980s, a supplemental sickness insurance scheme, supplied by private insurance companies, was increasingly used, e.g. in reimbursing costs of physician visits due to sickness in the private health-care sector. We shall study the variation of the proportion of privately insured persons in various income groups using data from the Finnish Health Security (FHS) Survey. The survey was conducted in 1987 by the Social Insurance Institution of Finland (Kalimo *et al.* 1992).

The FHS Survey was intended to produce reliable information required for evaluation of health and social security. Regionally stratified one-stage cluster sampling was used. Both substantive matters and economy of data collection motivated the use of households as the units of data collection. Of a sample of 6998 households, a total of 5858 (84%) took part in the survey. All eligible members in the sample households formed the element-level sample, consisting of a total of 16 269 interviewed non-institutionalized persons. Unit nonresponse was concentrated in urban regions, especially large towns such as Helsinki. Because of the nonignorability of the nonresponse, poststratification was used for adjusting so that the poststrata were formed by region, sex and age groups.

Personal interviews were conducted household-wise, but the main interest was on person-level inferences. It is obvious that many characteristics concerning health, use of health services, and health behaviour, tend to be homogeneous within households. Due to this, the corresponding study variables can be positively intra-cluster correlated. Design-effect estimates of means and proportions of such variables were often greater than one but less than two. The largest design-effect estimate (1.7) was found for a binary variable INSUR describing access to private sickness insurance.

A subsample of 2071 persons and 878 households living in the Helsinki Metropolitan Area, being one of the 35 strata, is considered in this case study. The estimated proportion of private sickness insured was relatively high, about 17% in the Helsinki Metropolitan Area, where the supply of private health-care services has also been high relative to other parts of the country. In rural areas this proportion was noticeably smaller.

Examining the association of INSUR with household incomes was seen to be relevant to the evaluation of the public sickness insurance scheme. The preliminary analysis, however, does not lend support to the hypothesis that having private sickness insurance depends on high incomes. Estimated INSUR proportions in three household-income categories (low, medium, high) are 15.2%, 17.3% and 18.1%, respectively. In a homogeneity test on these proportions, an observed value $X_P^2 = 2.15$ of the Pearson test statistic was obtained, with a p-value 0.342, clearly indicating nonsignificant variation. Further, a logit regression with INSUR as the response and household income as the quantitative predictor, with integer scores from 1 to 3, has a p-value 0.148, indicating a nonsignificant linear trend.

But being privately sickness-insured depends strongly on age. Private insurance appears to be a form of sickness insurance used especially for children. In the Helsinki Metropolitan Area, 43% of children were covered whereas the proportion for adults was only 9%. Moreover the need for visiting a physician due to a chronic or acute illness tends to increase the probability of being privately insured. Of those who had visited a doctor at least once in a given time interval, 27% had access to private sickness insurance. The proportion was 14% in the other group. Possible causal relationships (if any) can of course also work vice versa. Taking age of respondent and visiting a private physician as confounding factors can thus be informative when studying more closely the relationship of a household member being privately insured with income of the household.

An ANOVA-type logit model on the domain proportions of INSUR provides the simplest modelling approach for further studying the association. For simplicity, we choose the binary variables VISITS (visiting a private physician at least once during a fixed time interval), AGE (0–17-year-old child or over-17-year-old adult), and a three-category variable INCOME (household net income per OECD consumer unit, one-third parts) as the predictors in the ANOVA model. With these predictors, a total of 12 domains are produced. Because INCOME can also be taken as a quantitative predictor, we fit a logit ANCOVA model for these proportions to further examine the possible linear trend for household incomes.

Domain proportions of INSUR are displayed in Table 9.3. The proportions

Table 9.3   Unweighted and weighted proportion estimates $\hat{p}_j^U$ and $\hat{p}_j$ (%) of privately sickness-insured (INSUR) by VISITS, AGE and INCOME in the Helsinki Metropolitan Area (the FHS Survey).

Domain	VISITS	AGE	INCOME	$\hat{p}_j^U$	n_j	$\hat{p}_j$	$\hat{d}_j$	$\hat{n}_j$	m_j
1	None	Child	Low	27.6	145	29.0	1.7	140	86
2			Medium	33.3	135	33.6	1.7	125	93
3			High	41.3	75	41.2	1.3	69	57
4		Adult	Low	6.7	400	6.5	1.5	422	258
5			Medium	8.9	427	8.6	1.5	425	245
6			High	11.6	423	11.3	1.6	422	256
7	Some	Child	Low	60.5	43	60.3	1.4	44	33
8			Medium	74.4	39	75.2	1.4	37	30
9			High	75.6	41	75.4	1.3	41	35
10		Adult	Low	12.6	103	12.9	1.3	110	92
11			Medium	12.5	88	11.4	1.0	87	83
12			High	11.2	152	10.5	1.3	149	127
Total sample				17.2	2071	16.8	1.8	2071	878

INSUR	Access to private sickness insurance (binary response)
VISITS	Visiting to a private physician at least once in a given time interval
AGE	Age (Children 0–17 years / Adults 18–years and above)
INCOME	Household net income 1986/87 per OECD consumer unit (one-third parts)

$\hat{p}_j^U = n_{j1}/n_j$, and the domain sample sums n_{j1} of INSUR and the domain sample sizes n_j, are the original unweighted quantities used under the IID option. Under the other options, the proportions $\hat{p}_j = \hat{n}_{j1}/\hat{n}_j$ are used, which are reweighted for the unit nonresponse. The proportion estimators are thus consistent ratio estimators where $\hat{n}_{j1}$ and $\hat{n}_j$ are weighted domain sample sums and weighted domain sample sizes, respectively. The design-effect estimates $\hat{d}_j$ are for the weighted proportion estimates $\hat{p}_j$. The number of sample clusters m_j, i.e. households covered by each domain, is also displayed.

With VISITS and AGE fixed, the INSUR proportions increase with increasing income, except in the last three income groups. The proportions tend to be larger on average in the second VISITS group and in the first AGE group. The largest proportions are for children with at least one doctor's visit. The design-effect estimates indicate a slight clustering effect; their average is 1.4.

Methods

A logit ANOVA model is first fitted by the WLS method to the INSUR proportions $\hat{p}_j$ and $\hat{p}_j^U$ with VISITS, AGE and INCOME as the qualitative predictors.

Then, a logit ANCOVA model is fitted by the PML method for the same table, but the predictor INCOME is taken as quantitative with scores from 1 to 3. We use the WLS and PML methods under the DES, SRS and IID analysis options. Under the IID option, all design complexities are ignored, and only the weighting is accounted for under the SRS option. Under the DES option, the extra-binomial variation and the correlations between separate proportion estimates are allowed in addition.

There are obvious reasons for supporting the DES analysis option. The response variable INSUR appears positively intra-cluster correlated in such a way that if a household member, especially a child, is insured, then the others tend to be as well. This clustering effect is indicated in the design-effect estimate 1.8 of the overall INSUR proportion, and in the domain design effects which clearly indicate extra-binomial variation.

There is another important issue concerning the intra-cluster correlations with respect to the domain structure. VISITS and AGE obviously constitute cross-classes in that they cut across the clusters, i.e. the households. INCOME constitutes segregated classes because it is a household-level predictor. These predictors together thus produce a domain structure which is of a mixed-classes type. This causes pair-wise correlations between separate domain proportions $\hat{p}_j$. Not all proportions are allowed to be correlated, but only those corresponding to the respective INCOME groups, i.e. every third domain. So, in addition to the extra-binomial variation, positive covariances can be expected between the proportion estimates in these domains, also supporting the use of the DES analysis option.

The structure of the intra-cluster correlation is reflected in the 12×12 design-based covariance-matrix estimate $\hat{\mathbf{V}}_{des}$ of domain proportions $\hat{p}_j$. This estimate is displayed in Figure 9.1, where the corresponding binomial estimate $\hat{\mathbf{V}}_{bin}$ is shown for comparison. The estimate $\hat{\mathbf{V}}_{des}$, obtained by the linearization method, appears quite stable due to the large number of degrees of freedom, $f = m - H = 877$, and the condition number of $\hat{\mathbf{V}}_{des}$ is not large (37.4). It can thus be expected that the WLS and PML methods work adequately under the DES option. Because the variance estimates on the diagonal of $\hat{\mathbf{V}}_{des}$ are larger than the corresponding binomial variance estimates, liberal test results can be expected under the SRS option, relative to those obtained under the DES option.

As was shown in Chapter 8, the vector of domain proportion estimates and its covariance-matrix estimate, depending on the analysis option considered, are required for logit modelling with the WLS and PML methods. In the WLS analysis, equations (8.5) to (8.13) in Section 8.4 were used, and in the PML analysis, equations (8.24) to (8.27). Under the DES option, the estimates $\hat{p}_j$ and $\hat{\mathbf{V}}_{des}(\hat{\mathbf{p}})$ were used. Under the SRS option, the estimate $\hat{\mathbf{V}}_{bin}(\hat{\mathbf{p}})$ was used in addition to $\hat{p}_j$. Under the IID option, the estimates $\hat{p}_j^U$ and $\hat{\mathbf{V}}_{bin}(\hat{\mathbf{p}}^U)$ were used. The DES analyses were executed with the SUDAAN procedure CATAN for the WLS method and with the LOGISTIC procedure for the PML method, both

Design-based   Binomial

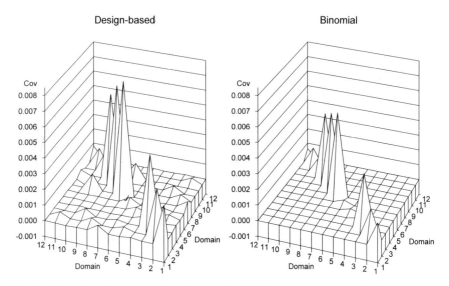

Figure 9.1   Covariance-matrix estimates for INSUR proportions $\hat{p}_j$. The design-based estimate $\hat{V}_{des}$ and the binomial estimate $\hat{V}_{bin}$ (the FHS Survey).

with the SUDAAN design option WR for the with-replacement assumption. The SAS procedure CATMOD was used under the other analysis options.

Results

Let us first consider the test results for the logit ANOVA model. We wish to study the dependence of being privately insured on incomes of the household with adjustment of the confounding effects of visiting a doctor and age of respondent. In addition to the corresponding main effects, possible interactions should be examined as well. Thus, the relevant saturated logit model is of the

Table 9.4   Wald-test results of goodness of fit of the logit ANOVA model, and of significance of the INCOME effect and the INCOME contrast 'low versus high', under the IID, SRS and DES analysis options (the FHS Survey).

Option	Model fit			Significance of INCOME effect			Significance of contrast low vs. high
	X^2	df	p-value	$X^2(\mathbf{b})$	df	p-value	p-value
IID	3.61	6	0.7290	9.31	2	0.0095	0.0023
SRS	4.52	6	0.6063	7.95	2	0.0188	0.0048
DES	4.23	6	0.6450	4.35	2	0.1138	0.0372

form $\log(P/(1-P)) = V + A + I + V*A + V*I + A*I + V*A*I$ where V refers to VISITS, A refers to AGE, I refers to INCOME, and P stands for the domain proportions of being privately insured. Note that in this expression, all the predictors are taken to be qualitative.

An ANOVA model with all the main effects, and an interaction of VISITS and AGE, appeared to fit reasonably well and could not be further reduced. Results on goodness of fit of the model are displayed in the leftmost part of Table 9.4, including the observed values of the Wald statistics X_{iid}^2, X_{bin}^2 and X_{des}^2. There is no need for F-corrections for unstability because of the large number of sample clusters. The reduced ANOVA model fits well according to the test results, under any of the analysis options.

The main interest in the analysis is the importance of the INCOME effect in the ANOVA model as a predictor of being privately insured. The Wald-test results under the selected analysis options, using the statistic $X^2(\mathbf{b})$, are given in the middle part of the table. The test results indicate that under the IID and SRS options, the INCOME effect clearly remains significant. The most liberal test, significant at the 1% level, is under the IID option. Under the SRS option, the test is significant at the 5% level. In both of these tests the clustering effect is ignored. But the INCOME effect turns out to be nonsignificant as soon as the extra-binomial variation and the correlations of the domain proportions are accounted for using the DES option. Then, the INCOME effect becomes nonsignificant even at the 10% level.

For more detailed inferences, we separately test the hypothesis that the model parameters for the low and high INCOME groups were equal. The test results for the corresponding contrast 'low versus high' are given in the rightmost part of Table 9.4. All the tests indicate significant difference at least at the 5% level, and the pattern of the p-value follow that of the previous tests. We next calculate the corresponding adjusted odds ratios and their 95% confidence intervals using the estimated model coefficients and their standard errors. Under the IID option, the odds ratio with its confidence interval is 0.61 (0.45, 0.84) for the first INCOME group, and 0.81 (0.59, 1.10) for the second group. Under the DES option it is 0.64 (0.42, 0.97) for the first INCOME group, and 0.82 (0.54, 1.24) for the second. Under both IID and DES options, the adjusted odds ratios for the first INCOME group differ significantly (at the 5% level) from one, which is the odds ratio for the highest INCOME group.

The results from the logit ANOVA model give some support to the conclusion that the access to private sickness insurance might not be equally likely in the two extreme income groups, although the overall effect of household incomes appeared nonsignificant when the clustering effect was accounted for. It is thus reasonable to model the variation further so that the possible linear trend in the proportions in the INCOME groups, adjusted for the confounding factors, can be tested more explicitly. This is carried out by a logit ANCOVA model, where INCOME is taken as a quantitative predictor so that

Table 9.5 Estimation and test results on the regression coefficient b_4 for INCOME in a logit ANCOVA model fitted by the PML method under the IID and DES options (the FHS Survey).

Option	$\hat{b}_4$	$\hat{d}(\hat{b}_4)$	s.e. $(\hat{b}_4)$	t-test	p-value
IID	0.246	1.00	0.081	3.02	0.0026
DES	0.229	1.77	0.109	2.10	0.0357

integer scores from 1 to 3 are assigned to the classes. Hence we increase the use of the information inherent in the variable INCOME.

A logit ANCOVA model is fitted by the PML method. A model with identical model terms as in the previous ANOVA model appears reasonable for further examination. Let us consider more closely the test results on the regression coefficient b_4 for INCOME in this model. The results obtained under the DES and IID options are given in Table 9.5. In fact, the IID results are based on the ML method, because the weighting is ignored. In the table, the t-test results under both the IID option and the DES option indicate significant deviation from zero (at least at the 5% level) for the regression coefficient of INCOME. Here also, the test under the IID option is liberal relative to the DES test. The test result under the SRS option would be intermediate. Note also that the estimates $\hat{b}_4$ somewhat differ; under the SRS option, an equal estimate to the DES counterpart would have been obtained.

Summary

We studied whether the access to private sickness insurance depends on household incomes when the confounding effects of visiting a private physician and age of respondent are adjusted for. For the analysis, the data were arranged in a multidimensional table of domain proportions. The proportions indicated slight clustering effects. Logit ANOVA modelling provided the simplest approach to studying the variation of the proportions. The effect of household incomes appeared significant when the clustering effects were ignored, but lost its significance when these effects were accounted for. In the test of a contrast, and in the odds ratio estimates, some evidence, however, was present on differences between the extreme income groups with respect to the coverage of private sickness insurance, thus supporting the need for further modelling. A logit ANCOVA model, where a linear trend on household incomes was more explicitly tested, provided results giving more evidence of having access to private insurance depending on high incomes. This result indicates that a private insurance scheme, as a supplement to a public insurance scheme, can involve inequality with respect to the access to, and use of, health-care services.

In the preceding analysis, the variable describing access to a private sickness insurance scheme was the binary response. This was used mainly for illustrative purposes; the intra-cluster correlation of that variable was relatively strong. It would also be reasonable to take the variable describing use of health services as the response with the insurance variable as one of the predictors. Then, a different view of the problem would be possible.

Methodological Conclusion

Positive intra-cluster correlation of a response variable can severely distort the test results in a multivariate analysis even if the correlations were relatively weak, as in the case demonstrated. In both logit ANOVA and ANCOVA modelling, ignoring the clustering effects resulted in overly liberal tests relative to those where the clustering effects were properly accounted for. This was because the standard errors of model coefficients were underestimated by ignoring the clustering effects. Hence the results indicate a warning against relying on results from standard analyses when working with data from a clustered design. For the nuisance approach, which appeared to be relevant in the analysis considered, the design-based methods provide a safe and easily manageable approach for modelling intra-cluster correlated responses. There also, the results should be carefully compared with alternative model formulations in order to reach valid inferences on the subject matter.

9.3   MATHEMATICS ACHIEVEMENTS IN EDUCATIONAL SURVEY

Logit regression on categorical data with a binary response and two predictors is used on a study problem concerning pupils' cognitive achievements in mathematics. Based on data from a clustered design with schools as clusters, the analysis situation is complex in two respects. The response variable is strongly intra-cluster correlated, and this is a property that should be taken into account in the analysis. Moreover, there are a small number of sample clusters available in the data set, resulting in instability of variance and covariance estimation and, thus, also of the test statistics. We demonstrate two approaches to taking into account these complexities. First, a logit model is fitted by the pseudolikelihood (PML) method with certain adjustments for the instability. This modelling takes place under the nuisance or aggregated approach which is the main strategy used in this book. Secondly, by emphasizing that the hierarchical structure of the population is intrinsically interesting, we apply the disaggregated approach by fitting a multi-level logit model. The results are also compared to those from a standard IID analysis where all design complexities are ignored.

The Study Problem and the Data

In the sampling design for an educational survey it is natural to utilize the existing administrative and functional structures of the school system. There, the schools can be taken as basic units, which are grouped by areas of school administration, such as school districts, or according to other administrative criteria, such as municipalities or provinces. On the other hand, the teaching is organized in teaching groups, which often coincide with school classes composed of the pupils and the teacher. In educational surveys, a school class is often taken as the unit of data collection, because of economy and other practical reasons. Further, school classes are readily stratified by grades, by school districts and by other administrative groupings. There is thus a natural hierarchy in the population, which is a property that is utilized both in the sampling design and in the modelling procedures for this case study.

The data are from the Second National Assessment of the Comprehensive School (the SNACS Survey), carried out in 1990 by the Institute for Educational Research at the University of Jyväskylä. These data consist of a sample of 6th graders, whose achievements in mathematics were measured. The population of clusters consisted of 4126 schools, from which a sample of 53 schools was drawn. This produced a sample of 1071 pupils from the population of 60 934 pupils. The pupil-level sample size was thus not fixed in advance.

More specifically, the sampling design used is an application of one-stage stratified cluster sampling from a national register of schools. The register includes auxiliary information on the geographical location of the schools, and on the number of teaching groups, pupils and teachers at each grade of the schools. Regional stratification of the population of the schools was used, and the stratum-wise samples were proportionally allocated. The unit in the register is thus a school, which is also the sampling unit in sampling of the school classes. The schools were sampled in each stratum with selection probabilities proportional to size, measured as the number of pupils. If there was only one target teaching group in a sampled school, it was taken, and if there were more, one of them was randomly selected. The composition of the population and the sample is displayed in Table 9.6.

We restrict the analysis to two strata, namely the second and third. Moreover, three teaching groups were excluded due to missing data, and thus, there are 40 sample clusters and 743 pupils in the available data set to be analysed.

The data consist of measurements at three levels: (1) the school level, (2) the teacher level, and (3) the pupil level. For simplicity, only the second- and third-level variables are used here. From these variables, a binary response

Table 9.6 The sample and population composition in the SNACS Survey.

Area		Pupils		Classes	
		Sample	Population	Sample	Population
1	Metropolitan	112	8 311	4	315
2	Southern	526	26 236	27	1573
3	Central	312	16 696	16	1405
4	Northern	121	9 691	6	833
Total		1071	60 934	53	4126

variable and two predictor variables are selected. The response variable (ACHIEV) measures whether or not a pupil has attained the desired level of knowledge in mathematics, as evaluated by the teacher. The variation in this response variable is studied by two originally quantitative variables, the one measuring in years the teaching experience of a teacher (TYRSEX), and the other measuring in minutes the time spent by a pupil on homework in mathematics per evening (STIME). Further, the predictors are categorized into three-class variables for the analysis, as is presented in Table 9.7. Note that the first predictor is a teaching-group-level variable, and the second is a pupil-level variable. Thus, the nine domains formed by these predictors constitute mixed classes. In the table, weighted proportions of the pupils who have reached the desired level in mathematics, and both unweighted and weighted sample

Table 9.7 Weighted domain proportion estimates $\hat{p}_j$ of achievements in mathematics (ACHIEV) with their estimated standard errors and design effects, and unweighted and weighted sample sizes, and the number of sample clusters, by TYRSEX and STIME (the SNACS Survey).

Domain	TYRSEX	STIME	$\hat{p}_j$	s.e.$(\hat{p}_j)$	$\hat{d}_j$	n_j	$\hat{n}_j$	m_j
1	1–10	0–14	0.877	0.032	0.5	57	56	8
2		15–30	0.789	0.058	2.5	121	120	10
3		31–	0.731	0.122	1.4	19	19	6
4	11–20	0–14	0.932	0.030	0.8	59	60	12
5		15–30	0.852	0.036	1.6	156	156	13
6		31–	0.592	0.105	1.4	32	32	9
7	21–	0–14	0.905	0.044	1.4	61	62	14
8		15–30	0.901	0.019	0.7	180	180	17
9		31–	0.673	0.082	1.8	58	58	15
Total sample			0.838	0.021	2.4	743	743	40

ACHIEV	Achievements in mathematics (binary response)
TYRSEX	Teaching experience of a teacher (in years)
STIME	Time spent by a pupil for homework per evening (in minutes)

sizes, are given, the weighted sample sizes being those obtained using appropriate weighting of the observations to compensate for varying element
inclusion probabilities. Design-effect estimates of proportion estimators are
also displayed in the table as well as the number of sample clusters (teaching
groups) covered by each domain.

There is a slight variation in the domain proportions varying in the range
$0.592 \leq \text{ACHIEV} \leq 0.932$; the overall proportion is 0.838. A relatively strong
clustering effect is indicated by the overall design-effect estimate 2.4. Moreover, it can be seen that the domain design-effect estimates vary considerably,
the smallest 0.5 and the largest 2.5. This is due to the property of the domains
that they constitute mixed classes which causes certain domains to cover
only a very small number of teaching groups. These findings, together with the
fact that the total number of sample clusters is small, indicate the presence of
an instability problem.

Methods

The pupil-level analyses are carried out under three approaches. The simplest
way is to assume that the pupil data constitute a simple random sample drawn
directly from the population of pupils. Analysis using this assumption falls
under the IID-based analysis option. An analysis under the full design-based
DES option takes into account the stratification and clustering of the sampling
design. An attempt can be made in the modelling procedure under this option
to account for the clustering effect which is taken as a nuisance. The source
of the clustering effect is in that the school classes can be expected to be to
some degree homogeneous with respect to learning achievements. Clustering
effect is indicated in the domain design-effect estimates and also causes
correlations between proportion estimates from separate domains. In the third
approach, the hierarchical structure of the population is incorporated in the
modelling procedure. In the structure, the schools constitute the highest level,
the school classes are nested within the schools, and the lowest level is the
pupil level in the school classes. Thus, there are three possible levels, of which
the two highest coincide in this case, so we actually have two levels of
hierarchy.

Let us consider first the analysis methodology under the nuisance approach. We denote by p_j the probability for a pupil in domain j achieving the
desired level in mathematics. By taking the predictors as continuous variables
with scores of 1 to 3 in both variables, a logistic regression model fitted on the
data in Table 9.7 is of the form

$$\text{logit}(p_j) = b_0(\text{INTERCEPT}) + b_1(\text{STIME})_j + b_2(\text{TYRSEX})_j$$

where b_0, b_1 and b_2 are the model coefficients to be estimated, and the index
$j \, (= 1, ..., 9)$ refers to domain.

Under the IID analysis option, the clustering and weighting are ignored and the coefficients of the logit regression model can be estimated by the standard ML method. Under the DES option, the weighting and clustering are accounted for by using the PML method in the estimation and the corresponding design-based test statistics in the testing procedures. Special attention is given to the statistical significance of the model coefficients, which guides the selection of the variables in an acceptable model. The tests on model coefficients are based on Wald statistics. As shown in Chapter 8, these statistics can be unreliable if instability is present. Adjustment for the instability is based on certain smoothing of a covariance-matrix estimate and degrees-of-freedom F-corrections to Wald statistics.

In multi-level modelling it is emphasized that there are two natural levels in the data: the pupil level as level 1 and the teaching groups (or schools) as level 2. An essential feature of a multi-level model is that random variation is allowed at each level of the model. Let us denote by p_{jk} the probability for a pupil in domain j of a teaching group k achieving the desired level in mathematics. The former model is thus modified to include the teaching-group level-2. The relevant multi-level logit model takes the form

$$\text{logit}\,(p_{jk}) = b_0(\text{INTERCEPT}) + b_1(\text{STIME})_{jk} + b_2(\text{TYRSEX})_k + u_k,$$

where the index $j \,(= 1,\ldots,9)$ refers to domain and $k \,(= 1,\ldots,40)$ refers to teaching group. The variable TYRSEX is a level-2 variable and thus has a constant value in a teaching group. The variable STIME is a level-1 variable for which the values can vary between pupils in a teaching group. The variate u_k is a level-2 residual which estimates deviation of a teaching group from the estimate $\hat{b}_0$ of the common intercept term. The variance between the level-2 units is $V(u_k) = \sigma_u^2$.

The observed value $\hat{p}_{jk}$ possesses residual variation on level-1, and we can write

$$\hat{p}_{jk} = p_{jk} + e_{jk}z_{jk},$$

where

$$V(e_{jk}z_{jk}) = \sigma_e^2 p_{jk}(1 - p_{jk})/n_{jk},$$

and the extra variable z_{jk} is

$$z_{jk} = (p_{jk}(1 - p_{jk})/n_{jk})^{0.5}.$$

If only the binomial variation is to be modelled, the level-1 variance term σ_e^2 is constrained to be 1. If extra-binomial variation is allowed on level-1, then σ_e^2 is freely estimated (Goldstein 1991; Paterson 1991).

Table 9.8 Logistic regression on ACHIEV proportions under the IID analysis option ignoring the design complexities.

Variable	Beta coefficient	Standard error	t-test	p-value	Design effect
Intercept	2.912	0.427	6.82	0.0000	1
STIME	−0.894	0.174	−5.14	0.0000	1
TYRSEX	0.254	0.127	2.00	0.0455	1

Results

Let us first consider the results under the IID-based approach. We thus use the unweighted proportion estimates and ignore clustering. The IID-based estimation and test results, calculated with SPSS, are given in Table 9.8. Recall that the observed value of a t-test statistic is the signed square root of the value of the corresponding Wald test statistic.

The test results indicate that all the predictors are significant at least at the 5% level. So, from this analysis it can be concluded that both TYRSEX and STIME are important predictors for the variation of learning achievements.

In the previous analysis, the clustering and weighting were ignored. Using the DES option under the nuisance approach, these complexities can be accounted for. We next fit the model by the PML method, and use the design-based Wald statistics in the testing on the model coefficients. The results are in Table 9.9. Because the SUDAAN procedure LOGISTIC was used, a t-test statistic is actually based on an F-corrected Wald test statistic from equation (8.19). Therefore, a certain correction for instability is included in the tests on the model coefficients.

The test results on model coefficients indicate that when the clustering effects are accounted for, the test on the coefficient of TYRSEX becomes nonsignificant. Thus, the IID-based test obviously is overly liberal.

The problem in the design-based analysis in this case is the small number of sample clusters (40). Thus, there are $f = 38$ degrees of freedom for the estimation of the 9×9 design-based covariance matrix of the domain proportions. Therefore, the estimate can be unstable. Although in the previous t-tests

Table 9.9 Logistic regression on ACHIEV proportions under the DES analysis option (the SUDAAN procedure LOGISTIC).

Variable	Beta coefficient	Standard error	t-test	p-value	Design effect
Intercept	2.899	0.578	5.02	0.0000	1.83
STIME	−0.906	0.211	−4.29	0.0001	1.47
TYRSEX	0.271	0.181	1.50	0.1426	2.03

Table 9.10 Logistic regression on ACHIEV proportions under the DES analysis option (the PC CARP program).

Variable	Beta coefficient	Standard error	t-test	p-value	Design effect
Intercept	2.899	0.597	4.86	0.0000	1.95
STIME	−0.906	0.219	−4.14	0.0002	1.58
TYRSEX	0.271	0.186	1.46	0.1533	2.14

a correction was present for the liberality of the Wald statistic due to the instability, we try to further reduce this effect.

A correction aiming at stabilizing the covariance-matrix estimate of the model coefficients (Morel 1989), implemented in the PC CARP program, is used next in the testing procedure for the model coefficients. The technique is different when compared to the previous F-correction; a similarity is that the impact of the correction on the t-test results increases with decreasing f. The test results are displayed in Table 9.10.

The results do not differ noticeably from those in the previous table. A slight difference can be found in the standard-error estimates, which are somewhat larger than those in Table 9.9 due to a more efficient stabilizing adjustment. The conclusion of nonsignificance of TYRSEX still remains in effect.

Let us finally turn to the results from multi-level logit modelling. We take a different view of the data. In the previous models, under the nuisance approach, we tried to clean out the effects of clustering and instability from the estimation and test results. In multi-level modelling, we model the hierarchical structure of the data: the pupils constitute the first level and the teaching groups constitute the second level. The multi-level logit model is fitted with the ML3 program (Prosser *et al.* 1991). The estimates for the fixed effects, and the corresponding t-test results, are displayed in Table 9.11.

Of the model coefficients, the intercept term and STIME indicate significant estimates but TYRSEX is nonsignificant as in the previous DES analyses. The level-2 variation is estimated as $\hat{\sigma}_u^2 = 0.42$ (with a standard-error estimate 0.189), which is significant. Thus, the teaching groups differ from each other

Table 9.11 Logistic regression on ACHIEV proportions with the multi-level modelling approach (the ML3 program).

Variable	Beta coefficient	Standard error	t-test	p-value	Design effect
Intercept	2.941	0.538	5.47	0.0000	1.59
STIME	−0.927	0.179	−5.18	0.0000	1.06
TYRSEX	0.254	0.188	1.35	0.1767	2.19

with respect to the teacher's evaluation on pupils learning achievements in mathematics. The level-2 residuals u_k, which can be taken as teaching-group effects, vary with $-0.993 \leq u_k \leq 0.850$. If one wishes to estimate the predicted values for each teaching group, the estimate of the intercept term has to be corrected by the factor u_k.

In the previous model, the level-1 variation was fixed to be binomial, in which case $\sigma_e^2 = 1$. To examine if there is extra-binomial variation, we let σ_e^2 to be freely estimated. The estimate turns out to be $\hat{\sigma}_e^2 = 0.941$ (with a standard-error estimate 0.050). The level-1 variance does not significantly differ from one, thus not supporting a conclusion of extra-binomial variation in the proportion estimates. Moreover, the estimated coefficients would be close to those previously obtained; the standard-error estimate for TYRSEX, however, would be slightly larger, leading to the most conservative test.

Conclusions

The results obtained are relevant from both methodological and subject-matter points of view. Design-based analyses under the nuisance approach, and multi-level modelling, provided similar results, and the same predictors appear in the accepted model, although the orientation of the modelling approaches to the structure of the population was different. In both approaches, the clustering effects could be accounted for. The extra contribution of a multi-level logit model was that an estimate of the differences between the teaching groups was provided. Undesirable effects of instability on the design-based analyses under the nuisance approach could be successfully decreased by using appropriately adjusted test statistics.

The results from the design-based analyses and from multi-level modelling indicate that the time pupils spend on their homework seems to predict the teacher's evaluation of their mathematics achievements. The sign of the corresponding estimate is negative, which may indicate that pupils spending a lot of time on homework may have learning difficulties in mathematics. The teacher's experience does not seem to have any explanatory power on the teacher's evaluation. In multi-level modelling it was noted that there were certain differences between teaching groups though they were only slight. It is interesting that the results from the design-based analyses are similar to those obtained from a PML logit regression of comparable data from the year 1979 (Pahkinen and Kupari 1991).

A standard IID-based analysis resulted in a model with more parameters than in the models from the other two approaches. In the IID analysis, teacher's experience appeared as a significant predictor for teacher's evaluation. In this analysis, the clustering effects were ignored. Because these effects were noticeable, the IID-based results are misleading, and methods that properly account for the clustering effects would be preferable.

Appendices

APPENDIX 1.   SOFTWARE REVIEW FOR SURVEY ANALYSIS

Four commercially or publicly available software products for survey analysis, OSIRIS IV, PC CARP, SUDAAN, and WesVarPC, are briefly described in this appendix.

OSIRIS IV

The OSIRIS IV program package for survey analysis includes programs for sample selection, imputation and statistical analysis (Institute for Social Research 1992).

The program SAMPLE can be used for sample selection from a frame population data set with basic sampling techniques, including simple random sampling (with replacement or without replacement), systematic sampling, and sampling with varying element inclusion probabilities (covering stratified sampling with non-proportionate allocation). Replicated subsamples can also be drawn using these schemes.

The program IMPUTE can be used for imputation of missing variable values with various imputation techniques. The techniques include mean imputation using overall or class means, hot-deck imputation using sequential imputation, random overall imputation, random imputation within classes, or simple random selection without replacement, and regression imputation using predicted values from OLS regression. In addition, multiple imputation can be used with the random imputation techniques.

Three specific sampling models are available for variance estimation in stratified multi-stage sampling designs: a *paired selection model* (with exactly two sampling-error computing units – SECUs – in each stratum, where SECU usually refers to the primary sampling unit), a *multiple selection model* (with two or more SECUs in each stratum), and a *successive differences model* (intended for a systematic sample of the clusters). With-replacement sampling of clusters is assumed in all sampling models and, for variance estimation, a possibly multi-stage design is reduced to a one-stage design.

The established OSIRIS program PSALMS (Sampling Error Analysis) is used for variance estimation of ratio means and totals, calculated for population subgroups if desired, in which case variances of differences of ratios can be estimated. Of the new programs, PSTOTAL (Sampling Errors for Estimated Totals) is used for variance estimation of subgroup totals, PSRATIO (Sampling Errors for Estimated Ratios) is used for variance estimation of subgroup means, and PSTABLE (Sampling Errors for Contingency Tables) is used for variance and covariance estimation of subgroup proportions. All these programs use the linearization method in the calculation of design-based variance estimates. The corresponding design-effect estimates are provided.

The program REPERR (Repeated Replication Sampling Error Analysis) is used for variance estimation of various nonlinear estimators with sample re-use methods. The techniques of balanced half-samples and jackknife are available. In addition to coefficients of a linear regression model, the methods can be used for variance estimation of ratio means and proportions, correlation coefficients, and partial and multiple correlations.

OSIRIS runs on mainframes with MTS, MVS or CMS operating systems. A UNIX version will be available.

PC CARP

Several analysis procedures are implemented in the PC CARP program for the analysis of complex survey data from multi-stage stratified samples (Fuller *et al.* 1989). The procedures include estimates and standard errors for simple descriptive parameters such as totals, means, ratios, and the difference of ratios. Analytical methods for more complex procedures such as the test of independence in a two-way table and quantile statistics are also available. Weighted regression equations can also be estimated. As a supplement, options for logistic regression and poststratified estimates are included. A subprogram PRE CARP, designed for imputation of missing variable values using the hot-deck technique, can also be called. Finite-population correction terms for multi-stage designs can be accounted for. PC CARP also has an option for collapsing strata, when only one observation is available within a stratum.

The program requires a complete data set as input with no missing values for any variables desired in the analysis (PRE CARP can be used for obtaining a complete data set). A separate data set, containing the sampling rate for each stratum, can be supplied to PC CARP.

The options for *basic survey estimation* include the estimation of population totals, ratios, the difference of ratios, means and proportions with their associated covariance-matrix estimates, corresponding to stratum and subpopulation analyses. Estimation of standard errors of ratio-type estimates is based on the linearization method. Design-effect estimates are produced for all these statistics. For future enhancements in survey estimation, Fuller and Hidiroglou (1992) have developed computational algorithms to accommodate auxiliary information in the estimation of totals, means, ratios, and differences of ratios for subpopulations. The computations use a transformation of the data which incorporates the auxiliary data via regression. This transformation has the property that regression estimates and their associated variances are produced, taking into account the auxiliary data. It is planned to incorporate some of these algorithms into PC CARP (Hidiroglou *et al.* 1993).

In *univariate analysis*, statistics describing the distribution of a study variable are provided in the current version of PC CARP. Estimates of overall or subpopulation means, variances, cumulative distribution function, quantiles and interquartile range with the corresponding standard-error estimates are computed.

The *analysis of a two-way table* is defined by choosing two classification variables and a response variable. Standard errors are computed for cell and marginal totals and proportions; the corresponding covariance matrices can also be obtained. A test for the hypothesis of proportionality, reducing to the test of independence if the response variable is with constant values of one, is carried out. It is based on a Satterthwaite adjusted Pearson chi-squared statistic. To protect against effects of possible instability, the test results are given in terms of degrees-of-freedom-corrected F-statistics.

The option for *regression analysis* uses weighted least squares estimation of the model coefficients. Iteratively reweighted least squares estimation, corresponding to the pseudolikelihood method, is used for *logistic regression*, where a polytomous response variable can be used in addition to a binary variable. The associated standard errors are estimated by the linearization method. Once more, all standard errors associated with estimated parameters reflect the sampling design. In logistic regression, a correction for possible instability is performed on the covariance matrix of the estimated model coefficients. In both the linear and logistic regression, corrected F test statistics with appropriate degrees of freedom are used. *Poststratified estimates* (with POST CARP) can be obtained for population totals, ratios and the difference of ratios, and for subpopulation totals, means, proportions and ratios.

PC CARP is menu-driven, runs on DOS operating systems and is programmed in Fortran. A batch-processing capability, and versions for other platforms including UNIX, are under current development.

PC CARP has been used in Chapters 2 to 4 for the calculation of various descriptive statistics and their associated variance estimates under a number of different sampling designs of varying complexity. PC CARP has been used for more analytical purposes in Sections 9.1 and 9.3.

SUDAAN

The SUDAAN software (Software for the Statistical Analysis of Correlated Data; Shah *et al.* 1995) is a comprehensive program package for the analysis of complex surveys. For continuous study variables, SUDAAN includes procedures for the estimation of population and subpopulation totals, ratios and the difference of ratios, proportions, means and geometric means, and medians and other quantiles, with the corresponding standard-error estimates reflecting the sampling design. Poststratified estimates can be computed for ratio estimates. Frequency and percentage distributions for multi-way tabulations can be produced for categorical variables. The methods also include linear and logistic regression, log-linear models, and proportional hazards models for survival data. Estimation of standard errors associated with estimated parameters is based on the linearization method. SUDAAN produces design-effect estimates for most statistics calculated.

There are several *design options* available covering the most common sampling designs. Under the WR option, a stratified multi-stage design is approximated by a stratified one-stage design (assuming sampling with replacement at the first stage and sampling with or without replacement at subsequent stages, with equal or unequal probabilities of selection at all stages). Under the WOR option, a multi-stage variance estimator is used (assuming sampling without replacement at the first stage and sampling with or without replacement at subsequent stages, with equal probabilities of selection at all stages). Under the UNEQWOR option also a multi-stage variance estimator is used (assuming sampling without replacement with unequal probabilities of selection at the first stage, and sampling with equal probabilities and with or without replacement at subsequent stages). In addition to these basic sampling-design options, there are two simpler design options: the STRWR option (as a special case of the WR option) and the STRWOR option (as a special case of the WOR option). The SRS option is available, corresponding to an analysis assuming simple random sampling with replacement. The procedure DESGCHK checks the specified design against the input data set.

The analysis methods are organized into several special-purpose analysis procedures with which the design complexities can be accounted for by using an appropriate design option. The CROSSTAB procedure, aimed primarily for *descriptive analyses of categorical variables*, produces weighted frequency and percentage distributions for one-way and multi-way tables. A test for an independence hypothesis can be obtained for a two-way table, based on either an *F*-corrected Wald statistic or a test of no interaction in a corresponding log-linear model. The estimation of odds ratios and relative risks can be requested as well as a test of association by the Cochran–Mantel–Haenszel test. The RATIO procedure is used for the *estimation of ratios* and *differences of ratios* with their associated standard errors for population subgroups. General linear contrasts of the ratio estimates can also be specified, and orthogonal polynomial contrasts can be requested for an ordinal subgroup variable. Direct standardization is available, and poststratified estimates can be computed for ratio estimates. The DESCRIPT procedure produces *descriptive statistics for continuous variables*. Estimates of totals, means, geometric means, and quantiles can be requested with the corresponding standard-error estimates. Contrast statistics, standardized estimates and poststratified estimates can also be obtained with DESCRIPT. The PRINTAB procedure permits custom printing of tables and statistics saved in SUDAAN data sets, and the procedure RECORDS allows printing of records from data files and conversion of an input file of one type to another.

The REGRESS procedure fits *linear models* by weighted least squares estimation. Predictor variables can be continuous or categorical, and interaction terms can be introduced in the model. Several test statistics are available for goodness of fit of a model and for model coefficients, including the design-based Wald test statistic with certain degrees-of-freedom F-corrections, and Satterthwaite adjusted standard test statistics, also with appropriate F-corrections. The LOGISTIC procedure is used for *logistic regression* of a binary response variable, based on the pseudolikelihood method. The estimation of odds ratios with their associated confidence limits is possible with the procedure. The CATAN procedure is used for *linear, logit* and *log-linear models* for frequency data with the weighted least squares method. The SURVIVAL procedure is for *survival analysis* with Cox's proportional hazards model or the discrete proportional hazards model. In all these procedures, similar testing options are available to those in the REGRESS procedure. The procedure HYPTEST is designed to perform further testing of user-defined hypotheses using parameter estimates and their associated covariance matrix estimates from other procedures as input.

Currently, stand-alone versions of SUDAAN are available for PCs under Windows and DOS, for Sun SPARCstations under SunOS, for DEC VAX mainframes and workstations under VMS and ULTRIX, and for IBM mainframes under MVS. The corresponding SAS-callable versions are available for DEC VAX mainframes and workstations under VMS, and for IBM mainframes under MVS. The procedures of these SUDAAN versions can be called in a similar way to standard SAS procedures. Thus, for example, SUDAAN jobs can be integrated with SAS by using SAS data sets as input, and full advantage can be taken of the data-management capabilities available in SAS.

Of the plans for the future (Shah 1993), it can be mentioned that in addition to general improvements and additional statistics in existing procedures, certain new analysis procedures for complex surveys are under consideration. These include, for example, a procedure to fit a logistic model to a polytomous response (currently available in Version 7), a procedure to fit a generalized linear model, a procedure to compute variance components, and a procedure to account for nonresponse and imputation.

SUDAAN has been used in Chapters 5–9 for variance and covariance estimation of ratio estimates, for analysis of two-way tables, and for multivariate survey analysis of data sets from stratified cluster sampling.

WesVarPC

WesVarPC is a software package that computes estimates and the corresponding variance estimates for data collected using complex sampling schemes including multistage, stratified, and unequal probability samples (A User's Guide to WesVarPC, Version 2.0, 1996). Descriptive parameters such as totals, means, ratios and differences of ratios, and more complex functions of totals, can be estimated for the entire population and for population subgroups. For means and proportions, the corresponding design-effect estimates are provided. The available model-assisted estimation capabilities include ratio estimation and poststratification.

Log-odds ratios for multiway tables can be estimated by WesVarPC. A test of independence for a two-way table uses first-order and second-order Rao–Scott corrections to the standard Pearson chi-square statistic. Linear regression for a continuous response and logistic regression for a binary response can be performed for a data set from complex sampling. Pseudolikelihood estimation is used in logistic regression. Testing procedures in regression analyses are based on a degrees-of-freedom F-corrected Wald statistic.

Variance estimation in WesVarPC is based exclusively on replication methods. Replicate weights required by this method can be produced in the program for several common sample designs and adjustment processes. The basic BRR and stratified jackknife (JK2) replication schemes assume a design with exactly two sample clusters per stratum. The leave-one-group-out jackknife replication scheme (JK1) can be used when many clusters have been selected. The program can produce replicate weights for these designs which incorporate post-stratification. For other schemes, for example if raking is used, the necessary replicate weights should be created before entering WesVarPC.

WesVarPC runs under Windows 3.1 and Windows 95 on IBM PC compatibles. ASCII and SAS data files can be imported for processing. The software is basically menu-driven, supplemented with a batch processing facility based on coding batch request files in C. Limited formatting and recoding capabilities are supported. There is an option to output generated tables in an Excel platform to produce high-level table formats. Future enhancements include estimating medians and conducting survival analyses.

A user-friendly property of WesVarPC is that the software and the accompanying manual can be obtained by downloading from Westat's site on the World Wide Web (http://www.westat.com).

Other Software

In addition to OSIRIS, PC CARP, SUDAAN and WesVarPC, there are other more specialized programs useful for the analysis of complex surveys. For example, EV CARP is a program for regression analysis of complex survey data containing measurement error in the explanatory variables. Some programs have been developed outside the survey sampling framework such as the MLn program and its predecessor ML3 program for multi-level analysis (Prosser *et al.* 1991, Rasbash and Woodhouse 1995). ML3 was used for two-level logistic regression in Section 9.3. In addition, the MECOSA program can also be used for multivariate analysis of complex surveys.

APPENDIX 2.   FINITE-POPULATION PROPERTIES OF THE ESTIMATORS FOR A TOTAL, RATIO AND MEDIAN: A MONTE CARLO STUDY

In Chapter 2 three basic estimators, $\hat{t}$ (for a total), $\hat{r}$ (for a ratio) and $\hat{m}$ (for a median), were introduced. Statistical properties of these estimators such as *design unbiasness* and *finite population consistency* are demonstrated by simulation methods in this Appendix.

A method of estimation is called *unbiased* if the average value of the estimate, taken over all possible samples of given size n, is exactly equal to the true population value. Further, a method of estimation is called *consistent* if the estimate becomes exactly equal to the population value when $n = N$, that is when the sample consists of the whole population (Cochran 1977, pp 21–22). In Särndal (1992, p. 168) this type of consistency is defined as *finite population consistency.*

The behaviour of total, ratio and median estimators are examined by Monte Carlo methods by simulating 1000 samples with SRS from a finite population. The target population is the *Province'91* population of size $N = 32$. Two study variables, UE91 and LAB91, are chosen for the analysis. All the three estimators are design-based estimators under the SRS design. Varying-size samples are selected; sample sizes vary from $n = 1$ to the population size $N = 32$.

$\hat{t}_{mc}$

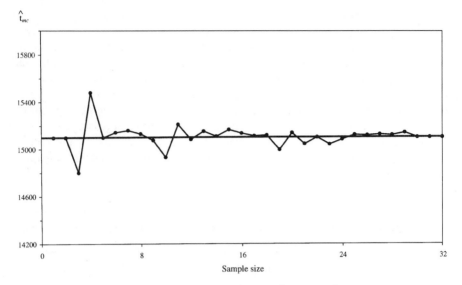

Figure A.1 Unbiased and consistent estimator $\hat{t} = N \times \sum_{k=1}^{n} y_k/n$ for the total $T = 15098$. Means $\hat{t}_{mc}$ of 1000 Monte Carlo estimates under different sample sizes.

s.e. $(\hat{t}_{mc})$

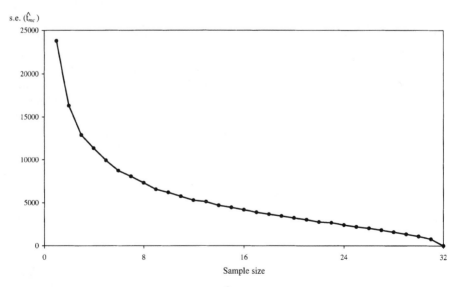

Figure A.2   Standard errors of the means $\hat{t}_{mc}$ of Monte Carlo estimates for the total.

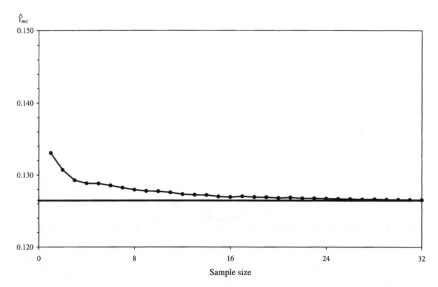

Figure A.3   Somewhat biased but consistent estimator $\hat{r} = \sum_{k=1}^{n} y_k / \sum_{k=1}^{n} x_k$ for the ratio $R = 0.1265$. Means $\hat{r}_{mc}$ of 1000 Monte Carlo estimates under different sample sizes.

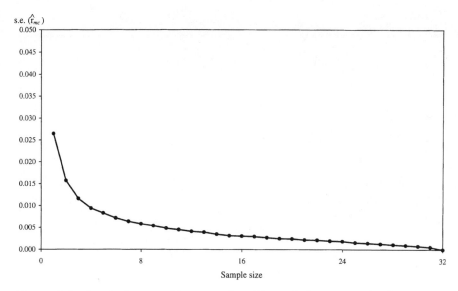

Figure A.4   Standard errors of the means $\hat{r}_{mc}$ of Monte Carlo estimates for the ratio.

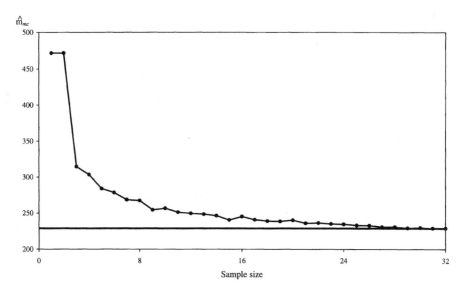

Figure A.5 Heavily biased but consistent estimator $\hat{m}_{mc} = \frac{1}{2}\left[y_{(n/2)} + y_{(n/2+1)}\right]$ for the median $M = 229$. Means $\hat{m}_{mc}$ of 1000 Monte Carlo estimates under different sample sizes.

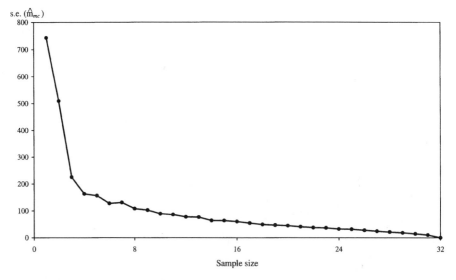

Figure A.6 Standard errors of the means $\hat{m}_{mc}$ of Monte Carlo estimates for the median.

APPENDIX 3. SAS MACRO FOR BOOTSTRAP VARIANCE APPROXIMATION

The SAS program outlined in this appendix can be used for bootstrap approximation of the variance of a ratio estimator from stratified cluster samples. For simplicity, we assume that an equal number (≥ 2) of sample clusters is drawn from each stratum. The MFH Survey sampling design (see Section 5.1) is of this type and is chosen for demonstration. The appropriate formulae for bootstrap variance estimation are given in Section 5.4. Results in Example 5.4 are obtained by using this SAS application.

The program is written in SAS language and is called as a SAS macro. In the macro, we use the SAS data step in the generation of bootstrap samples and in the processing of SAS data sets. In addition, SAS procedures APPEND, CHART, MEANS, PRINT, SORT and TRANSPOSE are used. The macro has two main parts. The appropriate cluster-level data set is first constructed. The second part includes subparts for bootstrap sample generation, for ratio estimation from the resulting bootstrap samples, and for bootstrap variance approximation. The steps are the following.

Step 1

Generate input data set. A cluster-level SAS data set MFH is obtained with 48 observations and 5 variables: the design identifiers STRATUM and CLUSTER, the variables SYSBP and CHRON for cluster-level sample sums of the response variables, and the variable X for cluster-level sample sizes. The data set can be obtained from an element-level data set by the SAS procedure SUMMARY; here, the cluster-level data set is given.

Step 2

Bootstrap samples and BOOT variance approximation. A SAS macro %BOOT is defined with seven macro variables. The variables SAMPLES and TIMES determine the number K of bootstrap samples to be generated (here 10, 100 or 1000), so that the product of SAMPLES and TIMES is the desired number K. In the simplest case, TIMES is set equal to one, but if intermediate simulation results are desired to be displayed, TIMES can be set equal to, say, 5 or 10 if $K = 1000$. The macro variable STRATA carries the number of strata ($H = 24$) and the variable CLUSTERS is for the number of sample clusters per stratum ($m_{h=2}$). The macro variables DSNAME, RESP and SEED are for the input data set name, for the response variable name, and for the initial seed for random number generation, respectively. First, the parent estimate $\hat{R}$ (the variable RHAT) is calculated from the data set PSUSUM which is derived from the input data set MFH. RHAT is saved to the corresponding data set.

The second step is further divided into three subparts:

Step 2.1   Weight matrix for bootstrap samples. A total of K independent bootstrap samples are generated by consecutively using the SAS random number function RANUNI which generates independent uniform random variates. The resulting SAS data set WGT, which includes the generated bootstrap samples, is reshaped to a weight matrix (data set W) by the SAS procedure TRANSPOSE. The data set W is merged with the data set PSUSUM to form a new SAS data set PSUSUMW. There are a total of K weight variables W1, W2, ... in the data set, each with a value 0, 1 or 2, so that the sum of the weights is two in each stratum and the sum of the weights is 48 for each weight variable. The data set PSUSUMW is printed using the procedure PRINT (with only the first ten weight variables).

Step 2.2   Ratio estimates from bootstrap samples. By using the weight variables in the data set PSUSUMW, ratio estimates $\hat{R}_k$ (the variable R in the SAS data set BOOT) are calculated from the bootstrap samples. The corresponding bootstrap estimate (the variable RBOOT) is then calculated as an average of the K estimates $\hat{R}_k$ and saved to the data set B. Data sets B and RHAT are merged with BOOT. Then, for the construction of the bootstrap variance estimates, the appropriate squared differences are calculated (the variables SR1 and SR2). The first ten observations are printed.

Step 2.3   Bootstrap variance approximation. The bootstrap variance approximations $\hat{v}_{1.boot}$ and $\hat{v}_{2.boot}$ (the variables VAR1 and VAR2 in the data set BOOTEST) with the corresponding standard-error estimates (the variables SE1 and SE2) are then calculated and printed together with the ratio estimates. Finally, a bootstrap histogram is produced from the estimates $\hat{R}_k$ by using the procedure CHART.

The statement %MEND BOOT closes the macro. The macro is called by using the statement
%BOOT(DSNAME,RESP,SEED, CLUSTERS, STRATA, SAMPLES, TIMES);
Here, the macro call for SYSBP is
%BOOT (MFH,SYSBP,876543,2,24,1000,1);
for $K = 1000$ bootstrap samples. The BOOT macro can also be used if the SAS macro facility is not available, by removing all macro statements and inserting appropriate numerical values in place of macro variables.

We next display the appropriate SAS code for the BOOT macro used for bootstrap approximation of the variance of a ratio estimator $\hat{R}$, which in the MFH Survey subsample of 30–64-year-old males can be either a proportion estimator $\hat{p}$ for the binary response CHRON or a mean estimator $\bar{y}$ for the continuous response SYSBP. For page limitations, results with the values $K = 10$, 100 and 1000 are displayed for SYSBP, and a full output is provided for the value $K = 10$.

The BOOT macro can be used for the estimation of the variance of a ratio estimator in designs where a constant number of $a \geq 2$ clusters is drawn from each stratum and the clusters are assumed to be drawn with replacement. The last assumption can be relaxed when sampling from a large population of clusters. The design can involve weighting of sample observations; in this case a cluster-level data set including weighted cluster sample sums obtained using rescaled element weights that sum up to the actual sample size, should be used as input. The response variable can be binary or continuous.

Input Data Generation

```
*   STEP 1. Generate input data set;
*   MFH Survey design, 30-64 years-old-males (n=2699);
*   H=24 strata, m=48 sample clusters;
*   The data set is on cluster level;
*   Responses: SYSBP (systolic blood pressure)
               CHRON (chronic morbidity);
*   Variable x: Count variable for cluster sample sizes;
```

```
data mfh;
input stratum cluster sysbp chron x;
lines;
```

stratum	cluster	sysbp	chron	x
1	1	29056	70	204
1	2	29417	74	210
2	1	3692	12	26
2	2	4564	14	30
3	1	7741	15	59
3	2	8585	16	63
4	1	6277	9	45
4	2	5668	14	43
5	1	2322	10	17
5	2	3960	16	30
6	1	3080	10	21
6	2	3252	6	22
7	1	3966	10	27
7	2	3261	4	24
8	1	4156	12	28
8	2	2852	6	20
9	1	6617	15	46
9	2	6616	23	48
10	1	10552	37	73
10	2	11032	25	77
11	1	8759	11	60
11	2	9876	25	72
12	1	9901	33	69
12	2	6828	24	47
13	1	8624	31	61
13	2	9390	27	66
14	1	6960	22	48
14	2	7130	20	49
15	1	6646	18	49
15	2	7094	22	49
16	1	9841	24	69
16	2	11786	37	83
17	1	6910	19	48
17	2	6446	23	45
18	1	10742	25	73
18	2	9026	29	61
19	1	9350	36	65
19	2	8912	34	62
20	1	3810	9	26
20	2	7098	22	51
21	1	6998	18	53
21	2	9970	34	69
22	1	11146	29	79
22	2	13215	41	94
23	1	6596	22	48
23	2	6002	18	41
24	1	3808	15	27
24	2	3148	7	22

```
;
run;
```

SAS Code for the BOOT Macro

```
%macro    boot (dsname, resp, seed, clusters, strata, samples,
times);

%let k=%eval (&samples*&times);
title1 "MFH Survey / Response &resp / Males aged 30-64 years";
title2 "Bootstrap variance approximation / &k bootstrap
       samples";
title3 "&strata strata, &clusters clusters in each stratum";
title4 'STEP 2.1. Cluster sample sums and weights (data
       PSUSUMW)';

* STEP 2. Bootstrap samples and BOOT variance approximation;

data psusum;
set &dsname(rename=(&resp=y) keep=stratum cluster &resp x);

proc means data=psusum noprint sum;
var y x;
output out=rhat sum=sumy sumx;

data rhat(keep=rhat);
set rhat;
rhat=sumy/sumx;
run;

* STEP 2.1. Weight matrix for bootstrap samples;

%do t=1 %to &times;
%let k= %eval (&samples * &t);

data wgt(keep=h i j w1-w&clusters);
length seed 8;
array w(&clusters) w1-w&clusters;
seed=symget('seed');
do i=1 to &samples;
do h=1 to &strata;
   do j=1 to &clusters;
     w(j)=0;
     end;
   do j=1 to &clusters;
     call ranuni(seed,u);
     ii=int(&clusters*u)+1;
     w(ii)+1;
     end;
   output;
   end;
end;
call symput('seed',seed); run;
```

```
proc sort data = wgt; by h i j; run;
proc transpose data=wgt out =w prefix =w;
by  h;
var w1 -w&clusters; run;

data psusumw;
merge psusum w(drop =_name_ h); run;

%if &t = 1 %then %do;
   proc print data = psusumw;
   var stratum cluster y x w1-w10;
   sumy x w1-w10;
   run;
%end;
```

```
* STEP 2.2. Ratio estimates from bootstrap samples;
```

```
%if &t = 1 %then %let name = boot; %else %let name = bootx;
data &name(keep = r);
retain y1-y&samples 0;
retain x1-x&samples 0;
set psusumw end=end;
array w(&samples) w1 -w&samples;
array sumx(&samples) x1 -x&samples;
array sumy(&samples) y1 -y&samples;
do i=1 to &samples;
   sumx(i)+w(i)*x;
   sumy(i)+w(i)*y;
   end;
if end then do;
   do i=1 to &samples;
   r=sumy(i)/sumx(i);
   output;
   end;
end;
run;
%if &t>1 %then %do;
   proc append data=bootx base=boot; run;
%end;

proc means data=boot noprint mean;
var r;
output out = b(drop = _type_ _freq_) mean = rboot; run;

data boot;
if _n_ =1 then merge b rhat;
set boot
%if &t>1 %then %do; (drop = rboot rhat) %end;
;
sr1=(r-rboot)**2;
sr2=(r-rhat)**2; run;
```

```
title2 "Bootstrap variance approximation / &k bootstrap
       samples";
title3 "&strata strata, &clusters clusters in each stratum";
title4 'STEP 2.2. Ratio estimates (data BOOT)';
proc print data=boot(obs=10); run;

* STEP 2.3. Bootstrap variance approximation;

proc means data = boot noprint sum;
var sr1 sr2;
id rhat rboot;
output out = bootest sum = var1 var2; run;

data bootest;
set bootest;
var1 = &clusters/(&clusters-1) *var1 / (&samples*&t);
se1 = sqrt(var1);
var2 = &clusters/(&clusters-1)*var2/(&samples*&t);
se2 = sqrt(var2); run;

title4 'STEP 2.3. Bootstrap estimates (data BOOTEST)';
proc print data = bootest(drop = _type_ _freq_) noobs;
format _all_ 12.7; run;

title4 'STEP 2.3. Bootstrap histogram';
proc chart data = boot;
hbar r / cfreq levels = 25; run;
%end;
%mend boot;

*Macro call;
*%boot(mfh,SYSBP,876543,2,24,1000,1);
*%boot(mfh,CHRON,876543,2,24,1000,1);
```

Output from STEP 2.1.   10 bootstrap samples.

MFH Survey / Response SYSBP / Males aged 30-64 years
Bootstrap variance approximation / 10 bootstrap samples
24 strata, 2 clusters in each stratum
STEP 2.1. Cluster sample sums and weights (data PSUSUMW)

OBS	STRATUM	CLUSTER	Y	X	W1	W2	W3	W4	W5	W6	W7	W8	W9	W10
1	1	1	29056	204	1	2	2	2	0	1	1	1	2	2
2	1	2	29417	210	1	0	0	0	2	1	1	1	0	0
3	2	1	3692	26	2	1	2	2	1	0	1	1	2	1
4	2	2	4564	30	0	1	0	0	1	2	1	1	0	1
5	3	1	7741	59	2	1	0	1	1	0	1	1	1	1
6	3	2	8585	63	0	1	2	1	1	2	1	1	1	1
7	4	1	6277	45	0	0	1	2	0	1	2	1	1	2
8	4	2	5668	43	2	2	1	0	2	1	0	1	1	0
9	5	1	2322	17	1	1	0	0	0	2	0	2	2	1
10	5	2	3960	30	1	1	2	2	2	0	2	0	0	1
11	6	1	3080	21	1	2	0	1	1	2	0	2	1	1
12	6	2	3252	22	1	0	2	1	1	0	2	0	1	1
13	7	1	3966	27	1	2	0	1	2	2	2	0	2	1
14	7	2	3261	24	1	0	2	1	0	0	0	2	0	1
15	8	1	4156	28	1	1	1	1	0	0	1	1	1	2
16	8	2	2852	20	1	1	1	1	2	2	1	1	1	0
17	9	1	6617	46	0	2	2	1	0	1	2	0	1	0
18	9	2	6616	48	2	0	0	1	2	1	0	2	1	2
19	10	1	10552	73	1	2	0	0	1	2	1	0	0	1
20	10	2	11032	77	1	0	2	2	1	0	1	2	2	1
21	11	1	8759	60	1	0	2	2	2	2	2	2	2	0
22	11	2	9876	72	1	2	0	0	0	0	0	0	0	2
23	12	1	9901	69	0	1	0	1	2	1	0	1	0	1
24	12	2	6828	47	2	1	2	1	0	1	2	1	2	1
25	13	1	8624	61	1	1	0	2	0	0	1	2	2	0
26	13	2	9390	66	1	1	2	0	2	2	1	0	0	2
27	14	1	6960	48	1	2	2	1	1	2	1	1	1	0
28	14	2	7130	49	1	0	0	1	1	0	1	1	1	2
29	15	1	6646	49	1	0	1	1	0	1	1	1	0	2
30	15	2	7094	49	1	2	1	1	2	1	1	1	2	0
31	16	1	9841	69	1	0	1	0	1	0	1	2	1	2
32	16	2	11786	83	1	2	1	2	1	2	1	0	1	0
33	17	1	6910	48	0	2	1	2	1	0	1	0	0	1
34	17	2	6446	45	2	0	1	0	1	2	1	2	2	1
35	18	1	10742	73	0	0	1	2	1	2	1	0	1	1
36	18	2	9026	61	2	2	1	0	1	0	1	2	1	1
37	19	1	9350	65	1	1	0	2	1	0	1	1	2	1
38	19	2	8912	62	1	1	2	0	1	2	1	1	0	1
39	20	1	3810	26	0	2	1	1	0	1	1	0	1	1
40	20	2	7098	51	2	0	1	1	2	1	1	2	1	1
41	21	1	6998	53	1	2	1	1	0	0	0	0	1	2
42	21	2	9970	69	1	0	1	1	2	2	2	2	1	0
43	22	1	11146	79	0	1	1	1	2	1	2	0	0	1
44	22	2	13215	94	2	1	1	1	0	1	0	2	2	1
45	23	1	6596	48	2	2	2	2	1	0	1	0	2	1
46	23	2	6002	41	0	0	0	0	1	2	1	2	0	1
47	24	1	3808	27	0	2	0	1	0	1	1	1	1	0
48	24	2	3148	22	2	0	2	1	2	1	1	1	1	2
			======	====	==	==	==	==	==	==	==	==	==	===
			382678	2699	48	48	48	48	48	48	48	48	48	48

Output from STEP 2.2. 10 bootstrap samples.

MFH Survey / Response SYSBP / Males aged 30-64 years
Bootstrap variance approximation / 10 bootstrap samples
24 strata, 2 clusters in each stratum
STEP 2.2. Ratio estimates (data BOOT)

OBS	RBOOT	RHAT	R	SR1	SR2
1	141.967	141.785	141.094	0.76223	0.47733
2	141.967	141.785	141.677	0.08432	0.01171
3	141.967	141.785	141.973	0.00003	0.03520
4	141.967	141.785	141.922	0.00209	0.01863
5	141.967	141.785	141.931	0.00130	0.02136
6	141.967	141.785	142.743	0.60243	0.91840
7	141.967	141.785	142.570	0.36359	0.61646
8	141.967	141.785	142.092	0.01550	0.09405
9	141.967	141.785	142.206	0.05696	0.17709
10	141.967	141.785	141.465	0.25262	0.10269

Output from STEP 2.3. 10 bootstrap samples.

MFH Survey / Response SYSBP / Males aged 30-64 years
Bootstrap variance approximation / 10 bootstrap samples
24 strata, 2 clusters in each stratum
STEP 2.3. Bootstrap estimates (data BOOTEST)

RHAT	RBOOT	VAR1
141.7851056	141.9672739	0.4282143

VAR2	SE1	SE2
0.4945849	0.6543809	0.7032673

Output from STEP 2.3 (continued). 10 bootstrap samples.

```
MFH Survey / Response SYSBP / Males aged 30-64 years
Bootstrap variance approximation / 10 bootstrap samples
24 strata, 2 clusters in each stratum
STEP 2.3. Bootstrap histogram


      R                                               Cum.
   Midpoint                                           Freq
           !
   140.96  !                                            0
   141.04  !                                            0
   141.12  !*******************                         1
   141.20  !                                            1
   141.28  !                                            1
   141.36  !                                            1
   141.44  !*******************                         2
   141.52  !                                            2
   141.60  !                                            2
   141.68  !*******************                         3
   141.76  !                                            3
   141.84  !                                            3
   141.92  !***************************************      5
   142.00  !*******************                         6
   142.08  !*******************                         7
   142.16  !                                            7
   142.24  !*******************                         8
   142.32  !                                            8
   142.40  !                                            8
   142.48  !                                            8
   142.56  !*******************                         9
   142.64  !                                            9
   142.72  !**************                             10
   142.80  !                                           10
   142.88  !                                           10
           -----------------+-------------------+
                            1                   2

                          Frequency
```

It should emphasized that estimation results with $K = 10$ bootstrap samples
are not valid for appropriate bootstrap variance approximation, because the
estimates are very unstable, as can be seen from the bootstrap histogram. But
these results can be compared with those obtained with larger values of K.
Variance approximation results with $K = 100$ and $K = 1000$ bootstrap samples
will be displayed.

Output from STEP 2.3.   100 bootstrap samples.

MFH Survey / Response SYSBP / Males aged 30-64 years
Bootstrap variance approximation / 100 bootstrap samples
24 strata, 2 clusters in each stratum
STEP 2.3. Bootstrap histogram (data BOOTEST)

RHAT	RBOOT	VAR1
141.7851056	141.7910487	0.2811686

VAR2	SE1	SE2
0.2812392	0.5302533	0.5303199

STEP 2.3. Bootstrap histogram

```
        R                                          Cum.
     Midpoint                                      Freq
            !
     140.88 ! 0
     140.96 ! **                                     1
     141.04 ! **                                     2
     141.12 ! ******                                 5
     141.20 ! ******                                 8
     141.28 ! ****                                  10
     141.36 ! **************                        17
     141.44 ! ***************                       24
     141.52 ! *********                             29
     141.60 ! **                                    30
     141.68 ! ****************************          46
     141.76 ! *********                             51
     141.84 ! ***************                       59
     141.92 ! ******************                    69
     142.00 ! ************                          75
     142.08 ! ******                                78
     142.16 ! **********                            84
     142.24 ! **********                            90
     142.32 ! ********                              94
     142.40 ! **                                    95
     142.48 ! ****                                  97
     142.56 ! **                                    98
     142.64 ! **                                    99
     142.72 ! **                                   100
     142.80 ! 100
            ---+---+---+---+---+---+---+
               2   4   6   8  10  12  14  16
                          Frequency
```

These bootstrap variance estimates are more appropriate than the previous ones and are not far from those obtained with other approximation methods (see Table 5.5). But the bootstrap histogram shows that the bootstrap estimates $\hat{R}_k$ are not well concentrated around the parent estimate $\hat{R} = 141.8$, although their mean is close to $\hat{R}$.

Output from STEP 2.3. 1000 bootstrap samples.

MFH Survey / Response SYSBP / Males aged 30–64 years
Bootstrap variance approximation / 100 bootstrap samples
24 strata, 2 clusters in each stratum
STEP 2.3. Bootstrap estimates (data BOOTEST)

RHAT	RBOOT	VAR1
141.7851056	141.7834897	0.2797674

VAR2	SE1	SE2
0.2797726	0.5289304	0.5289353

STEP 2.3. Bootstrap histogram

```
    R                                                              Cum.
 Midpoint                                                          Freq
          !
  140.6   !                                                           0
  140.7   !*                                                          2
  140.8   !**                                                         6
  140.9   !**                                                        12
  141.0   !****                                                      23
  141.1   !************                                              55
  141.2   !***********                                               85
  141.3   !******************                                       131
  141.4   !********************                                     188
  141.5   !**************************                               260
  141.6   !******************************                           344
  141.7   !*********************************************            454
  141.8   !*********************************************            564
  141.9   !***********************************************          679
  142.0   !***************************                              754
  142.1   !**************************                               830
  142.2   !***********************                                  893
  142.3   !******************                                       939
  142.4   !**********                                               963
  142.5   !*******                                                  981
  142.6   !*****                                                    993
  142.7   !**                                                       999
  142.8   !                                                         999
  142.9   !                                                        1000
  143.0   !                                                        1000
          ! ---+---+---+---+---+---+---+---+---+---+---+-
             10  20  30  40  50  60  70  80  90  100 110

                          Frequency
```

Approximation with $K = 1000$ bootstrap samples provides most appropriate variance estimates. The corresponding bootstrap histogram is quite symmetric and indicates good concentration of the bootstrap estimates $\hat{R}_k$ around the parent estimate $\hat{R} = 141.8$.

References

Aromaa A. *et al.* (1989) *Health, Functional Limitations and Need for Care in Finland. Basic Results from the Mini-Finland Health Survey* Helsinki, Turku: Publications of the Social Insurance Institution, Finland, AL:32. (In Finnish with English summary).

Barnett V. (1991) *Sample Survey Principles and Methods* Sixth Edition. London: Edward Arnold.

Barnett V. and Lewis T. (1984) *Outliers in Statistical Data* Second Edition. Chichester: Wiley.

Bean J. A. (1975) Distribution and properties of variance estimators for complex multi-stage probability samples *Vital and Health Statistics* Series 2, No. 65.

Biemer P. P., Groves R. M., Lyberg L. E., Mathiowetz N. A. and Sudman S. (eds) (1991) *Measurement Errors in Surveys* Chichester: Wiley.

Binder D. A. (1983) On the variances of asymptotically normal estimators from complex surveys *International Statistical Review* **51** 279–292.

Binder D. A. (1991) A framework for analyzing categorical survey data with non-response *Journal of Official Statistics* **7** 393–404.

Binder D. A. (1992) Fitting Cox's proportional hazards models from survey data *Biometrika* **79** 139–147.

Binder D. A., Gratton M., Hidiroglou M. A., Kumar S. and Rao J. N. K. (1984) Analysis of categorical data from surveys with complex designs: some Canadian experiences *Survey Methodology* **10** 141–156.

Binder D., Kovar J., Kumar S., Paton D. and Van Baaren A. (1987) Analytic uses of survey data: a review. In: MacNeill I. B. and Umphrey G. J. (eds) *Applied Probability, Stochastic Processes and Sampling Theory* Dordrecht: Reidel 243–264.

Breslow N. E. and Clayton D. G. (1993) Approximate inference in generalized linear mixed models *Journal of the American Statistical Association* **88** 9–25.

Brewer K. R. W. and Hanif M. (1983) *Sampling with Unequal Probabilities* New York: Springer.

Brier S. S. (1980) Analysis of contingency tables under cluster sampling *Biometrika* **67** 591–596.

Cochran W. G. (1977) *Sampling Techniques* Third Edition. New York: Wiley.

Cox B. G., Binder D. A., Chinnappa B. N., Christiansson A., Colledge M. J. and Kott P. S. (eds) (1995) *Business Survey Methods.* New York: Wiley.

Deville J.-C. and Särndal C. E. (1992) Calibration estimators in survey sampling *Journal of the Amercian Statistical Association* **87** 376–382.

Deville J.-C., Särndal C. E. and Sautory O. (1993) Generalized raking procedures in survey sampling *Journal of the American Statistical Association* **88** 1013–1020.

Diggle P. J., Liang K.-Y. and Zeger S. L. (1994) *Analysis of Longitudinal Data* Oxford: Oxford University Press.

Efron B. (1982) *The Jackknife, the Bootstrap and Other Resampling Plans* Philadelphia: Society for Industrial and Applied Mathematics.

Estevao V., Hidiroglou M. A. and Särndal C.-E. (1995) Methodological principles for a generalized estimation system at Statistics Canada *Journal of Official Statistics* **11** 181–204.

Fellegi I. P. (1980) Approximate tests of independence and goodness of fit based on stratified multistage samples *Journal of the American Statistical Association* **75** 261–268.

Frankel M. R. (1971) *Inference from Survey Samples* Ann Arbor: Institute for Social Research, The University of Michigan.

Frankel M. R. (1983) Sampling theory. In: Rossi P. H., Wright J. D. and Anderson A. B. (eds) *Handbook of Survey Research* New York: Academic Press, 21–67.

Freeman D. H. (1988) Sample survey analysis: analysis of variance and contingency tables. In: Krishnaiah P. R. and Rao C. R. (eds) *Handbook of Statistics 6. Sampling.* Amsterdam: North-Holland, 415–426.

Fuller W. A. (1987) Estimators of the factor model for survey data. In: MacNeill I. B. and Umphrey G. J. (eds) *Applied Probability, Stochastic Processes and Sampling Theory* Dordrecht: Reidel, 265–284.

Fuller W. A. and Hidiroglou M. A. (1992) *Using Auxiliary Information for a Number of Analyses Options in PC CARP* Ames, Iowa: Statistical Laboratory, Iowa State University.

Fuller W. A., Schnell D., Sullivan G., Kennedy W. J. and Park H. J. (1989) *PC CARP* Ames, Iowa: Statistical Laboratory, Iowa State University.

Glynn R. J., Laird N. M. and Rubin D. B. (1993) Multiple imputation in mixture models for nonignorable nonresponse with follow-ups *Journal of the American Statistical Association* **88** 984–993.

Goldstein H. (1987) *Multilevel Models in Educational and Social Research* London: Griffin.

Goldstein H. (1991) Nonlinear multilevel models, with an application to discrete response data *Biometrika* **78** 45–51.

Goldstein H. (1995) *Multilevel Statistical Models* London: Edward Arnold, New York: Halsted.

Goldstein H. and Rasbash J. (1992) Efficient computational procedures for the estimation of parameters in multilevel models based on iterative generalized least squares *Computational Statistics and Data Analysis* **13**, 63–71.

Goldstein H. and Silver R. (1989) Multilevel and multivariate models in survey analysis. In: Skinner C. J., Holt D. and Smith T. M. F. (eds) *Analysis of Complex Surveys* Chichester: Wiley, 221–235.

Grizzle J. E., Starmer C. F. and Koch G. G. (1969) Analysis of categorical data by linear models *Biometrics* **25** 489–504.

Groves R. M. (1989) *Survey Errors and Survey Costs* New York: Wiley.

Groves R. M., Biemer P. P., Lyberg L. E., Massey J. T., Nicholls II W. L. and Waksberg J. (eds) (1988) *Telephone Survey Methodology*, New York: Wiley.

Hedayat A. S. and Sinha B. K. (1991) *Finite Population Sampling* New York: Wiley.

Heliövaara M., Aromaa A., Klaukka T., Knekt P., Joukamaa M. and Impivaara O. (1993) Reliability and validity of interview data on chronic diseases *Journal of Clinical Epidemiology* **46** 181–191.

Hidiroglou M. A. and Paton D. G. (1987) Some experiences in computing estimates and their variances using data from complex survey designs. In: MacNeill I. B, Umphrey G. J. (eds) *Applied Probability, Stochastic Processes, and Sampling Theory* Dordrecht: Reidel, 285–308.

Hidiroglou M. A. and Rao J. N. K. (1987a) Chi-squared tests with categorical data from complex surveys: Part I *Journal of Official Statistics* **3**, 117–132.

Hidiroglou M. A. and Rao J. N. K. (1987b) Chi-squared tests with categorical data from complex surveys: Part II *Journal of Official Statistics* **3**, 133–140.

Hidiroglou M. A., Kennedy W. J. and Wang O. (1993) *Enhancements to PC CARP* Florence: Bulletin of the International Statistical Institute, 49th Session, Book 1, 559–560.

Holt D. and Ewings P. D. (1989) Logistic models for contingency tables. In: Skinner C. J., Holt D. and Smith T. M. F. (eds) *Analysis of Complex Surveys.* Chichester: Wiley, 261–284.

Holt D. and Smith T. M. F (1979) Post stratification *Journal of the Royal Statistical Society* A **142** 33–46.

Holt D., Scott A. J. and Ewings P. D. (1980) Chi-squared tests with survey data *Journal of the Royal Statistical Society* A **143**, 303–320.

Institute for Social Research (1992) *OSIRIS IV User's Manual* Eight Edition. Ann Arbor: The University of Michigan.

Judkins D. (1990) Fay's method for variance estimation, *Journal of Official Statistics* **6** 223–240.

Kalimo E., Karisto A., Klaukka T., Lehtonen R., Nyman K. and Raitasalo R. (1991) Occupational health services in Finland in 1985: a national survey. In: Rantanen J. and Lehtinen S. (eds) *New Trends and Developments in Occupational Health Services.* Amsterdam: Excerpta Medica, International Congress Series **890** 59–69.

Kalimo E., Klaukka T., Lehtonen R., Nyman K. and Raitasalo R. (1992) Health care development in Finland, 1960 to 2010 *World Health Forum* **13** 336–342.

Kalton G. (1977) Practical methods for estimating survey sampling errors. *Bull. Int. Stat. Inst* **47** *(Book 3)* 495–514.

Kalton G. (1983) *Introduction to Survey Sampling* Beverly Hills: Sage.

Keyfitz N. (1957) Estimates of sampling variance where two units are selected from each stratum *Journal of the American Statistical Association* **52** 503–510.

Kish L. (1965) *Survey Sampling* New York: Wiley.

Kish L. (1992) Weighting for unequal P_i *Journal of Official Statistics* **8** 183–200.

Kish L. (1995) Methods for design effects *Journal of Official Statistics* **11** 55–77.

Kish L. and Frankel M. R. (1970) Balanced repeated replications for standard errors *Journal of the American Statistical Association* **65** 1071–1094.

Kish L. and Frankel M. R. (1974) Inference from complex samples (With discussion) *Journal of the Royal Statistical Society* B **36** 1–37.

Koch G. G., Freeman D. H. and Freeman J. L. (1975) Strategies in the multivariate analysis of data from complex surveys *International Statistical Review* **43** 59–78.

Krewski D. and Rao J. N. K. (1981) Inference from stratified samples: properties of the linearization, jackknife and balanced repeated replication methods *Annals of Statistics* **9** 1010–1019.

Kumar S. and Singh A. C. (1987) On efficient estimation of unemployment rates from labour force survey data *Survey Methodology* **13** 75–83.

Laaksonen S. (1992) *Handling Household Survey Non-Response data* Helsinki: The Finnish Statistical Society, Statistical Research Reports 13.

Lee E. L., Forthofer R. N. and Lorimor R. J. (1989) *Analyzing Complex Survey Data.* Beverly Hills: Sage.

Lehtonen R. (1988) *The Execution of the National Occupational Health Care Survey* Helsinki: Publications of the Social Insurance Institution, Finland, M:64. (In Finnish with English summary).

Lehtonen R. (1990) *On Modified Wald Statistics. Their application to a Goodness of Fit Test of Logit Models under Complex Sampling Involving Ill-Conditioning* Helsinki: Publications of the Social Insurance Institution, Finland, M:74.

Lehtonen R. and Kuusela V. (1986) Statistical efficiency of the Mini-Finland Health Survey's sampling design. Part 5. In: Aromaa A., Heliövaara M., Impivaara O., Knekt P.

and Maatela J, (eds) *The Execution of the Mini-Finland Health Survey* Helsinki, Turku: Publications of the Social Insurance Institution, Finland, ML:65. (In Finnish with English summary).

Lessler J. T. and Kalsbeek W. D. (1992) *Nonsampling Error in Surveys* Chichester: Wiley.

Levy P. S. and Lemeshow S. (1991) *Sampling of Populations: Methods and Applications* New York: Wiley.

Liang K.-Y. and Zeger S. L. (1986) Longitudinal data analysis using generalized linear models *Biometrika* **73** 13–22.

Liang K. -Y., Zeger S. L. and Qaqish B. (1992) Multivariate regression analyses for categorical data. (With discussion) *Journal of the Royal Statistical Society* B **54** 3–40.

Little R. J. A. (1986) Survey nonresponse adjustments for estimates of means *International Statistical Review* **54**, 139–157.

Little R. J. A. (1991) Inference with survey weights *Journal of Official Statistics* **7**, 405–424.

Little R. J. A. (1993) Post-stratification: a modeler's perspective *Journal of the American Statistical Association* **88** 1001–1012.

Little R. J. A. and Rubin D. B. (1987) *Statistical Analysis with Missing Data* New York: Wiley.

McCarthy P. J. (1966) Replication. An approach to the analysis of data from complex surveys *Vital and Health Statistics* Series 2, No. 14.

McCarthy P. J. (1969) Pseudoreplication: further evaluation and application of the balanced half-sample technique *Vital and Health Statistics* Series 2, No. 31.

McCarthy P. J. and Snowden C. B. (1985) The bootstrap and finite population sampling *Vital and Health Statistics* Series 2, No. 95.

McCullagh P. and Nelder J. A. (1989) *Generalized Linear Models* Second Edition. London: Chapman and Hall.

Morel J. G. (1989) Logistic regression under complex survey designs *Survey Methodology* **15** 203–223.

Nathan G. (1988) Inference based on data from complex sample designs. In: Krishnaiah P. R. and Rao C. R. (eds) *Handbook of Statistics 6. Sampling.* Amsterdam: North-Holland, 247–266.

Nelder J. A. and Wedderburn R. W. M. (1972) Generalized linear models *Journal of the Royal Statistical Society* A **135** 370–384.

Pahkinen E. and Kupari P. (1991) Analysis of educational frequency data from a complex sample survey. Inferences on pupil level with teaching group as the primary sampling unit *Scandinavian Journal of Educational Research* **35** (3), 213–225.

Paterson, L. (1991) Multilevel logistic regression. In: Prosser R., Rasbash J. and Goldstein H. (eds) *Data Analysis with ML3* London: University of London, Institute of Education, pp. 5–18.

Pfeffermann D. (1993) The role of sampling weights when modeling survey data *International Statistical Review* **61** 317–337.

Pfeffermann D. and LaVange L. (1989) Regression models for stratified multi-stage cluster samples. In: Skinner C. J., Holt D. and Smith T. M. F. (eds) *Analysis of Complex Surveys* Chichester: Wiley, 237–260.

Plackett R. L. and Burman J. P. (1946) The design of optimum multifactorial experiments *Biometrika* **33** 305–325.

Prosser R., Rasbash J. and Goldstein H. (1991) *ML3 Software for Three-Level Analysis. User's Guide* London: Institute of Education, University of London.

Quenouille M. H. (1956) Notes on bias in estimation *Biometrika* **43** 353–360.

Rao J. N. K. (1985) Conditional inference in survey sampling *Survey Methodology* **11** 15–31.

Rao J. N. K. (1987) Analysis of categorical data from sample surveys. In: Puri M. L., Vilaplana J. P. and Wertz W. (eds) *New Perspectives in Theoretical and Applied Statistics.* New York: Wiley, 45–60.

Rao J. N. K. (1988) Variance estimation in sample surveys. In: Krishnaiah P. R. and Rao C. R. (eds) *Handbook of Statistics 6. Sampling* Amsterdam: North-Holland, pp. 427–447.

Rao J. N. K. and Scott A. J. (1981) The analysis of categorical data from complex sample surveys: chi-squared tests for goodness of fit and independence in two-way tables *Journal of the American Statistical Association* **76** 221–230.

Rao J. N. K. and Scott A. J. (1984) On chi-squared tests for multiway contingency tables with cell proportions estimated from survey data *Annals of Statistics* **12** 46–60.

Rao J. N. K. and Scott A. J. (1987) On simple adjustments to chi-square tests with sample survey data *Annals of Statistics* **15** 385–397.

Rao J. N. K. and Scott A. J. (1992) A simple method for the analysis of clustered binary data *Biometrics* **48** 577–585.

Rao J. N. K. and Thomas D. R. (1988) The analysis of cross-classified categorical data from complex sample surveys *Sociological Methodology* **18** 213–269.

Rao J. N. K. and Thomas D. R. (1989) Chi-squared tests for contingency tables. In: Skinner C. J., Holt D. and Smith T. M. F. (eds) *Analysis of Complex Surveys* Chichester: Wiley, 89–114.

Rao J. N. K. and Wu C. F. J. (1985) Inference from stratified samples: second-order analysis of three methods for nonlinear statistics *Journal of the American Statistical Association* **80** 620–630.

Rao J. N. K. and Wu C. F. J. (1988) Resampling inference with complex survey data *Journal of the American Statistical Association* **83** 209–241

Rao J. N. K., Hartley H. O. and Cochran W. G. (1962) A simple procedure of unequal probability sampling without replacement *Journal of the Royal Statistical Society* B **24** 482–491.

Rao J. N. K., Kumar S. and Roberts G. (1989) Analysis of sample survey data involving categorical response variables: methods and software (With discussion) *Survey Methodology* **15** 161–186.

Rao J. N. K., Wu C. F. J. and Yue K. (1992) Some recent work on resampling methods for complex surveys *Survey Methodology* **18** 209–217.

Rao J. N. K., Sutradhar B. C. and Yue K. (1993) Generalized least squares F test in regression analysis with two-stage cluster samples *Journal of the American Statistical Association* **88** 1388–1391.

Rasbash J. and Woodhouse G. (1995) *MLn Command Reference* London: Institute of Education, University of London.

Roberts G., Rao J. N. K. and Kumar S. (1987) Logistic regression analysis of sample survey data *Biometrika* **74** 1–12.

Rubin D. B. (1987) *Multiple Imputation for Nonresponse in Surveys* New York: Wiley.

Rust K. (1985) Variance estimation for complex estimators in sample surveys *Journal of Official Statistics* **4** 381–397.

Särndal C.-E., Swensson B. and Wretman J. (1992) *Model Assisted Survey Sampling* New York: Springer.

Satterthwaite F. E. (1946) An approximate distribution of estimates of variance components *Biometrics* **2** 110–114.

Scott A. J. (1986) Logistic regression with survey data *Proceedings of the Section on Survey Research Methods* American Statistical Association 25–30.

Scott A. J., Rao J. N. K. and Thomas D. R. (1990) Weighted least-squares and quasilikelihood estimation for categorical data under singular models *Linear Algebra and its Applications* **127** 427–447.

Shah B. V. (1993) *Survey Data Analysis Software (SUDAAN)* Florence: Bulletin of the International Statistical Institute, 49th Session, Book **2** 395–396.

Shah B. V., Barnwell B. G. and Bieler G. S. (1995) *SUDAAN User's Manual: Software for Analysis of Correlated Data. Release 6.40* Research Triangle Park, NC: Research Triangle Institute.

Singh A. C. (1985) *On Optimal Asymptotic Tests for Analysis of Categorical Data from Sample Surveys* Working Paper No. SSMD 86–002, Social Survey Methods Division, Statistics Canada.

Sitter R. R. (1992) A resampling procedure for complex survey data *Journal of the American Statistical Association* **87** 755–765.

Skinner C. J. (1986) Regression estimation and post-stratification in factor analysis *Psychometrika* **51** 347–356.

Skinner C. J., Holmes D. J. and Smith T. M. F. (1986) The effect of sample design on principal component analysis *Journal of the American Statistical Association* **81** 789–798.

Skinner C. J., Holt D. and Smith T. M. F. (eds) (1989) *Analysis of Complex Surveys* Chichester: Wiley.

Smith T. M. F. (1991) Post-stratification *The Statistician* **40** 315–323.

Smith T. M. F. (1993) Populations and selection: limitations of statistics *Journal of the Royal Statistical Society A* **156** 145–166.

Sudman S. (1976) *Applied Sampling* New York: Academic Press.

Tepping B. J. (1968) Variance estimation in complex surveys *Proceedings of the Social Statistics Section* American Statistical Association 11–18.

Thomas D. R. and Rao J. N. K. (1987) Small-sample comparisons of level and power for simple goodness-of-fit statistics under cluster sampling *Journal of the American Statistical Association* **82** 630–636.

Thompson S. K. (1992) *Sampling* New York: Wiley.

Valliant R. (1993) Poststratification and conditional variance estimation *Journal of the American Statistical Association* **88** 89–96.

Verma V., Scott C. and O'Muircheartaigh C. (1980) Sample designs and sampling errors for the World Fertility Survey *Journal of the Royal Statistical Society A* **143** 431–473.

Wald A. (1943) Tests of statistical hypotheses concerning several parameters when the number of observations is large *Transactions of the American Mathematical Society* **54** 426–482.

Williams D. A. (1982) Extra-binomial variation in logistic linear models? *Applied Statistics* **31** 144–148.

Wilson J. R. (1989) Chi-square tests for overdispersion with multiparameter estimates *Applied Statistics* **38** 441–453.

Wolter K. M. (1985) *Introduction to Variance Estimation* New York: Springer.

Woodruff R. S. (1971) A simple method for approximating the variance of a complicated estimate *Journal of the American Statistical Association* **66** 411–414.

Author Index

Subject Index